INTRODUCTION TO ELECTRONICS

Devices and Circuits —————————————————————

INTRODUCTION TO ELECTRONICS

Devices and Circuits ————————————

JOHN E. UFFENBECK

Hartnell College

Prentice-Hall, Inc., Englewood Cliffs, NJ 07632

Library of Congress Cataloging in Publication Data

Uffenbeck, John E.
    Introduction to electronics, devices and circuits.

    Includes index.
    1. Solid state electronics. I. Title.
TK7871.85.U37   621.381   81-11992
ISBN 0-13-481507-6   AACR2

Editorial/production supervision
and interior design by Karen Skrable
Manufacturing buyer: Gordon Osbourne
Cover design by Infield/D'Astolfo Associates

Printed in the United States of America

10 9 8 7 6 5 4 3 2 1

ISBN 0-13-481507-6

Prentice-Hall International, Inc., *London*
Prentice-Hall of Australia Pty. Limited, *Sydney*
Prentice-Hall of Canada, Ltd., *Toronto*
Prentice-Hall of India Private Limited, *New Delhi*
Prentice-Hall of Japan, Inc., *Tokyo*
Prentice-Hall of Southeast Asia Pte. Ltd., *Singapore*
Whitehall Books Limited, *Wellington, New Zealand*

to Mom and Dad

CONTENTS

FOUR—MEASURING AMPLIFIER PERFORMANCE *98*

FIVE—AC AND DC EQUIVALENT CIRCUITS *126*

SIX—SOME REAL CIRCUITS *154*

SEVEN—MULTISTAGE AMPLIFIERS AND COUPLING TECHNIQUES *193*

EIGHT—FIELD-EFFECT TRANSISTORS *221*

PREFACE

Consider the following facts:

(1) The *vacuum tube* was invented by Fleming in 1904.
(2) The *bipolar transistor* was invented at Bell Laboratories in 1947.
(3) The first *monolithic integrated circuit* was invented by Texas Instruments in 1958.
(4) Intel announced the first *computer-on-chip* in 1973.

Who knows what the next major breakthrough may be? Yet all of us studying electronics must begin at square one learning Ohm's Law and series and parallel circuits. By the time we get to transistors, we find they have been obsoleted by something called the *integrated circuit.*

These "multilegged bugs" are fast relegating the transistor to *support* roles in which the IC does all the thinking, and the transistor is used to *interface* with the real world.

With this perspective in mind, *Introduction to Electronics: Devices and Circuits* has been written. This book discusses solid state devices and their circuit applications, from the *pn* diode and bipolar and field-effect transistor, through the silicon controlled rectifier, TRIAC, and opto-coupler. It is suitable for a one- or two-semester course in linear applications of electronic devices. However, it is equally valid as a self-study text in this area. As such, numerous examples are presented in each chapter. In these examples, numbers are "plugged into" the various formulas to familiarize the reader with dealing in actual quantities. Detailed laboratory experiments are included at the end of each chapter and these should be worked out carefully to fully appreciate the capabilities of each device.

The reader is assumed to have a basic knowledge of dc and ac electronics and the necessary knowledge of mathematics does not go beyond anything required in these courses. A calculator is essential and particularly useful when working with decibels.

Basic semiconductor theory is presented in Chapter 1 so that the physical operation of each device can later be understood independent of any mathematics.

The devices and circuits in the title is meant literally, and several circuit applications are presented for each device. These circuits are chosen for their interesting nature in hopes the reader will be encouraged to try them out for him/herself. As an example, a *flashlight communicator* is presented in Chapter 9. This circuit is capable of transmitting voice communications using the light beam generated by an ordinary flashlight.

This text should bridge the gap between ac and dc electronics and the ever-increasing number of *linear integrated circuits* available today. Indeed, the book inevitably leads to the integrated circuit and the op-amp in particular. The last two chapters cover linear and non-linear applications of operational amplifiers and point out possibilities for the future as more complex circuit functions become integrated.

J. E. Uffenbeck

INTRODUCTION TO ELECTRONICS

Devices and Circuits ——————————————

ONE

SOLID STATE THEORY AND THE SEMICONDUCTOR DIODE

Any text on electronic devices must begin with a discussion of the atomic nature of materials. Once the electron-hole picture is grasped and understood, the whole world of solid state begins to open up. It is comforting to know that the complex integrated circuits common today still utilize the basic concepts used to explain the simple semiconductor diode.

1.1 WHAT IS SOLID STATE?

The term *solid state* is common in our electronic society today but few people can actually explain it. We have all heard of the solid-state transistor radio or the 100% solid-state television chassis, but what does this term *solid state* really mean?

To answer this question, let us examine the history of electronics. Before the advent of the transistor, the electronics world was dominated by the *vacuum tube*, a device dependent upon the principal of *thermionic emission.*

Thermionic Emission

If a wire, or *filament* as it is called in the vacuum tube, is heated to a sufficiently high temperature (for example, by passing a current through it), the electrons in the wire will become excited and their energy levels will increase. If the filament becomes hot enough, the electrons will actually break loose from the wire and form an electron "cloud" around it. This is referred to as *thermionic emission.* In a practical vacuum tube, a heater or filament is often used that in turn heats a specially coated cathode. The cathode then emits electrons by thermionic emission. Due to the high temperatures required for electron emission, it is necessary to place all of the tube elements in an evacuted cylinder (glass tube). If this is not done, the combination of extremely high temperature and oxygen will cause the cathode to burn up!

A *vacuum tube diode* is shown in Fig. 1-1 along with its IV curves. The plate or anode is connected to a positive potential so that the emitted electrons are attracted to it, and a flow of electrons occurs from cathode to plate. Notice that this electron flow can only occur in the direction shown. If the anode is made negative with respect to the cathode, the emitted electrons will be repelled by the plate and no current will flow. Accordingly, a common application of the diode is *rectification*, which is the conversion of alternating current (ac) to direct current (dc). We will study this in more detail later.

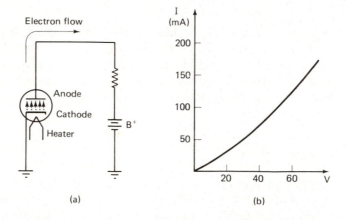

Figure 1-1   Vacuum tube diode. (a) basic circuit. (b) IV curve.

Fleming developed the vacuum tube diode in 1904, and in 1906 DeForest added a third element called the *grid*. This tube is referred to as a *triode*. This additional element allows a small grid bias voltage to control the number of electrons that arrive at the plate. Making the grid more negative repels electrons back towards the cathode, while making it less negative allows a greater number to reach the plate. In fact, if a small sine wave is superimposed on this negative grid voltage, the sine wave will control the plate current. As the sine wave increases (and the grid voltage becomes less negative), the plate current increases, with the opposite occurring for negative excursions of the input sine wave. As it turns out, the grid voltage variations required to cause large plate current variations are relatively small. For a typical triode, a 1-V change in grid potential will cause a 10 mA change in plate current. A typical triode amplifier circuit is shown in Fig. 1-2.

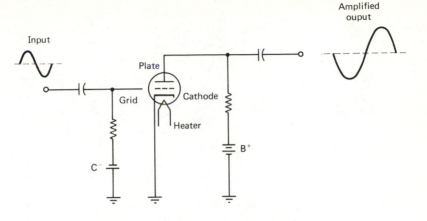

Figure 1-2   Triode vacuum tube amplifier.

Solid State

Prior to the 1950's, all electronic equipment was built using vacuum tube devices. Since that time solid-state devices such as the bipolar transistor, field-effect transistor, and integrated circuit have been developed and have systematically replaced the vacuum tube in most applications. In solid-state devices, the conduction process occurs at the *atomic level* much as conduction occurs in a wire. There is no red-hot filament to observe and verify that the device is working. Instead, current is controlled by chemical properties of the material not visible to the naked eye. To fully answer the question, "What is solid state?," we must examine this conduction process in more detail.

1.2   THE ATOMIC PICTURE

Recall that an atom consists of a *nucleus* made up of neutral particles called *neutrons* and positively-charged particles called *protons*. Revolving about this nucleus, much as the planets revolve about the sun, are the negatively-charged *electrons*.

There are 92 different elements that occur in nature and each differs from another in the size of the nucleus of its atoms and the number of electrons or protons associated with that atom, referred to as the *atomic number*. For example, hydrogen has only a single orbiting electron and single proton in its nucleus and thus an atomic number of 1. Silicon, the most common element used for solid-state devices, has an atomic number of 14. This is shown in Fig. 1-3.

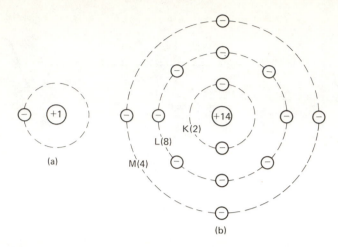

Figure 1-3   Models of the hydrogen and silicon atoms. (a) hydrogen, atomic number = 1. (b) silicon, atomic number = 14.

Regardless of the number of electrons a particular element may have, the electrons and protons are always equal in number. The significance of this is that all atoms are *charge neutral.* We will see that in the manufacture of solid-state devices, atoms may lose or gain electrons and no longer be charge neutral. These charged atoms are then referred to as *ions.*

Conduction of electricity (current), whether it be in a solid-state device, length of wire, or vacuum tube, is defined to be the *flow of charge.* One *ampere* of current is equal to one *coulomb* of charge flowing past a given point in one second. The charge on an electron (or proton) is quite small, 1.6×10^{-19} C, and thus one ampere of current requires 6.25×10^{18} electrons (1 C$/1.6 \times 10^{-19}$ C/electron) to flow past a given point in one second.

Although all elements have electrons, these electrons are not always available to be conductors of electricity. The electrons that a given atom has associated with it are distributed in orbits or *shells* about the nucleus. Each shell may contain only a specific number of electrons and no more as dictated from results of quantum mechanics. Table 1-1 summarizes the distribution of electrons in the various shells and subshells for silicon, copper, and neon. Notice that each main shell may contain only a specific number of electrons distributed over various *subshells* (s, p, d, f). Referring to Fig. 1-3, we see that each main shell corresponds to an orbit. For the case of silicon, three orbits are involved corresponding to the K, L, and M shells. Notice that the M shell is not completely filled.

In order for a given element to be a good conductor of electricity, we must be able to strip free its outermost electrons to use them as

TABLE 1-1 Electron Distribution for Silicon, Copper, and Neon

Main Shell	Subshells	Total Possible Electrons	Silicon	Copper	Neon
K	1s	2	2	2	2
L	2s	2	2	2	2
	2p	6	6	6	6
M	3s	2	2	2	
	3p	6	2	6	
	3d	10		10	
N	4s	2		1	
	4p	6			
	4d	10			
	4f	14			
		TOTAL	14	29	10

charge carriers. One method of doing this involves temperature. At absolute zero (–273°C) all electrons are contained in their various shells. As temperature is increased, the electrons tend to jump from orbit to orbit. The outermost electrons have a tendency to break free from their parent atoms and are then available to conduct electricity. The more complete the outside shell is, however, the less likely this is to happen. We all know that copper is an excellent conductor of electricity and from Table 1-1 we can see why this is so. Its outermost shell, the N shell, contains only one outside electron. At room temperature it is likely that every copper atom has lost this outermost electron. Thus, in a copper wire, millions and millions of *free* electrons are available to carry electricity and this material exhibits a low resistance.

On the other hand, consider neon. Its outermost shell (L shell) is completely filled. Neon is from a class of elements called the *inert gases* and would be considered an *insulator* (very high resistance). Silicon is between these two extremes with four outside electrons. For this reason it is called a *semiconductor*.

The Energy Band Diagram

Another way of representing the electron distribution for an atom is with an *energy band diagram*. This is shown in Fig. 1-4 for copper, silicon, and neon. Due to thermal agitation (temperature), electrons are continuously jumping from one orbit to another and as they do so, their energy levels will increase or decrease. The reason for this is that to jump to an orbit further from the nucleus, the electron must work against the pull from the positively-charged protons much as the engine of your car must work against gravity to climb a hill. Consequently, the

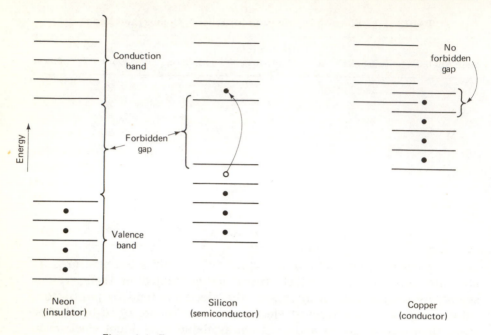

Figure 1-4 Energy band diagrams for neon, silicon, and copper.

outermost electrons will have acquired the maximum amount of energy (they have worked the hardest) and are the most likely to break free. In a sense, these electrons have "climbed the potential hill" and if they should receive sufficient additional energy, they will break free from the parent atom and be available to conduct electricity.

The energy band diagram of Fig. 1-4 plots increasing energy levels on the vertical scale, but you may also think of this as increasing orbit radius because the farther the electron orbit is from the nucleus, the greater its potential energy. Now we can observe an interesting result. Not just *any* energy level is allowed; that is to say, only certain orbits are allowed with a given number of electrons per orbit, as indicated in Table 1-1. To understand why this is true is beyond the scope of this text but is one of the most important concepts of solid-state physics.

Let us follow the path of one electron as it hops from one orbit to another due to some external energy source (heat, light). We might see this electron jump up several orbits and then back down and eventually reach the outermost orbit or energy level. All the while this electron has been contained in a band of energies called the *valence band*. Our electron now faces one final jump or gap. This is shown in Fig. 1-4 as the *forbidden gap*. This gap represents the amount of additional energy this electron must have to break free from the atom. When the electron somehow acquires this energy, it then resides in a new band of energy levels called the *conduction band*. We can think of the valence band as containing all electrons still held by their parent atoms, while the conduction band contains all *free* electrons.

All elements have corresponding energy band diagrams. The gap energy characterizes the element as an insulator or conductor. In Fig. 1-4, we can see that the gap for neon is large, meaning a very large external field will be required to get electrons into the conduction band. Hence, neon is an insulator and no conduction is possible. Copper, on the other hand, has the valence and conduction bands overlapping, which means essentially no external energy is necessary to have many free electrons. Copper is an excellent conductor. Silicon is in between these two cases and we might expect a limited amount of conduction to be possible.

Electrons and Holes

The materials we are most interested in for this text are semiconductors. In a semiconductor, electricity can actually be carried by two types of carriers: *electrons* and *holes*. Before you ask, "How can a hole carry electricity?" realize that conduction by holes still involves electrons as the actual charge carriers. Refer to Fig. 1-4 for the case of silicon. Notice that the electron that is now in the conduction band left behind a *vacant* energy level or *hole* at the top of the valence band. Another valence electron, with an energy level near this, may jump "into this hole" with the net result being that the hole moves to a still lower energy level. In this way electrons may move *up* in the energy diagram while holes are moving *down*. The hole is similar to a bubble in a liquid. It is easier to see the bubble moving upward even though it is actually the liquid moving downward.

Because the concept of the hole can be difficult to understand, let us consider a more physical representation than the energy band diagram affords. If we were to examine a semiconductor material such as silicon under a very high-powered microscope, we would see something like that shown in Fig. 1-5a. Each silicon atom shares its four outermost electrons with four adjacent silicon atoms, with the result being a regular geometric pattern as shown in Fig. 1-5a for three dimensions and in Fig. 1-5b for two dimensions. This sharing of electrons is referred to as *covalent bonding*. The result of this bonding is that although individual silicon atoms have only four outside electrons, these same atoms covalently bonded together with neighboring silicon atoms appear to have eight. This constitutes a full outside ring (3P subshell is full) and means these electrons are held tightly in place. Another way of saying this is that all of the electrons are contained within the valence band.

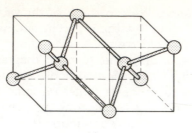

(a)

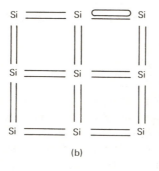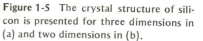

(b)

Figure 1-5 The crystal structure of silicon is presented for three dimensions in (a) and two dimensions in (b).

Current Flow in Semiconductors

In Fig. 1-5b the dashed lines represent the covalent bonds between outside electrons. At absolute zero all electrons are tightly held in place as described previously. As temperature is increased, a few electrons may momentarily break free from the crystal structure. This is shown in Fig 1-5b by the vacant spot in the upper right corner. It should be clear that we might also call this vacancy a *hole*. We might think of a hole as the missing electron in a covalent bond. If an electron should fall into this hole, then some place in the crystal another hole will appear where this electron came from. For pure silicon there should be one hole for every free electron.

Semiconductors, and silicon in particular, are unique in that electrons *and* holes contribute to the conduction process. Electrons in the conduction band are free to travel about the crystal, but electrons in the valence band may also move about, traveling from hole to hole. In this way, the *valence* electrons contribute to the current flow just as the *free* electrons do. Generally, it is easier to visualize conduction in the valence band as being due to holes rather than vacant energy levels. A hole can be thought of as a positive electron, and in a silicon crystal, if the electrons are traveling from left to right, the holes will be traveling from right to left.

Figure 1-6 shows a piece of pure silicon connected to a voltage source and the resulting action of the electron-hole pairs. The conduction electrons within the silicon bar are attracted to the positive potential and exit the bar at point B traveling back to the battery. In this way there is an electron flow in a *clockwise* direction around the loop.

Holes in the bar are attracted to the negative potential of point A and are traveling from right to left through the bar. At point A the incoming electrons "jump into" these holes (recombine), causing them to disappear. In order that the piece of silicon remain charge neutral (it would want to pick up a negative charge for every entering electron at point A), an electron must leave the bar at point B. To replenish the lost holes at point A, a covalent bond is broken at point B, generating a free electron and a hole. Thus holes are generated at point B, travel down the bar to point A, and are lost as the incoming electrons recombine with them.

The point of all this is the following:

1. Electrons travel in a clockwise direction through the bar and through the connecting wires.
2. Holes travel in a counter-clockwise direction but only within the semiconductor; no holes flow in the external connecting wires.
3. Only within the semiconductor itself can the current be broken into the electron and hole components.

Hopefully you now can see that a hole can also be a charge carrier and understand that a semiconductor has current components made up of electrons and holes. We will see in the following chapters that some solid-state devices have conduction due to only one of the two possible carrier types. For example, a *p*-channel FET or an *n*-channel MOSFET.

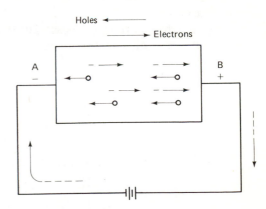

Figure 1-6 Hole and electron flow in a semiconductor. No holes flow in the external circuit.

1.3 DOPING OF SEMICONDUCTORS

One unique property of a semiconductor is that the current flow in such a material is due to both electrons and holes. This is quite different from other nonsemiconductor materials in which the flow of current is due only to conduction electrons. Of course, the hole current is actually an electron current, but it is due to valence band electrons only.

Pure or *intrinsic* semiconductors are those in which the number of electrons and holes are equal. By a process called *doping*, the number of free electrons can be made to exceed the number of holes or vice versa.

n-Type Silicon

If we were to place a pure silicon wafer (a wafer is an approximately 5-in. diameter, very thin and delicate piece of silicon or other semiconductor material) into an oven along with a gaseous form of *phosphorus* and heat this combination to $1000°C$, a chemical phenomenon known as *diffusion* would take place. The phosphorus atoms would diffuse into the silicon crystal structure and the outside electrons of the phosphorus atoms would form covalent bonds with the outside electrons of the silicon atoms.

The reason we have selected phosphorus is that it has 15 outside electrons, but more importantly, its outermost shell contains 5 electrons. It is sometimes referred to as a *pentavalent* impurity.

When phosphorus forms covalent bonds with silicon, one of its five outside electrons will remain as shown in Fig. 1-7. This extra electron is held very loosely compared to the covalent bonds of the other four. As a result, it is a very good assumption to assume that this electron will break free. That is to say, this electron will reside in the *conduction band*.

If, in this doping process, 1000 phosphorus atoms diffuse into the silicon crystal, then there will be 1000 free electrons due to this

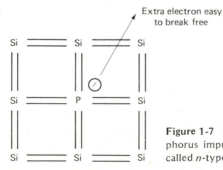

Extra electron easy to break free

Figure 1-7 Silicon doped with a phosphorus impurity. The resulting silicon is called *n*-type.

impurity (in addition to the free electrons normally present). The silicon is no longer pure (intrinsic) and the number of holes and electrons are no longer equal. This type of impurity is called a *donor* because it donates a free electron for each of its atoms.

Imagine now that we have doped the silicon such that the number of doped electrons greatly exceeds the intrinsically free electrons. If we were to pass a current through this material, we would observe two differences from the intrinsic case we studied before. First, there would be a larger current flow. This is because there are now more free electrons and the resistance has decreased accordingly. The second difference, which is actually the most important, is that the conduction process in the silicon would now be dominated by electrons. We could call this material *n*-type silicon to denote this.

p-Type Silicon

Of course we can also make *p*-type silicon. If we dope the pure silicon with a *trivalent* (three outside electrons) impurity such as *boron*, the situation illustrated in Fig. 1-8 results. Three of the four outside silicon electrons can form covalent bonds but the fourth covalent bond is incomplete. There is a strong probability that a nearby electron will "jump into" this hole, leaving another hole behind it. In this way, for each boron atom, an extra hole will be generated. Boron is referred to as an *acceptor* impurity (it accepts an electron).

p-type silicon will also have a lower resistance than intrinsic silicon and the conduction process will mainly be due to holes.

Table 1-2 summarizes some important properties for silicon, germanium, copper, and silicon dioxide (SiO_2), a common insulating material used in integrated circuit manufacture. Notice that the number of free electrons for the semiconductors is quite low compared to copper, a good conductor. Also note that there are very few free electrons in the insulator (SiO_2).

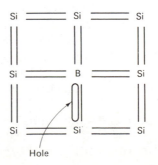

Figure 1-8 Silicon doped with a boron impurity. The resulting silicon is called *p*-type.

TABLE 1-2   Important Properties of Silicon, Germanium, Silicon
Dioxide and Copper. $T_A = 25^\circ C$

	Si	Ge	SiO_2	Cu
Atoms/cm^3	10^{22}	10^{22}	10^{22}	10^{22}
Free electrons/cm^3	10^{10}	10^{13}	10^1	10^{22}
Energy gap (electron volts)	1.1	0.67	8	—

Example 1-1 ──

Using a typical doping density of 10^{16} impurities/cm^3, compute the increase in the number of free electrons/cm^3 for silicon doped with phosphorus impurities.

Solution   If we assume all 10^{16} phosphorus atoms contribute one free electron each, then in one cm^3 of silicon there will be (referring to Table 1-2):

10^{10} intrinsic free electrons

10^{10} intrinsic holes

10^{16} doped free electrons

The doping process has contributed $10^{16}/10^{10} = 10^6 = 1$ million times as many free electrons/cm^3 as there are in pure silicon.

In Example 1-1 you should also notice that there are over a *million times* as many free electrons as holes after the doping. The conduction process would certainly be dominated by electrons and this material would be *n*-type.

Charge Neutrality

Earlier we mentioned that all atoms are charge neutral. Is this still true for a doped semiconductor? The answer to this question is yes, although it might seem that after doping a semiconductor *n*-type, it should take on a negative charge (or positive charge for *p*-type).

Each donor impurity that gave up one free electron had five outside electrons and five protons in its nucleus. Although one electron is now free and roaming about the crystal, its charge is cancelled by the net positive charge of the donor atom (now a donor ion, actually). The same is true for the acceptor impurity. When it accepts an electron, a positive hole appears where this electron had previously been. However, this positive charge is cancelled by the net negative charge of the acceptor ion.

The acceptor and donor ions differ from the electrons and holes in that they are "stuck" in the crystal and unable to move (immobile). Table 1-3 summarizes the types of charges found in a doped semiconductor.

TABLE 1-3 Mobile and Immobile Charges in a Semiconductor. For Charge Neutrality, $n_i = p_i$, $n = N_D$ and $p = N_A$.

Charge Carrier	Polarity	Symbol	Material
Intrinsic electron	(−)	n_i	n and p
Intrinsic hole	(+)	p_i	n and p
Doped electron	(−)	n	n type only
Doped holes	(+)	p	p type only
Donor ion	(+)	N_D	n type only
Acceptor ion	(−)	N_A	p type only

1.4 THE *pn* JUNCTION

The Equilibrium Condition

With the preceding sections as background, we are now prepared to discuss the *pn* junction or *semiconductor diode*.

If a voltage source is connected to the *p* or *n* material *individually*, a current will flow. In fact, it makes no difference which way the battery is connected, current flows as easily in either direction. This is shown in Fig. 1-9 for *p*-type silicon. In either case, the *p*- or *n*-type silicon appears as a resistor and there is no diode action.

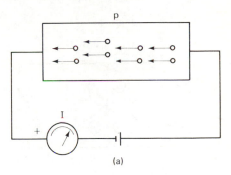

(a)

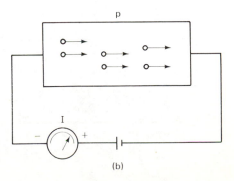

(b)

Figure 1-9 Current flow in *p*-type silicon. Current flows regardless of the battery polarity.

If, however, the *p* and *n* materials are joined together, a *pn* junction is formed. This is shown in Fig. 1-10. Visualize looking into the two pieces of silicon as they are first brought together. The following sequence of events will occur:

1. The holes and electrons are traveling randomly within the *p* and *n* materials respectively.
2. Because the motion of the two charges are random, they will tend to cross the junction and even their distribution throughout the *pn* crystal. This is shown in Fig. 1-10a.
3. Although we might expect this motion across the junction to continue until the holes and electrons have neutralized themselves on each side, it does not.
4. Each hole that crosses the junction leaves behind a negative acceptor ion.

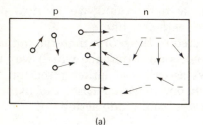

(a)

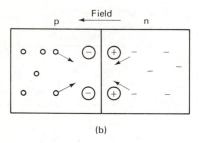

(b)

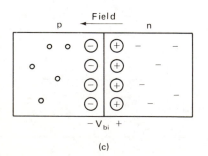

(c)

Figure 1-10 *PN* junction with no external bias. (a) random motion of electrons and holes causes them to cross the junction. (b) uncovered ions begin to exert a force on the electrons and holes as they cross the junction. (c) a built-in voltage develops and an equilibrium is established. No current flows.

5. Similarly, each electron leaves a positively-charged donor ion, as shown in Fig. 1-10b.

6. As more and more electrons and holes cross the junction, more and more of these ions become "uncovered"; that is, the charges of these ions are no longer cancelled by the free hole or electron. (Recall that these ions are immobile and stuck in the crystal.)

7. At the junction, a separation of charges now exists. Positive ions on the *n* side and negative ions on the *p* side. The junction area looks very similar to a capacitor (see Fig. 1-10c).

8. Eventually an equilibrium is reached in which the positive ions repel any holes attempting to cross the junction (like charges repel) while the negative ions repel the electrons.

9. All motion of electrons or holes across the junction *stops*. The junction area, now called the *depletion region*, is void or deplete of any carriers.

From the above description we conclude that a *pn* junction with no external bias has no current flowing and the bulk of the electrons remain on the *n* side as do the holes on the *p* side. The field that exists at the junction is referred to as *the built-in voltage* (V_{bi}). It is this voltage that prevents any current from flowing.

Reverse Bias

So far we have not applied an external voltage to the diode. In Fig. 1-11, a *reverse bias* is shown. The electrons in the *n* material are attracted

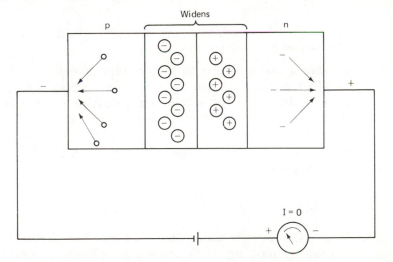

Figure 1-11 *PN* junction — reverse bias. No current flows.

to the positive source potential while the holes in the *p* material are attracted to the negative source potential. You should be able to see that this only adds to the built-in voltage, widening the depletion region. The ammeter in the circuit indicates no current flow.

Forward Bias

In Fig. 1-12 a *forward bias* is shown. This bias "pushes" the electrons and holes towards the junction. In so doing, the depletion region is made to shrink. If we increase the external voltage to the point at which it overcomes the built-in voltage, a substantial current will flow.

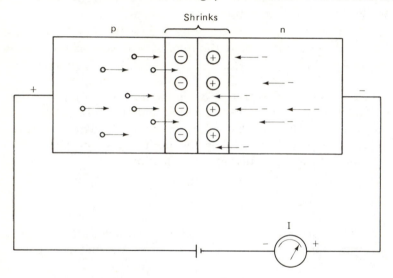

Figure 1-12 *PN* junction — forward bias. Electrons and holes freely cross the junction and a substantial current results.

We now have a device similar to the vacuum tube diode discussed previously in which conduction can occur in *one direction* only. However, no heater or glass envelope is required. In addition, the size of a typical semiconductor diode is about that of a one-half watt resistor. This is an inherent advantage of all solid-state devices. Because we are dealing with conduction processes at the atomic level, physical device sizes may be quite small.

IV Characteristics

Returning to the forward-biased diode, we may gain further insight into its operation by plotting its IV curves. An extremely useful tool for doing this is the *curve tracer* shown in Fig. 1-13. This instrument

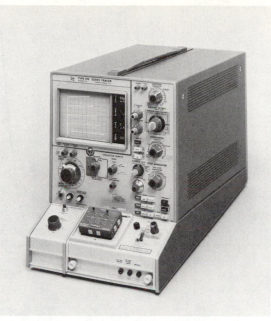

Figure 1-13 The curve tracer displays a device's current versus voltage character-istic display. (Courtesy Tektronix, Inc.)

yields a visual display of a device's current versus voltage characteristics. Because each device has its own unique pattern, the curve tracer is extremely useful for testing solid-state devices and spotting defective components quickly.

Of course we could also plot the diode's IV curve by using a power supply, ammeter, and voltmeter. In doing this for a silicon diode, we would find that current does not begin to flow until the built-in voltage of the diode has been overcome. Once this point is reached, the diode turns ON and the current rises in a nearly vertical fashion.

Different semiconductor materials will have different built-in voltages. This is clear when the IV curves of Fig. 1-14 are studied. A germanium diode begins to conduct with only 0.2 V of forward bias, while a silicon diode requires about 0.5 V and the light-emitting diode (LED) about 1.5 V.

The IV curve for a 100-Ω resistor is also shown in the figure. This curve is *linear* (straight line) and the current for a given voltage can be found from Ohm's Law, $I = V/R$, which is a linear equation. The Ohm's Law equation for a diode is more complex, but the IV curve suggests an *exponential* relationship.

Another interesting property of the diode is that over a wide range of current values, the voltage across the diode stays very nearly *constant*.

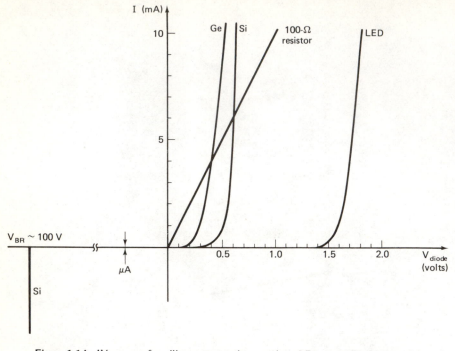

Figure 1-14 IV curves for silicon, germanium, and an LED. A 100-Ω resistor is shown for comparison.

In fact, for a typical silicon diode, the forward voltage drop across the diode will vary from 0.5 V to 0.7 V over a current range from $1 \, \mu A$ to 10,000 μA (10 mA). This is an increase in current of 10,000 times with a voltage change of only 0.2 V! For this reason, it is common to assume that a forward-biased silicon diode will have approximately a 0.6-V drop across it *independent* of its actual current. Of course Ohm's Law predicts $V = IR$ and as I increases, V must increase. This equation is also true for the diode, but its ON resistance is very small so that we may assume a nearly *constant* voltage drop over a wide range of currents. Throughout this text, unless otherwise indicated, we will assume the forward-biased diode drop for a silicon diode to be 0.6 V.

For other types of diodes this number is different. For example, for germanium diodes we might use 0.3 V, while for gallium arsenide diodes (commonly used for light-emitting diodes), we might use 1.6 V.

Figure 1-15 shows the schematic symbol used for the diode. The *p* side is called the *anode* while the *n* side is called the *cathode*. Most diodes are similar in size to resistors and usually have a band at one end to designate the cathode. Note that *conventional* current is shown (opposite to electron flow). Solid-state devices have their symbols drawn in such a way that the arrow part of the symbol points in the direction

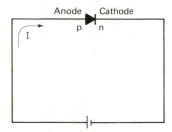

Figure 1-15 Schematic symbol for the semiconductor diode illustrating conventional current flow.

conventional current will flow when the device is forward-biased. We will use conventional current throughout this text.

Breakdown

Earlier we said that a reverse-biased diode conducts no current. Actually, a small current does flow. This current is usually in the micro-ampere range and is called the *leakage* current. It is due to *minority* carriers (holes in the *n* side or electrons in the *p* side) in the *n* and *p* materials.

Figure 1-16 shows the *pn* junction under reverse bias. The minority carriers that are shown are due to the intrinsic electrons and holes that are in the silicon crystal independent of the doping (for example, *n*-type silicon has many free electrons but there are also some holes due to thermal agitation breaking apart some covalent bonds).

The reverse bias appears as a forward bias to these minority carriers and they are free to cross the junction. The small leakage current of the diode is an indication that these carriers are present. As might be expected, this current is a strong function of *temperature* because an

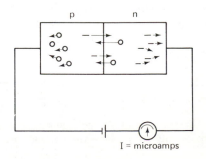

Figure 1-16 Minority carriers cross the junction under reverse bias.

increase in temperature will cause more electron-hole pairs to be generated. In fact, this is one of the main problems that plague all semiconductor devices. They are very sensitive to heat and their characteristics change accordingly. For silicon, the leakage current will approximately double for every $10°C$ increase in temperature.

If we continue to increase the reverse bias on the diode, the depletion region will continue to widen and the electric field will become very intense. A point will be reached where the diode is said to *breakdown* and a large current may now flow. Two different mechanisms may occur in breakdown.

1. *Zener breakdown.* If the field strength is sufficiently large, some covalent bonds may be torn apart, releasing electron-hole pairs with the resulting effect of a reverse current flow.
2. *Avalanche breakdown.* As the minority carriers accelerate across the junction, they may collide with the crystal and dislodge other electrons, which in turn may accelerate and knock other electrons free. The process continues in an avalanche fashion.

In most *pn* diodes, it is the avalanche process that causes breakdown. Once the diode does breakdown a substantial current can flow (limited by the external circuit), while the voltage across the diode remains essentially constant at the breakdown value. This is shown as the vertical line in the third quadrant in Fig. 1-14. It should be emphasized that breakdown is *not destructive* to the diode. However, the current that flows in breakdown must be limited to a safe value. Some diodes, called *zeners*, are meant to be operated in reverse bias and are manufactured to have specific breakdown voltages. The following examples will help you see how the diode is treated as a circuit element.

Example 1-2 ───

Determine the current, voltage, and power dissipation for the diode in Fig. 1-17a. Assume a silicon diode with a reverse breakdown of 100 V.

Solution   In Fig. 1-17a, the voltage source is such as to forward-bias the diode. Because the diode is silicon, we can assume a 0.6-V drop across the diode. By Kirchoff's Law, the voltage across the 1-kΩ resistor must be 10 V – 0.6 V = 9.4 V. Thus the diode current is: $I = V/R$ = 9.4 V/1 kΩ = 9.4 mA. The diode power dissipation is 9.4 mA $\times$ 0.6 V = 5.6 mW.

Example 1-3 ───

Repeat Example 1-2 for the circuit in Fig. 1-17b.

Solution   In this circuit the voltage source has been reversed. This means the diode is now reverse-biased. Because the voltage source is less than 100 V (the diode breakdown-voltage), the diode is not in breakdown. We can assume the diode current to be 0 A. This means the full source voltage is across the diode.

$$V_{\text{diode}} = 10 \text{ V} \qquad I_{\text{diode}} = 0 \text{ A} \qquad P_d = 0 \text{ W}$$

Example 1-4

Repeat Example 1-2 for the circuit in Fig. 1-17c.

Solution Again the diode is reverse-biased but this time a zener diode is used. (Note the difference in the diode symbol.) Because the source is greater than 5.1 V, the zener will break down and hold 5.1 V across itself. Thus, $V_{\text{diode}} = 5.1$ V. Again by Kirchoff's Law, $I_{\text{diode}} = (10 \text{ V} - 5.1 \text{ V})/1 \text{ k}\Omega = 4.9$ mA and $P_d = 4.9$ mA $\times$ 5.1 V = 25 mW.

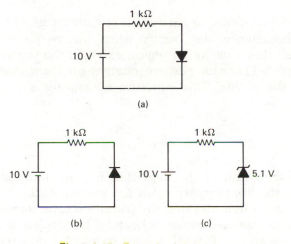

Figure 1-17 Examples 1-2 to 1-4.

Notice that the power dissipation of a forward-biased diode is quite low due to the low forward voltage. When the diode breaks down, the power dissipation may increase considerably. When purchasing a diode, the following are important considerations:

1. Type — Silicon, germanium.
2. Forward current (I_F) — Maximum current under forward bias.
3. Breakdown voltage (V_{BR} or *PIV*, peak inverse voltage) — Make sure this value is greater than any supply voltage in your circuit. Often a safety factor of 2 is used.
4. Zener voltage (V_Z) — For zener diodes this specifies the breakdown voltage.
5. Wattage rating (P_Z) — For zener diodes this is the maximum power the diode can safely dissipate in breakdown.

For all of the technology used in making a semiconductor diode, it is not an expensive component. For example, the 1N4002 is a 100-PIV, 1-A rectifier diode and sells for less than 10 cents!

1.5   OTHER SEMICONDUCTOR DIODES

In addition to the rectifier and zener diodes presented in the last section, there are a number of other semiconductor diodes each with their own unique properties.

The Varactor Diode

The *varactor* diode, or *varicap*, takes advantage of the *junction* or *transition* capacitance that occurs when the *pn* junction is reverse-biased. Recall that a depletion region exists in the vicinity of the junction (see Fig. 1-11), with positive charges on the *n* side and negative charges on the *p* side. This structure constitutes a capacitor and its value is given by

$$C_T = \frac{\epsilon A}{W} \tag{1-1}$$

where W is the width of the depletion region, A the area, and ϵ the dielectric constant or permittivity of the semiconductor.

Equation 1-1 indicates that the transition capacitance will *decrease* as the diode reverse bias *increases* because the depletion region widens under this condition. Typical values of C_T for commercial varactor diodes are 100 pF at 0-V and 10 pF at 20-V reverse bias.

The schematic symbol for the varactor diode is shown in Fig. 1-18. Because the capacitance values obtained are in the picofarad range, most applications involve high frequencies, for example, voltage-controlled tuning of CB radios or other high-frequency radio equipment. In this case the varactor diode can replace the bulkier *mechanical* tuning capacitor.

Figure 1-18 Schematic symbol for the varactor diode.

The Tunnel Diode

In 1958, Dr. Leo Esaki discovered that by doping a conventional *pn* diode to a much higher level than normal, a phenomenon known as *tunneling* would take place. The effect of this tunneling can be observed in the IV characteristic of the *Esaki* or *tunnel diode*, illustrated in Fig. 1-19a.

When the doping concentration is increased to a high level, the depletion region of the resulting diode is very narrow. In fact, it is possible for carriers to cross this barrier even though they do not have enough energy to overcome the *built-in* potential. As a result, a new current (the *tunnel* current) is seen for low values of forward bias when there is ordinarily no current flowing. As V_F is increased, the tunnel current decreases and the conventional diode current takes over. This is shown in the IV curves.

A particularly interesting property of the tunnel diode is the region on the curve between V_p and V_v. In this region the diode exhibits a *negative resistance*. This does not literally mean that the diode has a negative value for its resistance but rather that as voltage *increases*, the current *decreases*, and vice versa.

If the tunnel diode is biased (made to operate) in the middle of this negative resistance region, an *unstable* operating point will result. This is because a slight *decrease* in diode voltage will lead to an *increase* in diode current, which in turn leads to a still lower diode voltage. The

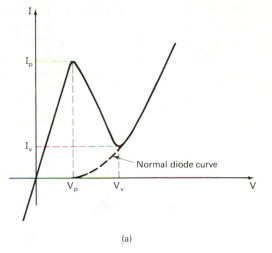

(a)

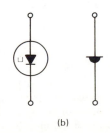

Figure 1-19 (a) IV characteristics for the tunnel diode. Typical values are I_p = .05A and V_v = .35 V. (b) two typical schematic symbols.

(b)

operating point will move to a point on the IV curve with a positive (stable) resistance value. A similar effect occurs for a slight *increase* in diode voltage.

This switching property of the tunnel diode makes it suitable for digital applications. Most logic circuits switch in *nanoseconds* (10^{-9} s), but because the tunnelling phenomenon occurs at the speed of light, tunnel diodes can switch in *picoseconds* (10^{-12} s).

Other applications of tunnel diodes include very-high-frequency oscillators and amplifiers. Two main disadvantages have kept the tunnel diode from widespread use. One is the low signal swing possible across the device and the other is a lack of *isolation* between the input and output circuits due to its two-terminal nature.

The Schottky Barrier Diode

The *Schottky* diode is unique in that a metal–semiconductor junction is used and conduction is due entirely to *majority* carriers.

The device structure and schematic symbol are illustrated in Fig. 1-20. When forward-biased, the *n*-type cathode injects electrons into the metal anode in a manner analogous to the way the cathode of a vacuum tube emits electrons into a vacuum. These carriers have a much higher velocity than those present in the metal anode and for this reason the diode is also often referred to as a *hot carrier* diode.

The properties of the Schottky diode depend to a great extent on the type of metal used for the anode. Typical characteristics include lower forward voltages for a given forward current (0.4 to 0.5 V for typical currents) and faster recovery (turn-off) times than those of

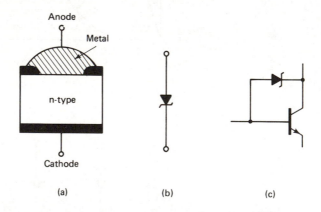

Figure 1-20 (a) The Schottky barrier diode, or hot carrier diode, consists of a metal–semiconductor junction, (b) the schematic symbol. (c) a Schottky-clamped transistor.

conventional diodes. These properties are taken advantage of in switch-ing power supply circuits in which the diode must turn on and off at a rapid rate (greater than 20,000 times/second).

Other applications include Schottky-clamped transistors (the 74S00 and 74LS00 logic families) in which a Schottky diode is con-nected between the transistor collector and base terminals to prevent saturation and dramatically reduce transistor switching times.

--------------------------------- KEY TERMS ---------------------------------

Solid State This term is commonly used to describe electronic devices in which the conduction process involves both holes and electrons and occurs within a semiconductor material at the atomic level.

Outermost Electrons These are the electrons that determine how an element will react chemically with other elements or bond with its own atoms. Silicon has four electrons and forms covalent bonds, resulting in a cube-like crystalline structure.

Valence Band Electrons that are held in covalent bonds of the silicon crystal are said to reside in the valence band of energies.

Conduction Band When an electron breaks free from its covalent bond and travels about the crystal, it is said to be in the conduction band of energies.

Hole A vacant spot in the covalent bond structure or an available energy level in the valence band is said to be a hole. Holes move in one direction as electrons move in the opposite direction.

Doping This is a technique in which pure silicon has impurities added to it in such a way that an excess number of free electrons or holes are generated in the crystal. The resulting silicon is then said to be n type or p type.

Forward and Reverse Bias A forward bias acts to force a current through the diode, while a reverse bias is in such a direction as to prevent current from flowing. A forward-biased diode is ON while a reverse-biased diode is OFF.

Breakdown With respect to a semiconductor diode, if the reverse bias is sufficiently large, the diode will break down and allow conduction. The external circuit must limit the current to prevent damage to the diode.

----------------------- QUESTIONS AND PROBLEMS -----------------------

1-1 How does conduction in a forward-biased vacuum tube diode differ from con-duction in a forward-biased semiconductor diode?

1-2 When is an electron considered to be in the valence band? The conduction band?

1-3 Why does $n_i = p_i$ for intrinsic silicon?

1-4 Specify the charge for each of the following:
(a) Intrinsic electron

(b) Acceptor ion

(c) Donor ion

(d) Proton

(e) Neutron

1-5   What is meant by an "uncovered ion"?

1-6   To dope silicon n-type, an acceptor impurity would be used. (True/False)

1-7   Conduction by holes is actually the same as conduction by electrons in the valence band. (True/False)

1-8   Once a diode has broken down, it cannot be used again. (True/False)

1-9   Why does the reverse-bias leakage current depend on temperature?

1-10 A certain donor impurity level is 10^{17} atoms/cm^3. If the silicon atoms are distributed 10^{22} atoms/cm^3, how many silicon atoms are there for each impurity atom?

1-11 Refer to Fig. 1-21.

(a) Calculate I_{diode}.

(b) If the voltage source is reversed, calculate I_{diode}, V_{diode} and P_d.

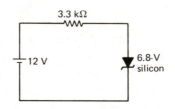

Figure 1-21

LABORATORY ASSIGNMENT 1:
DIODE FAMILIARIZATION

Objectives

1. Learning to use an ohmmeter to test a diode.
2. To observe differences in IV curves for silicon and germanium diodes.
3. Practice in using a curve tracer for testing a diode.
4. To understand the operation of a simple rectifier.

Introduction   In this laboratory assignment you will learn how to identify the cathode and anode ends of a diode, plot its IV curves, and observe one of its main applications, rectification.

Components Required

1 Silicon diode

1 Germanium diode

1 1-kΩ resistor

Part I: *Diode Identification*

STEP 1 Obtain a silicon diode and notice that one end has a band around it. This identifies the cathode lead.

STEP 2 Set the ohmmeter to the RX100 scale and measure the resistance with the cathode end connected to the negative lead and the anode connected to the positive lead. Repeat with the connections reversed and record the results.

$$R_{\text{forward}} = \underline{\hspace{2cm}}$$

$$R_{\text{reverse}} = \underline{\hspace{2cm}}$$

STEP 3 Repeat step 2 using the RX1000 scale.

$$R_{\text{forward}} = \underline{\hspace{2cm}}$$

$$R_{\text{reverse}} = \underline{\hspace{2cm}}$$

Question 1 How could an *unmarked* diode have its cathode and anode leads identified using an ohmmeter?

Question 2 How do you account for the differences in forward resistances measured in steps 2 and 3?

Part II: *IV Characteristics*

STEP 1 Set up the circuit of Fig. 1-22 using a silicon diode. Note that the voltmeter is only in the circuit when V_F is measured.

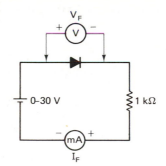

Figure 1-22 Part II, step 1.

STEP 2 Adjust the power supply until I_F is 100 μA. Connect the voltmeter across the diode and record V_F.

STEP 3 Repeat step 2 for various values of I_F through 25 mA.

STEP 4 Repeat steps 2 and 3 using a germanium diode.

STEP 5 Graph the data obtained for both diodes using linear graph paper.

Optional   Graph the data on semilog paper plotting I_F on the log scale and V_F on the horizontal.

Question 3   Based on your data, what approximate forward voltage would you assume for the silicon diode? The germanium diode?

STEP 6   Using a curve tracer, display the IV curves for the two diodes and compare with your graph (linear paper). You should be able to appreciate how much faster the IV curves can be generated. Using an oscilloscope camera, a "hardcopy" could also be obtained.

Part III: *Rectification*

STEP 1   Set up the circuit shown in Fig. 1-23.

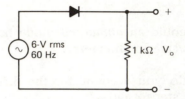

Figure 1-23   Part III, step 1.

STEP 2   Using a dual channel oscilloscope, observe and record the signal generator output on one channel and the output voltage across the 1-kΩ resistor on the other.

STEP 3   Repeat step 2 with the diode reversed.

Question 4   During which part of the a-c cycle does the diode conduct in steps 2 and 3?

TWO ―――――――――――――――――

CIRCUIT APPLICATIONS
OF DIODES

Although the semiconductor physics required to explain the solid-state diode is relatively complex, its operation in most circuits is straightforward. The diode allows current to flow in one direction only (indicated by the arrow in its symbol for conventional current flow) and, when conducting, has a typical voltage drop of 0.6 V for silicon.

When reverse-biased and therefore not conducting, the diode appears to be an open circuit with a voltage drop limited only by the source and the breakdown rating of the diode.

This chapter emphasizes several common applications of the diode and introduces the concept of *ideal* and *nonideal* diodes.

2.1 IDEAL AND NONIDEAL DIODES

An *ideal diode* is similar to a mechanical switch; zero resistance when closed (forward-biased) and infinite resistance when open (reverse-biased). Figure 2-1 compares an ideal diode curve with an actual diode curve. There are three main differences:

1. The silicon diode does not conduct until V = +0.6 V. The ideal diode conducts at 0 V (i.e., there is no offset voltage).
2. The ideal diode is a short circuit once it turns on (r_{on} = 0 Ω), while the actual diode has some ON resistance (its IV curve is not vertical).
3. The ideal diode is an open circuit for reverse-bias conditions, while an actual diode will break down if the reverse bias is sufficiently large.

Even though these differences exist, it is helpful to think of the

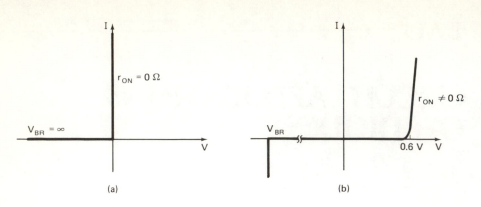

Figure 2-1 (a) Ideal diode. (b) silicon diode.

diode as ideal when analyzing many circuits. For example, consider the circuit shown in Fig. 2-2. Although this circuit may look complex at first glance, if we think of the diodes as ideal, its function is relatively simple.

When S_1 is in position 3, conventional current leaves the power source as shown. At point A it flows through L_3 *and* D_1 because it is in the forward direction for D_1 (D_1 appears as a shortcircuit). At point B the same thing occurs, with some current splitting to flow through L_2 and the remainder flowing through the forward-biased D_2. Finally, the remaining current flows through L_1 and back to the source. In effect, all three lamps are connected in parallel with the source.

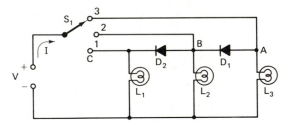

Figure 2-2 Current steering.

What happens if the switch is thrown to position 2? At point B, current from the power source cannot flow backwards through D_1 because it appears as an open circuit in this direction and only L_1 and L_2 are on.

Similarly, if S_1 is in position 1, only L_1 will light. In this application the diodes are functioning to "steer" the current to the desired lamps.

Example 2-1 ——————————————————————————

Determine which diodes are conducting for the circuit shown in Fig. 2-3.

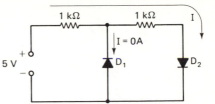

Figure 2-3 Example 2-1. Only D_2 conducts.

Solution Again, by tracing the path of conventional current flow, it can be seen that D_1 must be OFF but D_2 will be ON.

Although many circuits can be analyzed assuming ideal diodes, there are some cases in which this assumption can get you in trouble.

Example 2-2 ─────────────────────────

Determine which diodes are conducting in Fig. 2-4 and calculate the source current. Assume silicon diodes.

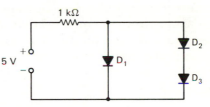

Figure 2-4 Example 2-2. Only D_1 conducts.

Solution At first glance we might assume all three diodes to be on. However, if we were to set this circuit up and take some measurements, we would find only D_1 to be on. The reason for this is that as D_1 begins to conduct, the voltage across it levels off at 0.6 V. Now in order for D_2 and D_3 to conduct (they are in series so they must conduct equally), a total of 1.2 V is required across this series branch. But D_1 is *clamping* the voltage across these series diodes to 0.6 V, which is insufficient to turn them on (in actuality D_2 and D_3 probably have about 0.3 V across each). The source current is then (5 V – 0.6 V)/1 kΩ = 4.4 mA.

Example 2-3 ─────────────────────────

Determine voltage V_X and current I for the circuit shown in Fig. 2-5. Assume a 6.2-V zener diode.

Solution All diodes are in the direction to be forward-biased except the zener diode. However, a quick check would indicate that the zener should be broken down because V_Z = 6.2V and the source voltage is 12 V (actually, we must consider the other diodes as well, meaning 12 V – 0.6 V – 0.6 V = 10.8 V is left to break the zener down). The voltage V_X is then: V_X = 6.2 V + 0.6 V = 6.8 V. The current can be found by applying Kirchoff's Law to find the voltage across the 1-kΩ resistor. On one side of the resistor is 12 V and on the other side is 6.2 V + 0.6 V + 0.6 V = 7.4 V. Then, $V_{1\,k\Omega}$ = 12 V – 7.4 V = 4.6 V and I = 4.6 V/1 kΩ = 4.6 mA.

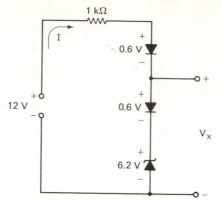

Figure 2-5 Example 2-3.

2.2 DIODE RECTIFIERS

Certainly one of the major circuit applications of the diode is *rectifica-tion*. Nearly all electronic equipment is powered by the 60-Hz 120-V rms ac main. However, most of that same equipment requires dc rather than ac and often at a substantially lower voltage. This section discusses the process of converting ac to dc.

The Half-Wave Rectifier

The conversion of ac to dc is not as difficult as it might seem. Fig-ure 2-6 illustrates a simple *half-wave rectifier* circuit used to convert the ac output of a signal generator to pulsating dc. As the generator output goes positive with respect to ground, the diode is forward-biased and conducts. Notice that V_o is 0.6 V below E_G due to the diode drop.

When E_G goes negative with respect to ground, the diode turns off and no current flows: $V_o = 0$ V. Because V_o never goes below 0 V, the output current must always be in the same direction (this is also obvious

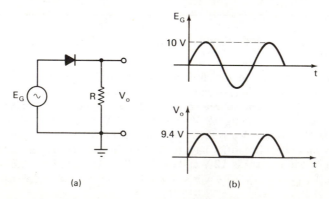

Figure 2-6 (a) Half-wave rectifier. (b) input and output waveforms.

when considering that the diode only conducts in one direction). By definition this is direct current.

It may already be obvious to you that the circuit in Fig. 2-6 does not produce a very useful dc. Half of the cycle time the output voltage is zero! Because the output voltage is not constant and appears in pulses, it is referred to as *pulsating dc*.

Figure 2-7a shows the schematic diagram of an unregulated power supply with half-wave rectifier and capacitive filter. In the waveform of Fig. 2-7b, the ac main is shown as a 60-Hz sine wave with peak value of

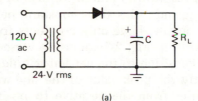

(a)

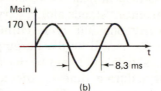

(b)

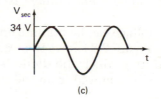

(c)

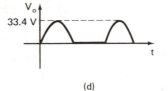

(d)

Figure 2-7 (a) Half-wave rectifier with capacitive filter. (b) the 120-V rms ac main. (c) 24-V rms secondary voltage. (d) V_O without capacitor. (e) V_O with capacitor.

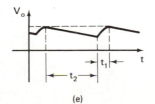

(e)

approximately 170 V (120 × $\sqrt{2}$). The transformer is used to step this voltage down to 24-V rms. The resulting waveform is shown in Fig. 2-7c. The waveform in Fig. 2-7d shows the output voltage *without* the filter capacitor. This is simply the pulsating dc we observed for the circuit in Fig. 2-6. However, if a capacitor is added across the load resistor as shown in Fig. 2-7a, the pulsating dc is smoothed out considerably, approaching a clean dc waveform. If the *time constant* of the load resistor and capacitor is sufficiently long, the output voltage will just barely decay before the next voltage pulse occurs, recharging the capacitor. This is shown in the waveform in Fig. 2-7e.

How long must this time constant be? To answer this question, first notice that the diode only conducts for a brief time, shown as t_1 in the waveform in Fig. 2-7e, and is reverse-biased for the remainder of the cycle. This is due to the capacitor holding the cathode end of the diode at nearly the peak value of the sine wave, while the anode end of the diode varies from the negative to positive peak of the secondary voltage. For example, when the transformer voltage is 30 V and the output voltage is held at near 34 V, the diode has a 4-V reverse bias. As a result, the capacitor "stands alone," trying to hold the output voltage constant for nearly one full cycle (time t_2 in Fig. 2-7e). During this time the load resistor will be discharging the capacitor with the result that the output voltage will begin to decrease.

In summary, during time t_2 the output voltage is decaying and during time t_1 it is charging. This charging and discharging of the filter capacitor causes an *ac ripple voltage* to appear, riding on the output dc level.

The ripple voltage can be minimized by selecting a sufficiently large filter capacitor. Figure 2-8 shows the output of the power supply during time t_2 (time during which the diode is reverse-biased). An amount of charge is going to be removed from the capacitor during this time, which can be found as

$$Q = I \times t_2 \qquad\qquad (2\text{-}1)$$

This equation assumes the capacitor will be discharged by a *constant* current instead of the usual *exponential* current. If the capacitor is not allowed to discharge by more than 10%, this equation will be reason-

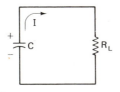

Figure 2-8 Equivalent circuit for the output of Fig. 2-7 during time t_2.

ably accurate. The charge may also be related to the size of the capacitor and the change in voltage (ripple) across it.

$$Q = C \times \Delta V \qquad (2\text{-}2)$$

where the symbol "Δ" is the Greek letter delta and represents the peak-peak ripple voltage across the capacitor. Equating Eq. 2-1 and 2-2

$$I \times t_2 = C \times \Delta V \qquad (2\text{-}3)$$

If we assume that t_1 is very short, which it is for light loads (load current not excessively large), then t_2 is approximately the same as the period of the ac waveform (1/60 s.). Equation 2-3 can then be rewritten to solve for C.

$$C = IT/\Delta V = I/(\Delta Vf) \qquad (2\text{-}4)$$

In this equation I is the dc load current, ΔV is the peak-peak ripple voltage, and f the frequency of the ripple.

Example 2-4

Determine suitable voltage and current specifications for the rectifier diode and required capacitor in Fig. 2-7a to allow 0.5 V p-p of ripple across a 500-Ω load resistor. Assume 1.5 Ω of dc resistance in the transformer secondary windings.

Solution When first energized, the capacitor appears to be a *short circuit* (it has not yet charged up), thus the full secondary voltage appears directly across the diode with only the transformer secondary resistance to limit this current. In this case

$$I = \frac{34\ \text{V} - 0.6\ \text{V}}{1.5\ \Omega} \cong 22\ \text{A}$$

Of course, this current quickly charges the capacitor and the diode need only withstand this *surge* for a few brief instants. For this reason, in addition to a maximum forward-current rating, most diodes also have a surge-current rating.

During time t_2 in Fig. 2-7e, the diode is *reverse-biased* with approximately +33.4 V on its cathode and as much as –34 V on its anode (at the negative half-cycle peak). This requires a PIV rating of 67.4 V.

The forward current through the diode is approximately

$$V_{\text{peak}}/R_{\text{L}} = 33.4\ \text{V}/500\ \Omega = 67\ \text{mA}$$

The 1N4002 has a PIV rating of 100 V, maximum forward current of 1 A, and a surge current rating of 30 A. This diode would be a good selection for this circuit.

Finally, C is found by applying Eq. 2-4.

$$C = \frac{67\ \text{mA}}{0.5\ \text{V} \times 60\ \text{Hz}} = 2233\ \mu\text{F}$$

The value of C obtained from this calculation is an approximate value and the closest standard value should be selected. Also, be sure to observe the *voltage rating* of the capacitor. Because the output voltage for this circuit is 33 V, a working voltage rating of about 50 V would provide a reasonable safety factor.

Full-Wave Rectifiers

The half-wave rectifier of the preceding section is not efficient because it forces the capacitor to hold the output voltage up for nearly a full cycle and wastes the negative half cycle completely (the diode is OFF). Figure 2-9a illustrates the schematic diagram of a *full-wave rectifier* using a center-tapped transformer.

This circuit is like two half-wave rectifiers, one operating on each half of the ac cycle. Recall that the phase of the ac waveform at the top-most winding of a transformer is always 180° out of phase (opposite) with respect to the bottom-most winding. Hence, during time t_1 (refer to Fig. 2-9b), D_1 is conducting just as it did for the half-wave case and during time t_2, D_2 is conducting. The resulting output voltage is the characteristic *double-hump* pattern shown in Fig. 2-9d.

Two significant differences result with this rectifier compared to the half-wave circuit:

1. A *center-tapped* transformer is required and the peak output voltage will be one half the peak full secondary voltage (only half of the secondary voltage is used for each half cycle).
2. The time between voltage pulses is now 1/120 s, allowing a smaller capacitor to be used to smooth the pulsating dc.

The ripple frequency of this circuit is 120 Hz because the pulses of voltage occur at a 120-Hz rate. This is compared to the half-wave circuit with a 60-Hz ripple frequency. Because the filter capacitor must hold the output voltage constant between these pulses, the higher the ripple frequency, the easier its job will be.

Example 2-5 ——————————————————————————————————

Refer to Fig. 2-9 and determine the appropriate diode specifications and filter capacitor size required to produce a dc output voltage with 0.5-V peak-peak of ac ripple. Calculate the approximate dc output voltage.

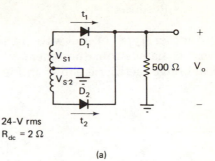

(a)

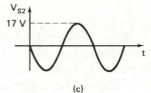

(b)

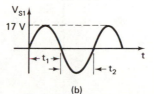

(c)

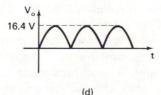

Figure 2-9 (a) Full wave center-tapped rectifier. (b) and (c) secondary voltages. (d) output voltage.

(d)

Solution In this case, the surge current is limited by one half the total secondary resistance.

$$I_s = \frac{17 \text{ V} - 0.6 \text{ V}}{1 \text{ }\Omega} = 16.4 \text{ A}$$

The approximate dc output voltage is 0.6 V below V_{pk} or 16.4 V, resulting in a forward diode current of 16.4 V/500 Ω = 33 mA. This means C is found as

$$C = \frac{33 \text{ mA}}{0.5 \text{ V} \times 120 \text{ Hz}} = 550 \text{ }\mu\text{F}$$

Finally, each diode has 16.4 V at the cathode terminal and −17 V (worst case) at the anode terminal. This is a 33.4-V PIV.

Another type of full-wave rectifier is called the *bridge*, which derives its name from the schematic representation of the circuit. A typical bridge rectifier is shown in Fig. 2-10.

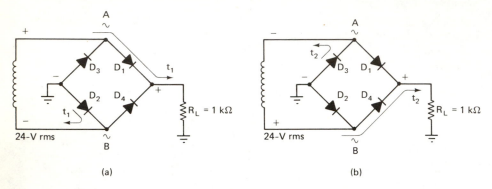

Figure 2-10 Full-wave bridge rectifier. (a) D_1 and D_2 conduct. (b) D_3 and D_4 conduct.

Notice that this circuit does not require a center-tapped transformer and therefore uses the full secondary voltage. Referring to Fig. 2-10a, during time t_1 and with the transformer polarity shown, conventional current leaves the transformer, passes through forward-biased diode D_1 (D_3 is reverse-biased), through the load resistor, and via the ground path through diode D_2 to the bottom transformer winding completing the circuit.

Figure 2-10b shows the current path during time t_2 when the bottom winding is positive with respect to the top winding. For this case, diodes D_3 and D_4 conduct while D_1 and D_2 are OFF. You should be able to see that the full secondary voltage (less two diode drops) is developed across the load each half cycle. This circuit provides full-wave (360°) conduction at the full secondary voltage.

Rather than using four discrete diodes, the bridge rectifier is commonly available in an encapsulated form as shown in Fig. 2-11.

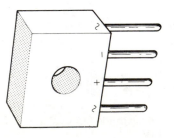

Figure 2-11 Encapsulated bridge rectifier.

Two of the four leads are marked with a sine wave symbol for connection to the transformer secondary. The other two leads are marked plus and minus. These four connections are also indicated on the schematic diagram in Fig. 2-10.

Example 2-6

Assume the transformer in Fig. 2-10 is rated at 24-V rms and sketch the expected output waveform across the load resistor and the waveforms at points A and B.

Solution   The peak secondary voltage will be $24 \times \sqrt{2} = 33.9$ V. The output voltage will be two diode drops below this or 32.7 V. During time t_1, point A will be 0.6 V above the output potential, but during time t_2 it will be at -0.6 V (D_3 is ON during time t_2, clamping point A at this level). Point B will be similar to point A but of the opposite phase. The waveforms are shown in Fig. 2-12.

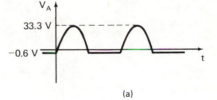

(a)

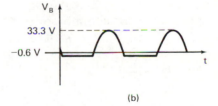

(b)

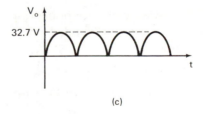

Figure 2-12   Bridge rectifier waveforms for Fig. 2-10. (a) point *A*. (b) point *B*. (c) output.

(c)

Ground Loops

As a final note, care must be taken when measuring the secondary voltage of either full-wave circuit. If the oscilloscope used for the measurements is at the same ground potential as the circuit (e.g., via the line cord three-prong plug), a potential short circuit can exist.

Figure 2-13 demonstrates this problem. Assume you want to measure the secondary voltage of the indicated transformer and so connect your oscilloscope leads to points A and B across the secondary as shown. If point B is the ground lead from the oscilloscope and the oscilloscope ground is the same as the circuit ground, then the bottom half of the transformer is actually *shorted* to the center tap and damage to the transformer and scope lead may result or a fuse may be blown.

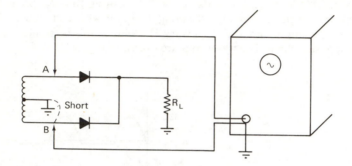

Figure 2-13 Potential short circuit when measuring $V_{\text{secondary}}$ in a full-wave rectifier.

The cure for this is to make sure the oscilloscope ground and circuit ground are *not* common or to first measure point A with respect to ground and then point B with respect to ground. As a general rule, it is good practice to have only *one* ground in a system and have all ground leads *common* to this point.

2.3   CLIPPING AND CLAMPING CIRCUITS

Another common circuit application of diodes is *wave-shaping*. Portions of waveforms may be clipped or clamped to specific dc levels by these special diode circuits.

Example 2-7 ——————————————————————————————————

Assume ideal diodes and sketch the output waveform for the circuit shown in Fig. 2-14a.

Solution   Until the input reaches +3 V, the ideal diode is reverse-biased and appears as an open circuit. As the equivalent circuit in Fig. 2-15a shows, V_o must equal 3 V. When V_i is greater than +3 V, the diode turns ON and appears as a short circuit, as shown in Fig. 2-15b. The output now follows the input until the input again drops below 3 V, at which time the output equals 3 V. The actual output waveform is shown in Fig. 2-14b.

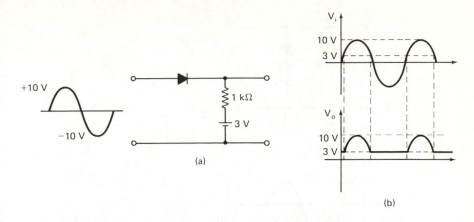

Figure 2-14 (a) Diode clipping circuit. (b) input and output waveforms assuming ideal diodes.

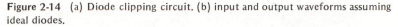

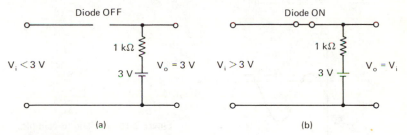

Figure 2-15 (a) For $V_i < 3$ V, the diode is OFF and an open circuit. (b) for $V_i > 3$ V, the diode is ON and a short circuit.

You should be able to see that this particular circuit clipped the input waveform below the source voltage of 3 V. Obviously, varying the source voltage would have the effect of changing the clipping level.

Clipping circuits generally consist of a diode, resistor, and voltage source. A *clamping* circuit is similar to the clipper but includes a capacitor with the previous components.

Example 2-8 —————————————————————————————

Sketch the output waveform for the clamping circuit in Fig. 2-16a.

Solution Initially the capacitor is uncharged and, until V_i exceeds 3V, the diode is OFF. As V_i increases beyond 3 V, the diode turns ON and a current flows through the capacitor and diode as shown in Fig. 2-17a. When $V_i = 10$ V, there will be 7 V across the capacitor and 3 V across the output. Note the polarity of charge on the capacitor. As V_i decreases, the capacitor would like to discharge but this would require the current to flow backwards through the diode, which it cannot do. As a result, the capacitor simply holds the 7-V charge as shown in Fig. 2-17b. In effect, the capacitor appears to be a 7-V source in series with the input and from Kirchoff's Law we can write that $V_o = V_i - 7$ V. The resulting waveform is shown in Fig. 2-16b.

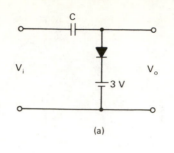

(a)

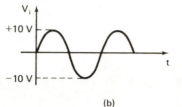

(b)

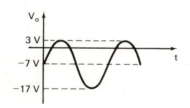

(c)

Figure 2-16 (a) Diode clamping circuit for Example 2-8. (b) sine-wave input. (c) level-shifted output. $V_o = V_i - 7$ V.

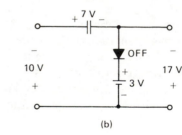

(a)   (b)

Figure 2-17 (a) Diode conducting $V_o = 10$ V $- 7$ V $= 3$ V. (b) diode OFF. $V_o =$ -10 V $- 7$ V $= -17$ V.

Note that the entire waveform has been level shifted down by 7 V. However, the shape of the waveform (and thus the peak-peak amplitude) has not been changed.

It is important that the load resistance seen by the clamping circuit be large for proper operation. If this is not the case, the capacitor will begin to discharge when the diode is OFF and proper circuit operation will not occur.

As a general result, a clipping circuit will change the shape of the input waveform, clipping off a portion above or below some dc level. A clamper, on the other hand, will not change the shape of the input waveform but will level-shift the waveform or clamp it about some new dc level.

An application of a circuit employing a clamper is shown in the following example.

Example 2-9

Determine the dc output voltage of the circuit in Fig. 2-18.

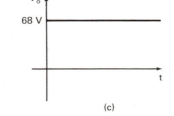

Figure 2-18 Voltage doubler for Example 2-9.

Solution In this circuit, C_1 and D_1 form a clamping circuit while D_2 acts as a rectifier and C_2 as a filter capacitor. The peak secondary voltage is approximately 34 V and when D_1 conducts, C_1 will charge to this value with the polarity as shown. The waveform across D_1 is a sine-wave level shifted *up* by 34 V as shown in Fig. 2-19b. Note that the peak value of this sine wave is now 68 V. D_2 allows current to

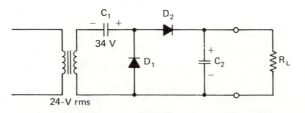

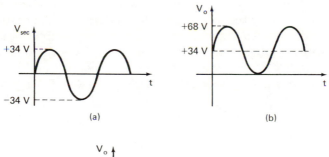

Figure 2-19 Waveforms for the circuit in Fig. 2-18. (a) secondary voltage of transformer. (b) voltage across D_1. (c) V_0 with C_2.

charge C_2 from the C_1-D_1 combination but does not allow C_2 to discharge back towards the transformer. As a result, C_2 charges to the peak value of 68 V and the resulting dc output voltage is 68 V.

This circuit functions as a *voltage doubler* because the dc output voltage is twice what would ordinarily be obtained from a normal rectifier circuit. Again, a key to proper operation is a long-time constant for the C_2-R_L combination.

2.4   ZENER DIODE CIRCUITS

The zener diode was discussed previously in Chapter 1. This diode is unique in that it is designed to be operated in its *breakdown region*. As mentioned in Chapter 1, this is not destructive provided the current through the diode is limited to a safe value.

Figure 2-20 indicates a typical zener diode circuit. The input voltage must be greater than V_Z to ensure that the diode operates in breakdown. If this is true, the output voltage will be approximately V_Z for a wide variation of load currents. In this respect the zener diode acts as a *voltage regulator*, maintaining a constant output voltage even as the load current varies.

Resistor R_s is necessary to absorb the difference in potential between V_i and V_Z. This resistor also protects the zener by limiting the maximum diode current, which occurs when R_L is disconnected. In this case $I_Z = I_T$. Notice that the minimum zener current occurs when R_L is minimum. In this case the load draws the majority of the current. If the load should attempt to draw *all* of the current, I_Z will become 0 A and the zener will no longer operate in breakdown. Now the output voltage will be determined by the voltage divider between R_L and R_S. This will result in very poor output voltage regulation. To avoid this problem, R_S

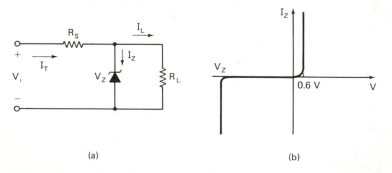

(a)   (b)

Figure 2-20   (a) Basic zener reference circuit. (b) zener diode characteristic IV curve.

should be selected to allow a minimum of approximately 10 mA of zener current under full load conditions (R_L minimum).

Example 2-10 ──────────────────────────────────────

Using the circuit of Fig. 2-20 as a guide, calculate component values and wattage ratings to produce a 10-V output. Assume V_i = 15 V and R_L = 100 Ω.

Solution Because V_o is to be 10 V, a 10-V zener diode is selected. The load current I_L is found as $I_L = V_Z/R_L$ = 10 V/100 Ω = 100 mA. To ensure that the zener operates in breakdown assume I_Z = 10 mA. Then $I_T = I_L + I_Z$ = 100 mA + 10 mA = 110 mA and $R_S = (V_i - V_Z)/I_T$ = (15 V - 10 V)/110 mA = 45 Ω. R_S will dissipate $(0.11$ A$)^2 \times 45$ Ω = 0.54 W and the zener must dissipate $(0.01$ A$)$ 10 V = 0.1 W. Selecting a 0.1-W zener could be dangerous, however. If the load should ever be disconnected, the zener would have to sink the total current (110 mA) and the power dissipation would jump to $(0.11$ A$)$ 10 V = 1.1 W! Therefore, a better choice would be a 2-W zener diode and a 1-W R_S resistor.

Example 2-11 ──────────────────────────────────────

Design a 5-V regulated power supply using a 6.3-V rms transformer with R_{dc} = 0.8 Ω, bridge rectifier, capacitive filter, and zener regulator. Assume a maximum load current of 250 mA and a maximum of 0.5-V peak-peak ripple across the filter capacitor. Determine all component values.

Solution Figure 2-21 indicates the circuit design. To determine component values, the individual currents must be found. I_L = 250 mA as required and I_Z should be at least 10 mA as discussed before. This means I_{RS} is 260 mA. This is the amount of current to be extracted from the capacitor. From Eq. 2-4

$$C = \frac{I}{(\Delta Vf)} = \frac{0.26 \text{ A}}{0.5 \text{ V} \times 120 \text{ Hz}} = 4333 \text{ } \mu\text{F}$$

The peak voltage across the capacitor should be 6.3 V $\times \sqrt{2}$ - 1.2 V = 7.7 V. A reasonable choice for the capacitor might be 5000 μF at 25 V.

R_S is found as R_S = (7.7 V - 5 V)/0.26 A = 10 Ω. $P_{RS} = (0.26$ A$)^2 \times 10$ Ω = 0.7 W. Choose 10 Ω and 1 W. Finally, a 5-V zener would be chosen. The wattage rating should be 5 V $\times$ 0.26 A (worst case) = 1.3 W. Choose 2 W.

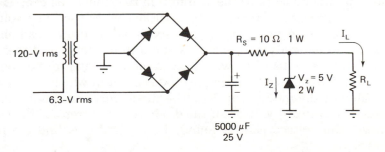

Figure 2-21 Zener-regulated power supply for Example 2-11.

With the bridge configuration, the surge current (and load current) flows through two series diodes and in this example is limited to

$$I_S = \frac{V_{pk} - 1.2 \text{ V}}{R_{dc}} = \frac{7.7 \text{ V}}{0.8 \text{ }\Omega} = 9.6 \text{ A}$$

Recalling the waveforms of Fig. 2-12, each diode in the bridge circuit must withstand a PIV equal to the peak secondary voltage. This is because the *conducting* diodes clamp the anodes of the nonconducting diodes at approximately 0 V (see Fig. 2-12). In this case the PIV is approximately 9 V.

The zener diode in Fig. 2-21 should maintain the output voltage constant at 5 V, independent of the load current. We can measure its ability to do this and express the result as a percentage using the following equation.

$$\% \text{ voltage regulation} = \frac{V_o \text{ (no load)} - V_o \text{ (full load)}}{V_o \text{ (full load)}} \times 100 \qquad (2\text{-}5)$$

For example, if the output of the circuit in Fig. 2-21 is 5.1 V with no load resistor (open circuit) but 4.8 V with a 250-mA load current, the regulation is $[(5.1 \text{ V} - 4.8 \text{ V})/4.8 \text{ V}] \times 100 = 6.25\%$. The lower this number the better.

Example 2-11 points out one of the main limitations of the simple zener regulator. The wattage rating of the zener diode must be relatively large to handle even moderate loads. Several circuits have been developed to solve this problem but one of the best solutions to date is the *three-terminal* regulator. These are available with current ratings as high as 10 A. Chapter 12 discusses three-terminal regulators in more detail.

2.5 DIODE LOAD LINES

In most circuit applications the assumption that the diode drop is 0.6 V is adequate. However, there may be occasions when the exact voltage is desired. In this case the *load line* method must be used. Load lines are also useful when studying transistors and a discussion about them now will be beneficial to us when we discuss transistors later in the book.

Figure 2-22 illustrates a simple diode circuit in which we desire the *exact* value of the diode drop V_F. The diode curve shown in Fig. 2-22 indicates the forward voltage of the diode versus current. If the current is known, the voltage can be found. But how do we find the proper current value?

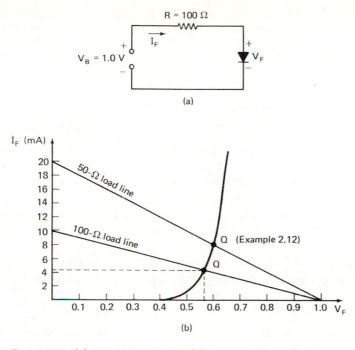

Figure 2-22   (a) Simple diode circuit. (b) characteristic diode curve and load lines.

Kirchoff's Law requires that the sum of V_R and V_F equal the battery voltage. In equation form

$$V_B = V_R + V_F \qquad (2\text{-}6)$$

and substituting $V_R = I_F R$

$$V_B = I_F R + V_F \qquad (2\text{-}7)$$

Equation 2-7 has two unknowns, I_F and V_F. If a value for I_F is chosen, the corresponding V_F value can be calculated. For example, when $I_F = 0$ A, $V_B = V_F = 1$ V. A *table* of values can be calculated as shown in Table 2-1.

All of these data points satisfy Kirchoff's Law, but which represents the diode operating point?

To answer this, recall that the correct $I_F V_F$ operating point must *also* lie on the diode curve. If all of the points indicated in Table 2-1 are

TABLE 2-1

I_F	$V_F = V_B - I_F R = 1\ \text{V} - (I_F \times 100\ \Omega)$
0	$V_F = 1$ V
5 mA	$= 0.5$ V
10 mA	$= 0$ V

located on the graph of Fig. 2-22, a straight line results. When this line intersects the diode curve, Kirchoff's Law *and* the diode curve are both satisfied and the operating point of the circuit has been found. This point is often called the *Q point.* Referring to Fig. 2-22b, the Q point is $I_F = 4.5$ mA and $V_F = 0.57$ V.

Because a straight line requires only two points for its definition, a simplified procedure for finding the load line is as follows:

1. The *X*-axis intercept occurs when $I_F = 0$ A. In this case, $V_F = V_B$.
2. The *Y*-axis intercept occurs when $V_F = 0$ V. In this case, $I_F = V_B/R$.

Example 2-12 ───

Assume the resistor in Fig. 2-22 is changed to 50 Ω. Use the load line method to calculate the exact diode drop.

Solution   The V_F-axis value is $V_F = V_B = 1.0$ V. The I_F axis value is $V_B/R = 1$ V/ 50 Ω = 20 mA. The 50-Ω load line is shown in Fig. 2-22b and the new Q point is $I_F = 8$ mA and $V_F = 0.6$ V.

You may have guessed that the main drawback to this method is that a diode curve must be available. This may require a curve tracer to generate the curve or data sheets from the manufacturer indicating the $I_F V_F$ characteristics. A second drawback is that results obtained graphically are subject to small errors due to the graphical technique.

─────────────────────── KEY TERMS ───────────────────────

Ideal Diode   Similar to a perfect switch. When forward-biased, it appears to be a short circuit and when reverse-biased, an open circuit.

Rectification   The process of converting an ac, signal to a pulsating dc voltage. A full-wave rectifier conducts on the positive *and* negative halves of the ac cycle and results in a ripple frequency of 120 Hz. A half-wave rectifier will conduct for only half the cycle and has a 60-Hz ripple frequency.

Ripple   This is the undesirable ac remaining after rectification and filtering in a power supply. It is usually measured in peak-peak volts.

Clipper   This circuit clips off a portion of the input waveform but causes no level shifting.

Clamper This circuit will shift the input waveform up or down by some dc value but will not change the waveform shape.

Zener Diode A diode meant to be operated in breakdown and often used as a voltage regulator.

Voltage Regulation The process of maintaining a constant output voltage as the load current varies.

Load Line A graphical technique for determining the exact operating point of a diode circuit.

─────────────── QUESTIONS AND PROBLEMS ───────────────

2-1 How can you tell if a diode is forward-biased? reverse-biased? operating in breakdown?

2-2 How does a silicon diode differ from a mechanical switch?

2-3 Indicate which diodes are conducting for the circuits in Fig. 2-23. Assume silicon diodes.

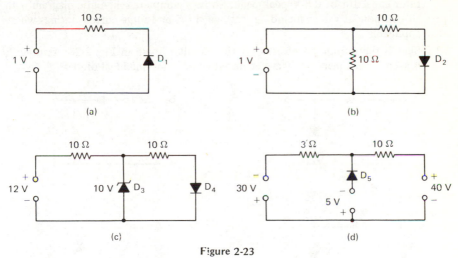

Figure 2-23

2-4 Refer to Fig. 2-3. Calculate the circuit if
(a) D_2 is an ideal diode.
(b) D_2 is a silicon diode.

2-5 Repeat Example 2-2 if D_1 is replaced with a 3.3-V zener diode connected opposite to the D_1 in Fig. 2-4.

2-6 Calculate the diode currents for the four circuits shown in Fig. 2-23. Assume ideal diodes.

2-7 Refer to Fig. 2-5. If a zener diode with a higher V_Z rating is substituted for the one shown, will the circuit current increase or decrease? Explain.

2-8 If the zener diode in Fig. 2-5 is connected opposite to that shown in the figure, recalculate I and V_X.

2-9 If the diode in Fig. 2-6 is reversed, redraw the output waveform.

2-10 Show the schematic diagram of a power supply using a half-wave rectifier, 12-V rms transformer, capacitive filter, and 1-k Ω load resistor. Determine a suitable value for the filter capacitor and PIV and surge current ratings for the diode, assuming $R_{dc} = 1.75\ \Omega$. Calculate the expected dc output voltage. Assume 0.3-V peak-peak of ripple.

2-11 You are given an 18-V rms center-tapped transformer. Determine the maximum output voltage for each of the following rectifiers assuming a capacitive filter and silicon diodes.
(a) Half-wave
(b) Full-wave center-tapped
(c) Bridge

2-12 Sketch the schematic diagram of a bridge rectifier circuit that will produce a negative output voltage.

2-13 Design a 12-V dc power supply using a 24-V rms center-tapped transformer with $R_{dc} = 1\ \Omega$, full-wave rectifier, capacitive filter, and zener regulator. Assume a 500-mA load current and maximum allowable ripple across the filter capacitor of 0.8-V peak-peak. Show a complete schematic diagram with all component values including wattage, PIV, and surge current ratings where applicable.

2-14 Sketch the output waveforms for the circuits shown in Fig. 2-24. Assume V_i is a 10-V peak-peak sine wave centered at 0 V. Assume ideal diodes.

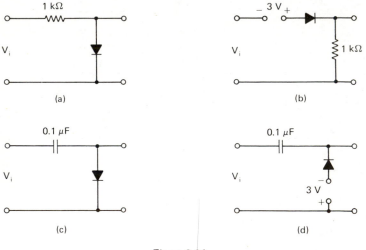

Figure 2-24

2-15 Show a circuit diagram with component values to obtain a regulated 6 V at 100 mA from a 12-V source.

2-16 *Line regulation* refers to how much the output voltage of a power supply changes due to changes in the input voltage. Assume the input voltage to your circuit in Problem 2-15 increases to 13 V and the output voltage rises to 6.1 V. Calculate the percent *line regulation* using a formula similar to Eq. 2-5.

2-17 The following data is collected on a particular power supply: V_o (open circuit) = 12.1 V, V_o (100-Ω load) = 11.89 V. Calculate the percent voltage regulation.

2-18 Calculate V_X and I for the circuit in Fig. 2-25. Assume silicon diodes.

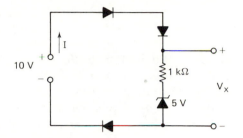

Figure 2-25

2-19 Sketch the output voltage for the circuit in Fig. 2-14 if the diode is reversed.

2-20 Repeat Problem 2-19 for the circuit in Fig. 2-16.

2-21 Verify that the Q point found graphically in Example 2-12 satisfies Kirchoff's Law for this circuit.

2-22 Refer to the diode curve and circuit in Fig. 2-22. If $R = 50$ Ω and $V_B = 0.8$ V, use the load line method to find I_F and V_F.

2-23 Repeat Problem 2-22 with the voltage source changed to 1.6 V. *Hint:* Locate two points that fit on the curve shown.

──────── LABORATORY ASSIGNMENT 2: DIODE APPLICATIONS ────────

Objectives

1. To gain experience with the diode in several typical applications.
2. To gain practical experience in identifying forward- and reverse-biased diodes.
3. To observe the difference between the ac and dc functions of the oscilloscope.

Introduction   In this laboratory assignment, you will study several typical diode applications, including clipper and clamper circuits, current steering, zener regulators, and a simple dc power supply. Diode load lines using the curve tracer will also be studied.

Components Required

4 Silicon diodes

1 0.1-μF capacitor

   10-, 47-, and 470-μF electrolytic capacitors

1 5-V zener diode

1 10-V zener diode

miscellaneous ¼-W resistors

12.6-V center-tapped ac source

Part I: *Current Steering*

STEP 1 Set up the circuit indicated in Fig. 2-26. Record the diode voltages for both polarities of the voltage source. In both cases maintain the positive voltmeter lead on the diode anode with the negative lead connected to the cathode (in this way a positive voltage indicates a *forward bias* while a *reverse bias* is indicated by a negative voltage).

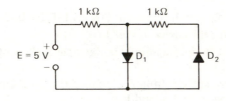

Figure 2-26 Current steering.

$$E = +5 \text{ V} \qquad V_{D_1} = \underline{\hspace{1.5cm}} \qquad V_{D_2} = \underline{\hspace{1.5cm}}$$
$$E = -5 \text{ V} \qquad V_{D_1} = \underline{\hspace{1.5cm}} \qquad V_{D_2} = \underline{\hspace{1.5cm}}$$

STEP 2 With $E = +5$ V, reverse diode D_2 and again measure V_{D_1} and V_{D_2}

$$V_{D_1} = \underline{\hspace{1.5cm}} \qquad V_{D_2} = \underline{\hspace{1.5cm}}$$

Question 1 Are both diodes forward-biased in step 2? Are they both ON? Explain.

Part II: *Clipping and Clamping Circuits*

STEP 1 Set up the circuit indicated in Fig. 2-27 and record the input and output waveforms. Do this with your oscilloscope input in the dc position.

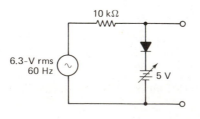

Figure 2-27 Clamping circuit.

STEP 2 Vary the dc voltage. What affect does this have on the output waveform?

STEP 3 Replace the 10-k Ω resistor with a 0.1-μF capacitor. Record the output waveform with the dc source adjusted to +5 V.

STEP 4 Vary the dc voltage and again observe the effect on the output waveform.

Question 2 In step 4, the oscilloscope *must* be in the dc position. Explain why this is so (if this is not clear, try step 3 with your oscilloscope in the ac position).

Part III: *Zener Diode*

STEP 1 Set up a circuit similar to Fig. 2-20a. Record V_o under open circuit conditions for E_i = +5 V, +10 V, +15 V, and +20 V. Use R_S = 1 k Ω and a 10-V zener diode.

V_i =	5 V	10 V	15 V	20 V
V_o =				

Question 3 Does the zener diode regulate *any* input voltage to its zener voltage? Explain.

Question 4 Accurately measure V_o with a digital voltmeter for the 15 and 20 V cases. What causes V_o to increase as V_i increases?

Part IV: *Power Supplies*

STEP 1 Set up a half-wave rectifier circuit such as that shown in Fig. 2-6a. Use a 12.6-V rms ac source and R = 33 k Ω. Accurately sketch the output waveform with the oscilloscope in the *dc position*.

STEP 2 Bridge the output of the rectifier with a 10-μF, 47-μF, and 470-μF filter capacitor. Measure the dc output voltage for each case and the peak-peak ripple voltage. (Note that the ripple voltage can be more accurately measured by using the *ac position* of the oscilloscope.)

STEP 3 Repeat steps 1 and 2 using a full-wave center-tapped circuit like that shown in Fig. 2-9a. Use R = 18 k Ω.

STEP 4 Set up a bridge-rectifier zener-regulated +5-V power supply circuit such as that shown in Fig. 2-21 using the 12.6-V rms ac source. Calculate a value for resistor R_S and the filter capacitor assuming a minimum zener diode current of 10 mA and a maximum load current of 50 mA.

STEP 5   With a 10-k Ω load resistor, measure the peak-peak ripple voltage across the capacitor and across the load resistor.

Question 5   How do you account for the difference in these two voltages? Does the zener diode regulator help reduce the ripple riding on the dc output?

STEP 6   Test the regulation of this circuit by measuring the dc output voltage for each of the following loads:

R_L =	100 Ω	330 Ω	1 k Ω	10 k Ω
V_o =				

STEP 7   Calculate the percent voltage regulation. Assume R_L = 10 k Ω corresponds to the *no load* condition and R_L = 100 Ω corresponds to the *full load* condition.

Part V: *Load Lines* (Optional)

STEP 1   Using the curve tracer, select a silicon diode and display its IV curves to a maximum current of 25 mA.

STEP 2   Estimate the *Q* point your diode will have for the circuit shown in Fig. 2-22 if the resistor is changed to 51 Ω. Do this on the curve tracer screen using a straight edge.

STEP 3   Set this circuit up and measure I_F and V_F. Compare with the data obtained from the curve tracer.

Question 6   What error sources could contribute to any discrepancies between the results found in steps 2 and 3?

THREE ────────────────────

THE BIPOLAR JUNCTION TRANSISTOR

The invention of the bipolar junction transistor by Bardeen, Shockley and Brattain (see Fig. 3-1) in 1947, while all were working for Bell Telephone Laboratories, stands as one of the landmark events in the chronology of electronic devices. This single component, more than any other, has revolutionized our modern society in terms of the reliability and miniaturization of electronics-related equipment. Certainly man walking on the moon would have been impossible without the transistor. Satellite communications, computers, and even the pocket radio would be impractical using the bulky vacuum tube. Even today's microcomputers and pocket calculators use *integrated circuits*, which are simply the logical progression from a single *discrete transistor* to the creation of a component with thousands of transistors on a single piece of silicon.

In this chapter we discuss the physics of how and why the bipolar transistor functions as it does. Its characteristic IV curve is discussed, and the load line concept developed for diodes is carried over from the last chapter to demonstrate how *amplification* can be achieved.

It is interesting to note that the inventors of the bipolar transistor were actually working on what is now called the *field-effect transistor*. When their experiments with the bipolar transistor proved successful, all work on the field-effect device was halted. It was not until 1958 that commercial field-effect transistors (FETs) became available. Most large scale integrated (LSI) circuits today use field-effect transistors.

3.1 BASIC TRANSISTOR ACTION

The bipolar junction transistor or *BJT* is similar to the *pn* diode discussed previously. Figure 3-2 illustrates the two transistor types and corresponding schematic symbols. Two junctions are now involved,

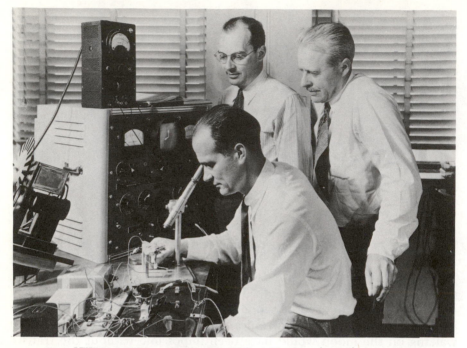

Figure 3-1 Nobel Prize winners, Drs. William Shockley (seated), John Bardeen (left), and Walter H. Brattain, at Bell Telephone Laboratories with apparatus used in their first investigations that led to the invention of the transistor. The trio received the 1956 Nobel Physics award for their invention, which was announced by Bell Laboratories in 1948. (Courtesy of Bell Laboratories)

referred to as the collector-base (J_1) and base-emitter (J_2) junctions. The transistor may be *npn* or *pnp*, with the base region always sandwiched between the collector and emitter as shown. We will see that electrons and holes both are required for operation and hence the term *bipolar* junction transistor.

Recall that when *p* and *n* material are joined, a barrier potential or *built-in voltage* forms at the junction. This causes an equilibrium condition in which the electrons remain in the *n* material and holes in the *p* material. An external bias must be applied to overcome this potential and cause current to flow across the junction. This is accomplished when a forward bias is applied (+V to *p* material, –V to *n* material).

Figure 3-3a is a pictorial of an *npn* transistor biased for amplification. A back-back diode analogy is also shown. The base terminal of the transistor is connected to a positive potential (V_{BB}), the collector to a larger positive potential (V_{CC}), and the emitter is at ground. Because the emitter is *common* to the base and collector circuits, this configuration is referred to as the *common emitter*. Its schematic representation is shown in Fig. 3-4.

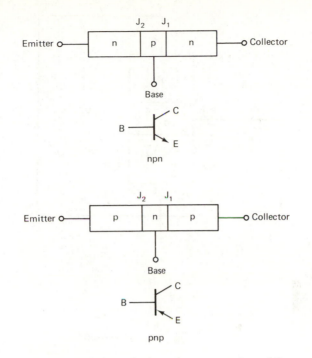

Figure 3-2 Pictorial and schematic representations of the two bipolar transistor types.

Let us make a number of observations about the biasing circuit in Fig. 3-3:

1. Due to the polarity of V_{BB}, the base-emitter junction is *forward-biased*. This means the emitter will be injecting electrons into the base (and the base holes into the emitter).

2. Because V_{CC} exceeds V_{BB}, the collector-base junction will be *reverse-biased*. This is illustrated in Fig. 3-3b, where the collector potential is shown to be 5 V but the base potential can only be 0.6 V due to its forward bias. As a result, the collector-base diode has 4.4 V of reverse bias. A *depletion region* exists in the vicinity of the collector-base junction.

3. Injected electrons from the emitter come "zipping" into the base and are attracted by the large positive potential at the collector. Accordingly, they flow through the base and across the reverse-biased collector-base junction into the collector.

4. In a typical transistor, the emitter is doped much more heavily than the base to cause a large number of electrons to be injected. The base width is also made very narrow to increase the likelihood that most injected electrons reach the collector.

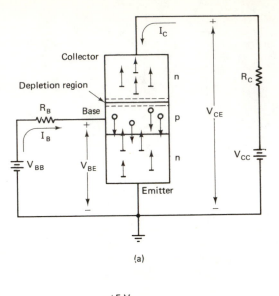

(a)

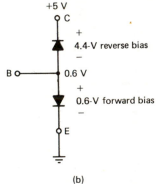

(b)

Figure 3-3 (a) *npn* transistor illustrating hole and electron flows. (b) two diode equivalent circuit.

5. Some injected electrons meet up with holes in the base and are lost ("fall into holes"). To maintain charge neutrality in the base, the V_{BB} potential must supply a *hole current* equal to the number of holes lost due to recombination.

6. Two current paths exist: the *base current* supplied by V_{BB} to replenish any holes lost due to recombination with injected electrons in the base (actually *electrons* exit at the base terminal, leaving a hole behind), and the *collector current* made up of those electrons reaching the collector attracted by V_{CC}. In a good transistor, the majority of injected electrons reach the collector without recombining with holes and the necessary base current is therefore very small. In a sense, the base current can be thought

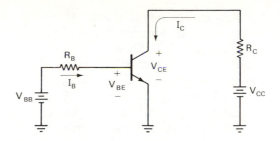

Figure 3-4   Common-emitter circuit. Schematic diagram of Fig. 3-3.

of as a "defect" current because a perfect transistor would not need any base current.

7. Probably the most significant result is that V_{BB} *controls* the amount of emitter injection by increasing or decreasing the base-emitter forward bias and therefore the amount of collector current. Particularly important is the fact that only a *small change* in V_{BB} will cause a *large change* in collector current. This is because the collector current is much larger than the base current. It is this property which allows amplification.

8. Finally, note that if V_{CC} should increase, a greater attractive force will be established and more of the injected electrons will be attracted to the collector.

The Transistor Symbols

Figure 3-2 indicated the *npn* and *pnp* transistor symbols. The arrow shown on the emitter indicates the direction of *conventional* emitter current flow when the base-emitter junction is forward-biased.

Two voltages are commonly referred to when discussing the transistor. They are the base-emitter voltage, V_{BE} (voltage at the base with respect to the emitter), and the collector-emitter voltage V_{CE} (voltage at the collector with respect to the emitter). Be careful not to confuse these voltages with V_{BB}, the base supply voltage, or V_{CC}, the collector supply voltage.

IV Curves

What characteristic IV curves might we expect for the transistor? Because it is a three-terminal device, there is more than one characteristic curve to consider. For the common emitter circuit in Fig. 3-4, we have an input side, I_B versus V_{BE}, and an output side, I_C versus V_{CE}.

The I_B versus V_{BE} curve is simply that of a diode. This is reasonable when considering the two diode analogy in Fig. 3-3b.

The I_C versus V_{CE} curve will take a bit more explaining. Assume $V_{CC} = 0$ V. No electrons are attracted to the collector, $I_C = 0A$ and $V_{CE} = 0$ V. Now as V_{CC} increases (and thus V_{CE}), the collector current will rise. However, this does not continue indefinitely. Eventually V_{CE} is large enough so that *all* available electrons are being attracted to the collector. At this point, I_C must become constant, and further increases in V_{CE} will *not* increase I_C.

We could repeat the previous experiment for a higher initial value of base current but the same result would occur. The final collector current would be higher in this case but it would still approach a constant value.

In summary we expect

1. For fixed V_{CC}: I_C will increase as I_B does (or V_{BB}).
2. For fixed V_{BB}: I_C will increase until all available electrons are collected, at which time I_C becomes constant.

Figure 3-5 is a typical set of common-emitter characteristic curves for an *npn* transistor. The I_B versus V_{BE} curve is that of a forward-biased diode, while the I_C versus V_{CE} set of curves is a "family" for various values of I_B. The characteristic for $I_B = 0.2$ mA is highlighted. For small values of V_{CE} (1–3 V), the collector current increases *linearly* as V_{CE} increases (region 1). But notice that eventually this curve folds over, leveling off near 2 mA of collector current when all injected electrons have been collected (region 2). This characteristic is repeated for all I_B values, with the only difference being higher initial base currents resulting in higher final collector currents.

Base-Width Modulation

You might notice that the I_C curves are not completely flat after they fold over but have a slight slope to them. This is due to an effect known as *base-width modulation*. As V_{CE} increases, the collector-base junction becomes increasingly more reverse-biased. Recall that a reverse-biased junction has a depletion region in which no carriers exist. As the reverse-bias increases, this depletion region increases. In the transistor this means the *effective* base width becomes smaller as the depletion region pushes into the base. There is now less likelihood of an injected electron "falling into a hole" due to the shorter base and therefore more electrons reach the collector. As a result, as V_{CE} increases, I_C also increases slightly.

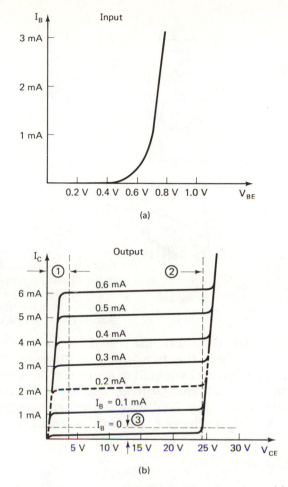

Figure 3-5 Common emitter characteristic curves. (a) I_B vs. I_{BE}. (b) I_C vs. V_{CE} and illustrating the three operating regions, 1 saturation, 2 active, 3 cutoff.

If the collector to emitter voltage is increased to a sufficiently high value, a *breakdown* voltage is eventually reached and a substantial current may flow. Under normal conditions the transistor should be operated with voltages less than this breakdown value.

3.2 PACKAGING

Modern transistors are available in a variety of packages. Figure 3-6 illustrates some of the more commonly available types. Often the *emitter* terminal can be identified as the pin closest to a metal tab on

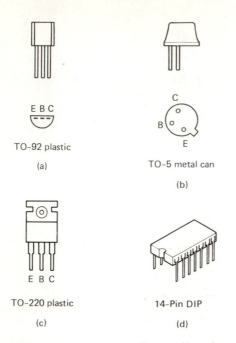

E B C

TO-92 plastic

(a)

C

B

E

TO-5 metal can

(b)

E B C

TO-220 plastic

(c)

14-Pin DIP

(d)

Figure 3-6 Transistors are available in a variety of package types. The integrated circuit package in (d) may contain 4–5 transistors with the leads connected to separate pins on the package.

the package perimeter. Generally, the size of the package is a reflection of the amount of power the transistor can dissipate. The TO–220 package in Fig. 3-6c has its own built in *heat sink* and is intended to be bolted to a larger metal frame to help conduct heat. The 14-pin DIP (dual-in-line package) in Fig. 3-6d may contain an *array* of transistors and is commonly used to achieve a high packing density.

3.3 CURRENT RELATIONSHIPS

The various currents in *npn* and *pnp* transistors are shown in Fig. 3-7. Notice that the *pnp* has the opposite carrier types when compared to the *npn*. For the *pnp* case, the emitter injects holes into the base, which are collected at the collector. The base current must now be made up of electrons to make up for those electrons in the base that are lost due to recombination with the injected holes.

For both types of transistor, the injected carrier travels *vertically* from emitter through the base and into the collector. The base current enters the base terminal and makes up for the carriers lost due to recombination. This maintains charge neutrality for the base.

Conventional current arrows are drawn for I_B, I_C, and I_E in Fig. 3-7. Assuming the transistor to be a node and applying Kirchoff's Current Law,

$$I_E = I_C + I_B \qquad\qquad (3\text{-}1)$$

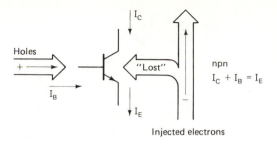

Injected electrons

(a)

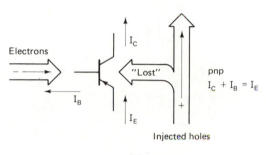

Injected holes

(b)

Figure 3-7 Direction of current flows. (a) the *npn* transistor. (b) the *pnp* transistor.

Note that this result is true for the *npn* or *pnp* transistor. Often I_B is very small, in which case I_E and I_C are approximately equal.

Figure 3-8 indicates schematically the proper biasing for amplification for *npn* and *pnp* amplifiers. Note that in both cases the base-emitter junction is forward-biased while the collector-base junction is reverse-biased. The current arrows indicate the *actual* direction of conventional current flow (opposite to electron flow).

The *pnp* collector curves are shown in Fig. 3-9. Note that these curves appear identical to the corresponding *npn* curves except that they are upside down. This is because by definition all transistor currents are defined as *entering* the transistor. For the *pnp* this means I_B and I_C must be negative (current leaves the transistor). Often the *pnp* curves will be displayed right-side up but with I_C and V_{CE} labelled negative.

α and β

Two parameters commonly used to characterize the transistor are α (alpha) and β (beta). α is defined as the ratio of I_C to I_E.

$$\alpha = \frac{I_C}{I_E} \qquad\qquad (3\text{-}2)$$

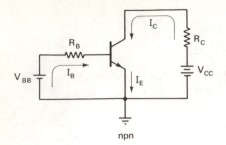

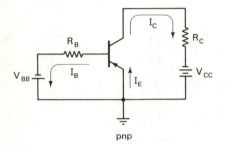

Figure 3-8 Conventional current flow in *npn* and *pnp* common-emitter amplifiers.

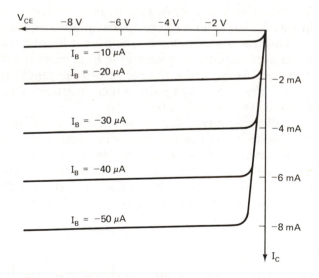

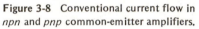

Figure 3-9 Typical common-emitter characteristic curves for a *pnp* transistor.

64

Typical values for α are 0.99 – 0.999, indicating that 99% of the emitter current gets to the collector. α is typically used when discussing common base circuits.

The common emitter circuit is probably the most prominently used today and β is most applicable for this circuit. It is defined as the ratio of I_C to I_B.

$$\beta = \frac{I_C}{I_B} \qquad (3\text{-}3)$$

Another symbol for β that is commonly used is H_{FE}. Typical values for β range from 50 to 300. This indicates how much larger I_C is compared to I_B. A perfect transistor would have $I_C = I_E$ and $I_B = 0$. For this cae, $\alpha = 1.0$ and $\beta = \infty$.

Figure 3-10 indicates the three possible amplifier configurations and applicable current relationships.

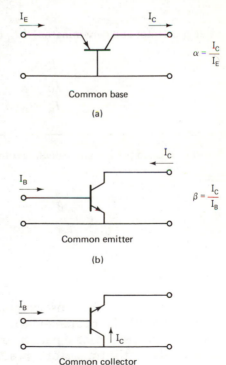

$$\alpha = \frac{I_C}{I_E}$$

Common base

(a)

$$\beta = \frac{I_C}{I_B}$$

Common emitter

(b)

Common collector

(c)

Figure 3-10 The three transistor configurations.

Example 3-1

A certain transistor has $I_B = 50\ \mu A$, and $I_C = 5$ mA. Find α and β.

Solution We may find β directly as $\beta = I_C/I_B = 5$ mA/50 $\mu A = 100$. To find α, first find I_E as $I_E = I_C + I_B = 5$ mA + 50 $\mu A = 5.05$ mA. Then, $\alpha = I_C/I_E = 5$ mA/5.05 mA = 0.99.

Example 3-2

A certain transistor has $I_C = 1$ mA and $\beta = 200$. Find I_B, I_E, and α.

Solution From Eq. 3-3, $I_B = I_C/\beta = 1$ mA/200 = 5 μA. $I_E = I_B + I_C = 5\ \mu A + 1$ mA = 1.005 mA, and $\alpha = I_C/I_E = 1$ mA/1.005 mA = 0.995.

Example 3-3

Find an equation for α in terms of β.

Solution $\alpha = I_C/I_E$ and $I_E = I_C + I_B$. Therefore

$$\alpha = \frac{I_C}{I_C + I_B}$$

Now divide numerator and denominator by I_B to obtain

$$\alpha = \frac{I_C/I_B}{\dfrac{I_C}{I_B} + \dfrac{I_B}{I_B}} = \frac{\beta}{\beta + 1}$$

Example 3-4

Use the result of Example 3-3 and calculate α from β directly in Example 3-2.

Solution

$$\alpha = \frac{\beta}{\beta + 1} = \frac{200}{201} = 0.995$$

3.4 TYPICAL DATA SHEETS

Although β (or H_{FE}) is the most common figure of merit for a transistor, there are several other parameters that are of importance and should be considered when selecting a particular transistor for an application.

Data sheets for the transistor types 2N2218A, 2N2219A, 2N2221A, and 2N2222A are included in Fig. 3-11. Table 3-1 is an explanation of some of the common specifications.

TABLE 3-1   Common Transistor Specifications

Symbol	Explanation	Test Circuit
(a) BV_{CEO}	Breakdown voltage from collector to emitter with base lead open.	
(b) BV_{CBO}	Same as (a) accept the emitter is open. This tests the collector-base diode.	
(c) BV_{EBO}	Same as (a) and (b) with the collector open. This tests the emitter-base diode.	
(d) I_{CEX}	This is the reverse leakage current when collector-base and base-emitter are reverse-biased.	
(e) I_{CBO}	This is the reverse leakage current of the base-collector diode.	Same as (b)
(f) I_{EBO}	This is the reverse leakage current of the base-emitter diode.	Same as (c)
(g) β (or H_{FE})	DC gain. I_C/I_B. Usually indicated at some specific I_C and V_{CE} value.	Same as Fig. 3-8
(h) $V_{CE(sat)}$	Collector to emitter voltage when the transistor is saturated. The specific I_C and I_B are usually indicated.	Same as Fig. 3-8
(i) f_T	Frequency at which the ac $\beta = 1$ (see Section 3.8, AC β)	

(a) I_{CEO} Open V_{CEO} + −

(b) I_{CBO} V_{CBO} + − Open

(c) Open V_{EBO} + − I_{EBO}

(d) I_{CEX}

2N2218A · 2N2219A · 2N2221A · 2N2222A
NPN HIGH SPEED SWITCHES

DIFFUSED SILICON PLANAR* EPITAXIAL TRANSISTORS

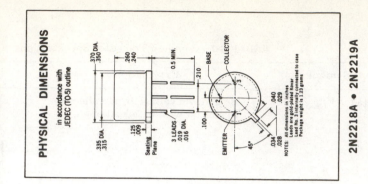

PHYSICAL DIMENSIONS

in accordance with
JEDEC (TO-5) outline

NOTES: All dimensions in inches
Leads are gold-plated Kovar
Lead No. 3 internally connected to case
Package weight is 1.20 grams

2N2218A · 2N2219A

GENERAL DESCRIPTION - These Fairchild devices are NPN silicon PLANAR epitaxial transistors designed for high-speed switching at collector currents up to 500 mA. They feature useful beta over a wide range of collector current, low leakage currents, and low saturation voltages.

ABSOLUTE MAXIMUM RATINGS [Note 1]

Maximum Temperatures

Storage Temperature	-65°C to +200°C
Operating Junction Temperature	+175°C Maximum

Maximum Power Dissipation

	2N2218A 2N2219A	2N2221A 2N2222A
Total Dissipation at 25°C Case Temperature (Notes 2 & 3)	3.0 Watts	1.8 Watt
at 25°C Ambient Temperature (Notes 2 & 3)	0.8 Watt	0.5 Watt

Maximum Voltages and Current

V_{CBO}	Collector to Base Voltage	75 Volts
V_{CEO}	Collector to Emitter Voltage (Note 4)	40 Volts
V_{EBO}	Emitter to Base Voltage	6.0 Volts
I_C	Collector Current	800 mA

Figure 3-11   (Courtesy Fairchild Camera and Instrument Corporation)

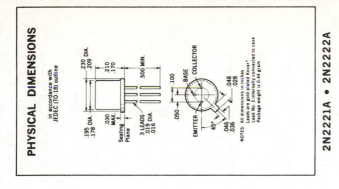

.230 DIA.
.209

.210
.170

.500 MIN.

.195 DIA.
.178

.030 MAX.

Seating Plane

3 LEADS
.019 DIA.
.016

.100

.050

BASE

COLLECTOR

EMITTER

45°

.048
.028

.046
.036

NOTES: All dimensions in inches
Leads are gold plated Kovar*
Lead No. 3 internally connected to case
Package weight is 0.44 gram

2N2221A • 2N2222A

FAIRCHILD
SEMICONDUCTOR
A DIVISION OF FAIRCHILD CAMERA AND INSTRUMENT CORPORATION

ELECTRICAL CHARACTERISTICS (25°C Free Air Temperature unless otherwise noted)

Symbol	Characteristic	2N2218A 2N2221A Min.	Max.	2N2219A 2N2222A Min.	Max.	Units	Test Conditions
h_{FE}	DC Current Gain	20		35			$I_C = 100\ \mu A$ $V_{CE} = 10\ V$
h_{FE}	DC Current Gain	25		50			$I_C = 1.0\ mA$ $V_{CE} = 10\ V$
h_{FE}	DC Pulse Current Gain (Note 5)	35		75			$I_C = 10\ mA$ $V_{CE} = 10\ V$
h_{FE}	DC Pulse Current Gain (Note 5)	40	120	100	300		$I_C = 150\ mA$ $V_{CE} = 10\ V$
h_{FE}	DC Pulse Current Gain (Note 5)	25		40			$I_C = 500\ mA$ $V_{CE} = 10\ V$
$h_{FE}(-55°C)$	DC Pulse Current Gain (Note 5)	15		35			$I_C = 10\ mA$ $V_{CE} = 10\ V$
h_{FE}	DC Pulse Current Gain (Note 5)	20		50			$I_C = 150\ mA$ $V_{CE} = 1.0\ V$
$V_{CE}(sat)$	Collector Saturation Voltage (Pulsed, Note 5)		0.3		0.3	Volts	$I_C = 150\ mA$ $I_B = 15\ mA$
$V_{CE}(sat)$	Collector Saturation Voltage (Pulsed, Note 5)		1.0		1.0	Volts	$I_C = 500\ mA$ $I_B = 50\ mA$
$V_{BE}(sat)$	Base Saturation Voltage (Pulsed, Note 5)	0.6	1.2	0.6	1.2	Volts	$I_C = 150\ mA$ $I_B = 15\ mA$
$V_{BE}(sat)$	Base Saturation Voltage (Pulsed, Note 5)		2.0		2.0	Volts	$I_C = 500\ mA$ $I_B = 50\ mA$
h_{fe}	High Frequency Current Gain (f = 100 MHz)	2.5		3.0			$I_C = 20\ mA$ $V_{CE} = 20\ V$
f_T	Gain-Bandwidth Product (f = 100 MHz)	250		300		MHz	$I_C = 20\ mA$ $V_{CE} = 20\ V$

*Planar is a patented Fairchild process.

313 FAIRCHILD DRIVE, MOUNTAIN VIEW, CALIFORNIA, (415) 962-5011, TWX: 910-379-6435

FAIRCHILD TRANSISTORS 2N2218A • 2N2219A • 2N2221A • 2N2222A

ELECTRICAL CHARACTERISTICS (25°C Free Air Temperature unless otherwise noted)

SYMBOL	CHARACTERISTIC	2N2218A 2N2221A		2N2219A 2N2222A		UNITS	TEST CONDITIONS
		MIN.	MAX.	MIN.	MAX.		
I_{CEX}	Collector Reverse Current		10		10	nA	$V_{EB} = 3.0$ V, $V_{CE} = 60$ V
I_{CBO}	Collector Reverse Current		10		10	nA	$I_E = 0$, $V_{CB} = 60$ V
I_{CBO} (+150°C)	Collector Reverse Current		10		10	µA	$I_E = 0$, $V_{CB} = 60$ V
I_{EBO}	Base Current		10		10	nA	$I_C = 0$, $V_{EB} = 3.0$ V
C_{obo}	Common Base, Open Circuit Output Capacitance (f = 100 kHz)		8.0		8.0	pF	$I_E = 0$, $V_{CB} = 10$ V
C_{ibo}	Common Base, Open Circuit Input Capacitance (f = 100 kHz)		25		25	pF	$I_C = 0$, $V_{EB} = 0.5$ V
$Re(h_{ie})$	Real Part of Common-Emitter High Frequency Input Impedance (f = 300 MHz)		60		60	Ohms	$I_C = 20$ mA, $V_{CE} = 20$ V
BV_{CBO}	Collector to Base Breakdown Voltage	75		75		Volts	$I_C = 10$ µA, $I_E = 0$
BV_{CEO}	Collector to Emitter Break-down Voltage (Notes 4 & 5)	40		40		Volts	$I_C = 10$ mA, $I_B = 0$
BV_{EBO}	Emitter to Base Breakdown Voltage	6.0		6.0		Volts	$I_C = 0$, $I_E = 10$ µA
I_{BL}	Base Current		20		20	nA	$V_{EB} = 3.0$ V, $V_{CE} = 60$ V
t_d	Turn-on Delay Time		10		10	ns	$I_{CS} = 150$ mA, $V_{CC} = 30$ V, $I_{B1} = 15$ mA, V_{BE}(off) = 0.5 V
t_r	Rise Time		25		25	ns	$I_{CS} = 150$ mA, $V_{CC} = 30$ V, $I_{B1} = 15$ mA, V_{BE}(off) = 0.5 V
t_s	Storage Time		225		225	ns	$I_{CS} = 150$ mA, $V_{CC} = 30$ V, $I_{B1} = 15$ mA, $I_{B2} = 15$ mA

SYMBOL	CHARACTERISTIC	2N2218A 2N2221A		2N2219A 2N2222A		UNITS	TEST CONDITIONS
		MIN.	MAX.	MIN.	MAX.		
t_f	Fall Time		60		60	ns	I_{CS} = 150 mA, V_{CC} = 30 V; I_{B1} = 15 mA, I_{B2} = 15 mA
τ_A	Active Region Time Constant		2.5		2.5	ns	I_C = 150 mA, V_{CE} = 30 V
$r_b'C_c$	Collector Base Time Constant (f = 31.8 MHz)		150		150	ps	I_C = 20 mA, V_{CE} = 20 V
NF	Noise Figure (f = 1.0 kHz)		4.0		4.0		I_C = 100 μA, V_{CE} = 10 V; R_g = 1.0 kΩ, BW = 1.0 Hz

SMALL SIGNAL CHARACTERISTICS (f = 1 kHz)

SYMBOL	CHARACTERISTIC	2N2218A 2N2221A		2N2219A 2N2222A		UNITS	TEST CONDITIONS
		MIN.	MAX.	MIN.	MAX.		
h_{ie}	Input Resistance	1.0	3.5	2.0	8.0	kΩ	I_C = 1.0 mA, V_{CB} = 10 V
		0.2	1.0	0.25	1.25	kΩ	I_C = 10 mA, V_{CB} = 10 V
h_{oe}	Output Conductance	3.0	15	5.0	35	μmhos	I_C = 1.0 mA, V_{CB} = 10 V
		10	100	25	200	μmhos	I_C = 10 mA, V_{CB} = 10 V
h_{re}	Voltage Feedback Ratio		500		800	$\times 10^{-6}$	I_C = 1.0 mA, V_{CB} = 10 V
			250		400	$\times 10^{-6}$	I_C = 10 mA, V_{CB} = 10 V
h_{fe}	Forward Current Transfer Ratio	30	150	50	300		I_C = 1.0 mA, V_{CB} = 10 V
		50	300	75	375		I_C = 10 mA, V_{CB} = 10 V

NOTES:

(1) These ratings are limiting values above which the serviceability of any individual semiconductor device may be impaired.

(2) These are steady state limits. The factory should be consulted on applications involving pulsed or low duty cycle operations.

(3) These ratings give a maximum junction temperature of 175°C and junction-to-case thermal resistance of 50°C/Watt (derating factor of 20 mW/°C); junction-to-ambient thermal resistance of 188°C/Watt (derating factor of 5.33 mW/°C) for the 2N2218A and 2N2219A. For the 2N2221A and 2N2222A, junction-to-case thermal resistance of 83.5°C/Watt (derating factor of 12 mW/°C; junction-to-ambient thermal resistance of 300°C/Watt (derating factor of 3.33 mW/°C).

(4) This rating refers to a high-current point where collector-to-emitter voltage is lowest.

(5) Pulse Conditions: length = 300 μs ; duty cycle = 1%.

Example 3-5

A certain one-stage transistor amplifier has the following requirements:

1. DC gain $\geqslant 30$
2. $I_C(\text{max}) \geqslant 500$ mA
3. $f_T \geqslant 275$ MHz
4. Power supply = 24 V

Select an appropriate transistor or transistors from the data sheets supplied.

Solution Referring to the data sheets, only the 2N2219A and 2N222A have $H_{FE} > 30$ for all test conditions. All transistor types meet the $I_C(\text{max})$ specification (data sheet indicates 800 mA absolute maximum). Only the 2N2219A and 2N2222A meet the f_T requirement. Finally, all transistors meet the power supply requirement of 24 V. Note that although V_{EBO} is 6 V minimum, this junction is normally forward-biased and therefore its reverse breakdown is not of concern. Normally V_{CEO} will be the minimum breakdown voltage to consider and should always be larger than the supply voltage. Either the 2N2219A or 2N2222A could be selected for this application.

Example 3-6

A student wishes to test the collector-base breakdown voltage of a 2N2222A transistor on a curve tracer. How should this be done?

Solution Usually the curve tracer will have three output connections labelled base (B), emitter (E), and collector (C). Connect the transistor to these terminals as shown in Fig. 3-12a. As the sweep control of the curve tracer is varied, the voltage between the collector and emitter terminals will vary accordingly. Notice that in this test the base terminal is not used. The vertical or current sensitivity should be set to 1 μA or 10 μA per division, while the horizontal should be set to 10 V per division. The data sheet specification for BV_{CBO} is 75 V minimum. Figure 3-12b indicates a typical curve tracer display for this test. BV_{CBO} is approximately 85 V.

3.5 REGIONS OF OPERATION

Depending on the specific voltages and currents (also called the bias point), a transistor is said to operate in one of three regions: *cutoff*, *saturation*, or the *active region.*

Cutoff

A transistor is said to be cutoff when I_B is reduced to 0 A. The collector current is then very small (microamperes of leakage current). Figure 3-13a indicates a transistor in cutoff. Notice that there is no base battery and thus $I_B = 0$ A. With no forward bias of the base-emitter

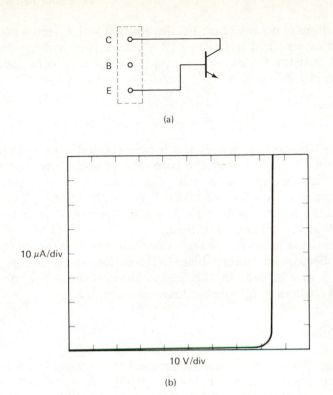

(a)

10 µA/div

10 V/div

(b)

Figure 3-12 (a) Curve tracer connections to measure BV_{CBO}.
(b) curve tracer display indicating an 85-V-breakdown.

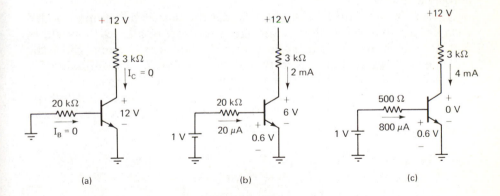

(a)

(b)

(c)

Figure 3-13 (a) Cutoff. (b) Active region; $\beta = 2\,\text{mA}/20\,\mu\text{A} = 100$.
(c) saturation region; $\beta = 4\,\text{mA}/800\,\mu\text{A} = 5$.

junction, there is no injection by the emitter and I_C must also be 0 A. The collector terminal indicates +12 V (there is no voltage drop across the 3-k Ω resistor because I_C = 0 A). Cutoff can also be identified as region 3 in Fig. 3-5.

Active

If the base-emitter junction is forward-biased, a base current flows as shown in Fig. 3-13b and the base-emitter voltage is approximately 0.6 V. In this example we assume silicon transistors and calculate the base current as (1 V – 0.6 V)/20 kΩ = 20 μA. If β is 100, then I_C = $\beta \times I_B$ = 100 $\times$ 20 μA = 2 mA. This 2 mA causes a 6-V drop across R_C, leaving 6 V at the collector terminal.

A transistor in the *active region* has the base-emitter forward-biased and the collector-base reverse-biased. The active region can be identified as region 2 in Fig. 3-5. In this region the transistor operates linearly with small changes in I_B causing larger changes in I_C.

Saturation

In Fig. 3-13c, the base resistor has been reduced to only 500 Ω. The base current is now (1 V – 0.6 V)/500 Ω = 800 μA. If I_C was $\beta \times I_B$, then I_C would be 100 $\times$ 0.8 mA = 80 mA! This would cause a 80 mA $\times$ 3 kΩ = 240 V drop across R_C! This is obviously not logical. You should be able to see that as I_B increases, so does I_C and so does the drop across R_C ($I_C \times R_C$). However, the drop across R_C cannot exceed the power supply voltage itself, in this case 12 V. When this occurs, I_C must equal 4 mA (12 V/3 k Ω = 4 mA) and the collector voltage will now be 0 V.

What is happening in this case is that the base is becoming flooded with carriers but the collector voltage is too low to attract them all. In fact, it is inevitable that as I_B increases, I_C will increase and cause the collector voltage to fall due to the IR drop across R_C. Eventually, V_C will be 0 V and any further increases in I_B will prove fruitless.

For this reason, β (or H_{FE}) is not really defined for saturation. Notice that β is only 5 (4 mA/800 μA) in Fig. 3-13c. Saturation can usually be spotted by any of the following:

1. Very low value for I_C/I_B
2. Very low collector voltage (< 1 V)
3. Forward-biased collector-base junction

Saturation can be identified as region 1 in Fig. 3-5. Note that in this region all of the I_B curves lie on top of each other. Then, as V_{CE} increases, they break out into their own individual curve in the active region.

3.6   THE FIXED-BIAS COMMON-EMITTER CIRCUIT

All of the circuits we have discussed so far have required two power supplies: V_{CC} and V_{BB}. Figure 3-14 illustrates a circuit in which only *one* power supply is used. Because the base is biased by the fixed resistor R_B to V_{CC}, this circuit is called the *fixed-bias* common-emitter amplifier. The base-emitter junction is forward-biased and the voltage at the base is approximately 0.6 V for a silicon transistor. The base current can then be found from

$$I_B = (V_{CC} - 0.6 \text{ V})/R_B \tag{3-4}$$

The collector current can be found using the β relationship

$$I_C = \beta \times I_B \tag{3-5}$$

and finally V_{CE} is found from

$$V_{CE} = V_{CC} - I_C R_C \tag{3-6}$$

Example 3-7 ───

Calculate I_B, I_C, and V_{CE} for the circuit in Fig. 3-14 if R_B = 285 k Ω, R_C = 1 k Ω, β = 100, and V_{CC} = 12 V.

Solution   From Eq. 3-4, I_B = (12 V – 0.6 V)/285 k Ω = 40 μA. Then I_C = $\beta \times I_B$ = 100 × 40 μA = 4 mA. And V_{CE} is found using Eq. 3-6; V_{CE} = 12 – (4 mA × 1 k Ω) = 8 V.

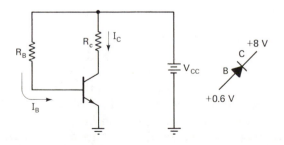

Figure 3-14   Fixed-bias common-emitter amplifier and equivalent base-collector diode, illustrating the reverse bias of this junction.

Example 3-8 ───

Design a fixed-bias common-emitter circuit using a *pnp* transistor. Choose a bias point such that V_{CE} = -6 V and V_{CC} = -12 V. Assume I_C = 2 mA and β = 150.

Solution The circuit is shown in Fig. 3-15. Notice that this circuit is identical to Fig. 3-14 but that the power supply polarity is reversed.

Because V_{CE} = -6 V, there must also be 6 V dropped across R_C. R_C can be found as $R_C = V_{RC}/I_C$ = 6 V/2 mA = 3 k Ω. R_B can be found if I_B is known. $I_B = I_C/\beta$ = 2 mA/150 = 13.3 μA. Then R_B = (12 V – 0.6 V)/13.3 μA = 857 k Ω.

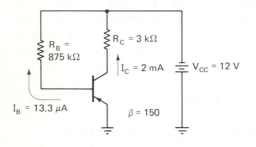

Figure 3-15 Biasing for the *pnp* amplifier in Example 3-8.

Notice in both of these examples that the base resistor is quite large. This is due to the small base current required and the fact that the relatively large V_{CC} must be dropped down to the small base-emitter diode drop (0.6 V).

The bias point of both of these last two circuits is in the *active region*. We can verify this by noting that in both cases the base-emitter junction is forward-biased while the base-collector junction is reverse-biased. For Example 3-7, V_B = 0.6 V and $V_C = V_{CE}$ = 8 V. This means the collector-base diode has 8 V – 0.6 V = 7.4 V in a reverse direction across it (collector more positive than the base). This is shown on the equivalent base-collector diode in Fig. 3-14.

3.7 THE DC LOAD LINE

One of the main applications of the transistor is as an amplifier. Up to now we have seen how the bias currents and voltages may be calculated and have defined the various regions of transistor operation. The *dc load line* will give us another viewpoint on the transistor's operation. In later chapters, we develop equations to predict how much gain an amplifier has. However, an equation can be difficult to visualize even though it may give accurate results. The load line method of analysis is not precise but does allow a better understanding of why the transistor can amplify and where the optimum bias point for a particular circuit should be chosen.

Finding the Q Point

The circuit shown in Fig. 3-16 has four unknowns: I_B, V_{BE}, I_C, and V_{CE}. We desire to find these unknowns by using a load line drawn on the appropriate IV curves. Because the base-emitter junction of the transistor is a forward-biased diode, the input load line will be drawn using a technique similar to that used in the previous chapter on diodes.

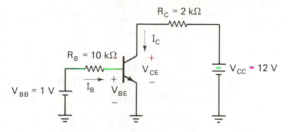

Figure 3-16 Common-emitter circuit illustrating the four unknowns: I_B, I_C, V_{BE}, and V_{CE}.

The equation for the *base* circuit is

$$I_B R_B + V_{BE} = V_{BB} \qquad (3\text{-}7)$$

A table of values for I_B and V_{BE} is shown in Fig. 3-17 along with the resulting load line. This establishes I_B and V_{BE} for the circuit in Fig. 3-16.

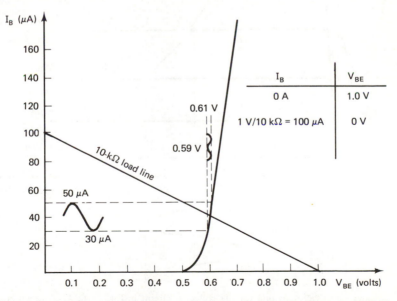

Figure 3-17 DC load line for the base circuit in Fig. 3-16. Also shown is a 0.02-V peak-peak sine wave causing a 20-μA peak-peak variation in base current.

The *collector* circuit of Fig. 3-16 can be described by the equation

$$I_{\mathrm{C}} R_{\mathrm{C}} + V_{\mathrm{CE}} = V_{\mathrm{CC}} \qquad (3\text{-}8)$$

The Q point of this circuit must satisfy this equation and lie on the characteristic curves shown in Fig. 3-18. Again, a table of values can be made and two points for the load line located. In this case, when $I_{\mathrm{C}} = 0$ A, $V_{\mathrm{CE}} = V_{\mathrm{CC}}$, and when $V_{\mathrm{CE}} = 0$ V, $I_{\mathrm{C}} = V_{\mathrm{CC}}/R_{\mathrm{C}}$.

However, you might notice that any load line drawn on the characteristic curves in Fig. 3-18 will intersect these curves in many different places (unlike the diode curve in Fig. 3-17). How do we select the proper Q point? As you may have guessed, we must choose the intersection with the proper value of I_{B} determined from the *first* load line in Fig. 3-17. Referring to this figure, we find $I_{\mathrm{B}} = 40$ μA. Now

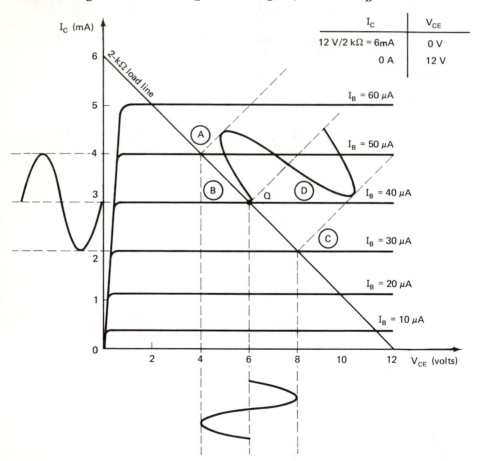

Figure 3-18 DC load line for the collector circuit in Fig. 3-16. Also shown is a 20-μA peak-peak base current and resulting collector current and voltage variations.

referring to Fig. 3-18, the output Q point is found where its load line intersects the $I_B = 40\,\mu A$ curve. This results in $I_C = 3\,mA$ and $V_{CE} = 6\,V$.

In summary, to find the Q point of a transistor amplifier:

1. Sketch the input load line on the base-emitter diode curve and obtain I_B and V_{BE}.
2. Sketch the output load line on the collector set of curves and obtain I_C and V_{CE} at the I_B value obtained in (1).

One value of the load line is that we can see by inspection if the bias point is safely within the active region and not threatening to go into saturation or cutoff.

Example 3-9 ————————————————————————————————————

Refer to Fig. 3-18 and indicate in which direction the Q point will move (A,B,C, or D) as each of the following is varied:

1. R_B increases
2. V_{CC} increases
3. R_C decreases

Solution   (1) If R_B increases, the base current will decrease. The Q point of the collector circuit will have to move down the load line towards point C. (2) If V_{CC} increases, the load line will move to the right but remain parallel with the existing line (this is because R_C is not changing). In this case the Q point moves towards point D. (3) Finally, if R_C decreases, the load line will pivot about the point $V_{CE} = 12\,V$ and $I_C = 0\,A$, rotating upwards (becoming more vertical). This causes the Q point to move towards point D.

Example 3-10 ———————————————————————————————————

Determine the value of β at the Q point in Fig. 3-18.

Solution   At the Q point, $I_C = 3\,mA$ and $I_B = 40\,\mu A$. Thus, $\beta = I_C/I_B = 3\,mA/0.04\,mA = 75$.

Identifying the Operating Regions

If the base current is increased, we learned that I_C will increase and the transistor will eventually saturate. In Fig. 3-18, the Q point moves along the load line in direction A as I_B increases. As this occurs, V_{CE} is decreasing and I_C increasing. Eventually the Q point cannot go any further to the left, V_{CE} is quite small, and the transistor is *saturated*.

Decreasing I_B has the opposite effect. The Q point moves *down* the load line approaching the point $I_C = 0\,A$ and $V_{CE} = V_{CC}$. In this case the transistor is in *cutoff*. Q points in between these two extremes result in bias points in the active region. However, we will see that some bias points are better than others.

3.8 AMPLIFICATION

You may be growing inpatient to see exactly how the transistor can amplify but we are about to discuss it. Figure 3-19 is the circuit from Fig. 3-16 but with an ac input applied to the base of the transistor through a *coupling* capacitor. The ac output will be taken at the collector terminal through another coupling capacitor.

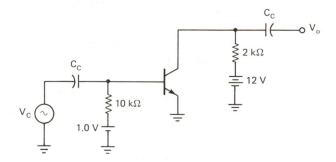

Figure 3-19 Common-emitter amplifier. The ac signal is applied through a coupling capacitor.

Assume that a 20-mV peak-peak sine wave is applied to the base of the transistor. Becaue the base Q point was previously found to be 0.6 V, this ac input will cause the base-emitter voltage to swing 0.01 V above and below this value (20 mV peak-peak). As Fig. 3-17 illustrates, as the base-emitter voltage varies between 0.59 V and 0.61 V, the base current will vary from 30 to 50 μA about the 40-μA operating point.

Now referring to Fig. 3-18, as I_B varies about the Q point, the collector current and voltage will also vary. For the specific case shown, V_{CE} will vary from 4 V to 8 V and I_C from 2 mA to 4 mA. These points are found by following I_B up and down the load line between 30 and 50 μA. Note that, in all cases, these variations are set up due to the input waveform at the base terminal, in this case a sine wave.

Various waveforms for the amplifier are illustrated in Fig. 3-20. Lower case letters are used to designate *instantaneous* values of the input and output signals. The 0.02-V peak-peak input signal causes the base current to vary as a sine wave, which in turn causes the collector current to vary similarly. Note that the waveforms for V_{CE} and V_o are 180° out of phase with respect to the input and base waveforms. This is because the collector voltage *decreases* as I_C *increases* due to the R_C resistor ($V_{CE} = V_{CC} - I_C R_C$). Also note that all waveforms except V_i

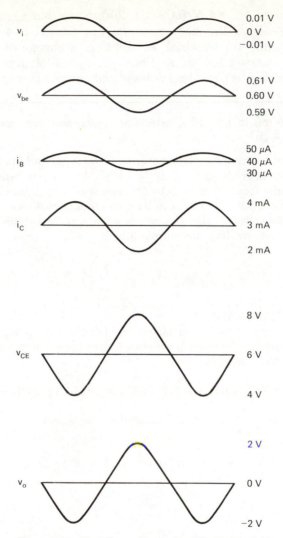

Figure 3-20 Wave forms for the amplifier in Fig. 3-19.

and V_o are riding on their respective dc bias levels. Due to the output coupling capacitor, there is no dc in the output sine wave.

In summary, the common emitter amplifier takes the ac input signal, amplifies it, phase shifts it by 180°, adds a dc level, and finally provides an amplified *ac* output via the output coupling capacitor.

The voltage gain of the amplifier can be found by taking the ratio of V_o to V_i.

$$A_V = V_o/V_i \qquad\qquad (3\text{-}9)$$

and for this case: A_V = 4 V/0.02 V = 200. Note that a small change of 0.02-V peak-peak at the base terminal has caused a 4-V peak-peak change at the collector terminal. In addition, a change of 20-μA peak-peak in base current led to a 2-mA peak-peak change in collector current. This circuit is providing voltage and current amplification.

Example 3-11

Using the waveforms in Fig. 3-20, determine the current and power gain of the amplifier in Fig. 3-19.

Solution   At first glance there may appear to be no input current. However, note that the base current is biased to 40 μA by the 1-V V_{BB} supply. When the ac input is applied, I_B varies from 30 to 50 μA. This variation in I_B is caused by V_i and we could consider the input current (i_i) to be a 20-μA peak-peak sine wave. Similarly, on the output the collector current varies as a 2-mA peak-peak sine wave. Taking the ratio of these two results in the current gain

$$A_I = \frac{i_o}{i_i} \tag{3-10}$$

or A_I = 2 mA/0.02 mA = 100.

The power gain may be calculated by determining the input and output powers. The power on the input is supplied by the generator and is calculated as $V_i \times I_i$ where V_i and I_i must be rms.

$$P_i = V_i I_i = (0.01 \text{ V} \times 0.707)(10 \text{ } \mu\text{A} \times 0.707) = 0.05 \text{ } \mu\text{W}$$

The power out of the amplifier is dissipated in the collector biasing resistor and may be found as

$$P_o = V_o I_o = (2 \text{ V} \times 0.707)(1 \text{ mA} \times 0.707) = 1000 \text{ } \mu\text{W}$$

The power gain G is

$$G = \frac{P_o}{P_i} \tag{3-11}$$

or G = 1000 μW/0.05 μW = 20,000.

Example 3-12

Show that the power gain of an amplifier is just the product of $A_V \times A_I$.

Solution   Power gain is $(V_o I_o)/(V_i I_i)$. This can be rewritten as

$$G = V_o/V_i \times I_o/I_i = A_V \times A_I \tag{3-12}$$

Notice that V and I do *not* have to be rms to calculate voltage, current, *or* power gain. This is because any conversion factor applied to V_o would also have to be applied to V_i and therefore would cancel when taking the ratio. Be sure, however, not to mix units, for example, taking the ratio of rms volts to peak volts. Applying Eq. 3-12, $G = A_V \times A_I = 200 \times 100 = 20,000$, which checks with Example 3-11.

AC β

When we take the ratio of the dc collector current to the dc base current (I_C/I_B), we are calculating the dc β. There is also an ac β that is defined:

$$\beta_{ac} = \Delta I_C/\Delta I_B \tag{3-13}$$

where Δ means "a small change in." The ac β says that a *small* change in the base current will result in a correspondingly *larger* change in the collector current. Usually the ac and dc values of β are nearly the same for a given operating point.

Example 3-13 ───────────────────────────────

Using the curves in Fig. 3-18, calculate β_{ac}.

Solution   From Eq. 3-13

$$\beta_{ac} = \frac{\Delta I_C}{\Delta I_B} = \frac{4 \text{ mA} - 2 \text{ mA}}{50 \,\mu\text{A} - 30 \,\mu\text{A}} = \frac{2 \text{ mA}}{20 \,\mu\text{A}} = 100$$

This compares to 75 for the dc β calculated in Example 3-10 for this operating point.

Predicting Signal Swing

The load line allows us to determine the output signal variations (i_C and v_{CE}) for a given i_B variation. It can also be used to predict the *maximum possible* output signal variations before distortion occurs.

Refer to Fig. 3-21. The bias point is shown to be $I_B = 60 \,\mu\text{A}$, $I_C = 6$ mA, and $V_{CE} = 6$ V. As the base current varies, corresponding variations in I_C and V_{CE} occur as shown. However, V_{CE} is limited to 12 V (V_{CC}) and 0 V by the power supply. When the base current increases, I_C increases and V_{CE} decreases, pushing the operating point up the load line towards saturation. The opposite occurs when I_B decreases and the operating point moves down the load line towards cutoff. The maximum output signal occurs when V_{CE} swings from *saturation to cutoff*, or V_{CC} volts peak-peak.

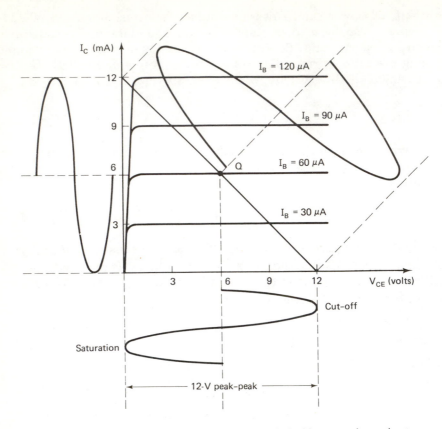

Figure 3-21 The maximum signal swing possible is limited by saturation and cut-off to V_{CC} volts.

If the Q point is exactly centered between V_{CC} and 0 V, the maximum possible signal swing will be V_{CC} peak-peak, or 12 V for our example. This is illustrated in Fig. 3-21. In reality, the transistor curves tend to become nonlinear (a constant change in I_B does *not* cause a constant change in I_C and V_{CE}) as saturation and cutoff are approached, meaning the maximum *undistorted* signal swing is even less than V_{CC}.

If the input is sufficiently large, the amplifier may be overdriven, causing flat-topping (a severe form of distortion). This occurs when the Q point is pushed all the way to the top (or bottom) of the load line. The output voltage then remains constant at either 0 V (saturation) or V_{CC} (cutoff). This is illustrated in Fig. 3-22. The output no longer is a sine wave. In fact, if the input is large enough, the output may appear to be a *square* wave. Because square waves are not pleasing to the ear, your stereo starts making "painful sounds" when the volume is set too high.

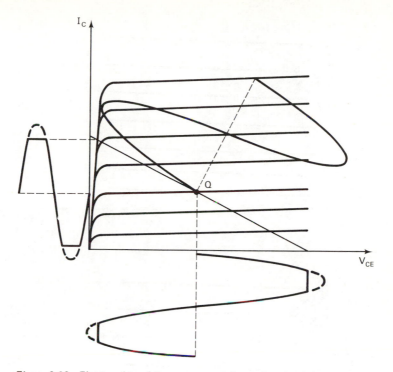

Figure 3-22 Flat-topping of the output waveforms due to too large an input signal.

What happens if the Q point is not biased exactly in the center of the active region? First of all, the amplifier will certainly still amplify. However, the signal swing will be restricted. With the Q point biased off of center, the output will be able to swing further in one direction than the other. Figures 3-23a and b illustrate this problem. In Fig. 3-23a, the bias point is closer to cutoff and the output will become distorted if V_o exceeds 4-V peak. In Fig. 3-23b, the opposite problem occurs; the bias point is too close to saturation and again the maximum undistorted output signal swing is 4-V peak.

3.9 POWER DISSIPATION

We have now seen that there are several factors involved in selecting an operating point for a transistor amplifier. The power supply voltage itself should not exceed the breakdown voltages of the transistor, and the Q point should be in the active region and preferably centered between V_{cc} and 0 V for *maximum* signal swing.

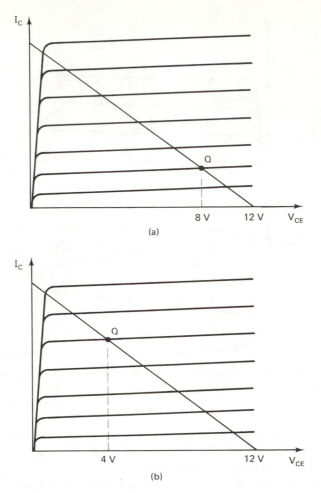

Figure 3-23 The output signal is limited to approximately 4-V peak in both cases due to the Q point not being centered in the active region.

In addition to these considerations, all transistors have a maximum *junction temperature* that must not be exceeded or destruction of the transistor may result. This temperature is affected by the collector current, collector-emitter voltage, package type, and heat sink used (if any). For example, a certain transistor without a heat sink may be able to dissipate 500 mW of heat. However, if a heat sink is added, this may increase to 5 W. In both cases the junction temperature is the same, but with the heat sink in place, more heat can be dissipated to the surrounding air.

For a given package and heat sink, a maximum power dissipation will result. The power dissipated at a given operating point is calculated as

$$P = I_C \times V_{CE} \tag{3-14}$$

When selecting a Q point and load line, care must be taken that the power calculated in Eq. 3-14 does not exceed the limit established by the package and heat sink.

Example 3-14 ————————————————————————

Determine the base and collector Q points for the *pnp* amplifier circuit in Fig. 3-24a using the indicated characteristic curves. Assume a TO-92 plastic package is used with a 200-mW maximum power dissipation. Verify that the operating point will not exceed this value.

Solution We could determine the operating point by using the method described in Section 3-6, but to gain further experience with load lines we will obtain our results graphically.

A problem arises attempting to draw the base load line. V_{CC} = 20 V, which is off the graph for V_{BE}. However, two points are needed to determine a line, so we will have to pick two points that are *on* the curves in Fig. 3-24b.

$I_B = (V_{CC} - V_{BE})/R_B$	V_{BE}
20 V/194 kΩ = 103 μA	0 V
19 V/194 kΩ = 98 μA	1 V

The load line is drawn connecting these two points. The resulting base Q point is I_B = 100 μA, V_{BE} = -0.65 V. Note that I_B and V_{BE} are negative due to the *pnp* transistor.

The collector load line can be found in the standard way:

I_C	V_{CE}
0 A	20 V
20 V/680 Ω = 29.4 mA	0 V

The collector Q point is I_C = -15 mA, V_{CE} = -10 V.

A graphical representation of the power dissipation may be obtained by plotting all points where $I_C \times V_{CE}$ = 200 mW. A table of values is prepared below.

$-I_C$	$-V_{CE}$	P
8 mA	25 V	200 mW
12 mA	16.7 V	200 mW
15 mA	13.3 V	200 mW
20 mA	10 V	200 mW
25 mA	8 V	200 mW

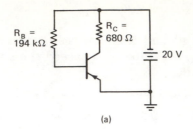

(a)

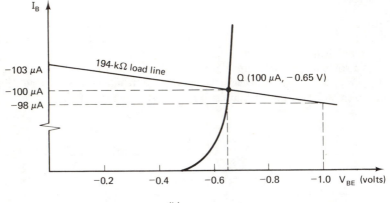

(b)

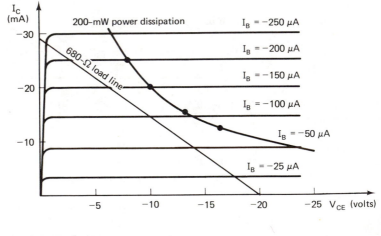

(c)

Figure 3-24 (a) Common emitter *pnp* amplifier. (b) base characteristics and load line. (c) collector characteristics and load line. The *Q* point lies beneath the constant 200 mW power dissipation curve.

When these data points are sketched on the collector curves in Fig. 3-24c, a *parabola* results. All points beneath the parabola have power dissipations less than 200 mW. As can be seen, our Q point and load line are safely within this region.

3.10 THE TRANSISTOR AS A SWITCH

All of the transistor circuits discussed to this point have been amplifiers. The Q point has been carefully chosen to be in the *active* region and within the maximum power dissipation of the device.

Of course the transistor is not always used as an amplifier. In digital applications, *logic-1* and *logic-0* output voltages are required. By biasing the transistor at cut off ($V_o = V_{CC}$) or saturation ($V_o = 0$ V), a *logic switch* can be designed.

Example 3-15 ───

Design a logic inverter that will convert an input from a console switch to standard logic levels (0 V and 5 V) across a 560-Ω load. The input voltage from the switch is either 0 V or 2 V through a 1-k Ω resistor. Assume $V_{CC} = 12$ V and a 2N2221A transistor.

Solution   The circuit is shown in Fig. 3-25a and its logic symbol in Fig. 3-25b.

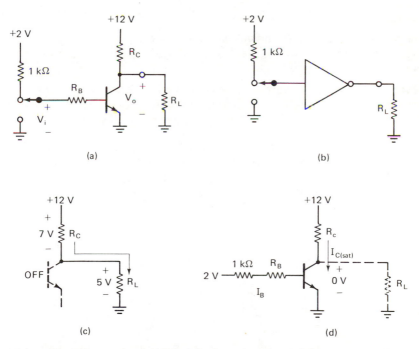

Figure 3-25   The transistor switch. (a) schematic diagram. (b) logic symbol. (c) case 1: the transistor is cut off. (d) case 2: the transistor is saturated.

Note that a common-emitter circuit is used. However, these are no coupling capacitors, as no ac signals are involved.

Case 1 $V_i = 0$ V. See Fig. 3-25c.

The transistor will be cut off and $I_C = 0$ A. This means R_C and R_L are in series. To determine the value of R_C, first find the load current: $I_L = V_o/R_L = 5$ V$/560$ $\Omega = 8.9$ mA. This same current flows through R_C and we may therefore find R_C as $R_C = V_{RC}/I_L = (12$ V $- 5$ V$)/8.9$ mA $= 787$ Ω.

Case 2 $V_i = 2$ V. See Fig. 3-25d.

In this case, the transistor must be saturated to ensure $V_o = 0$ V. In saturation, the full supply voltage will drop across R_C, resulting in $I_C = V_{CC}/R_C = 12$ V$/787$ $\Omega = 15.2$ mA. Note that the saturated collector-emitter has "shorted-out" the load resistor so that it is not important for this case.

To determine the value of R_B, the base current must be known. Referring to the 2N2221A data sheet, β has a minimum value of 35. For this worst case condition, the base current would have to be $I_B = I_C/\beta = 15.2$ mA$/35 = 434$ μA. To be doubly certain that the transistor saturates, we will double this I_B value to 868 μA. This will allow for resistor tolerances, power supply variations, and temperature, all working against us to prevent saturation. Now R_B can be calculated as $R_B = (2$ V $- 0.6$ V$)/I_B = 1.4$ V$/868$ μA $= 1.6$ kΩ. Because the switch already contains 1 kΩ, we choose $R_B = 600$ Ω.

Example 3-16

Design a transistor LED (light-emitting diode) driver. Assume a 2-V input should turn the LED ON and the $V_{LED} = 1.2$ V and $I_{LED} = 20$ mA. $V_{CC} = +5$ V, $\beta = 50$ minimum and $V_{CE(sat)} = 0$ V.

Solution The circuit is shown in Fig. 3-26. When $V_i = 2$ V, the transistor should saturate, allowing 20 mA to flow through the LED. R_C can be calculated once its voltage is known. $V_{RC} = 5$ V $- 1.2$ V $- V_{CE(sat)} = 3.8$ V. $R_C = V_{RC}/I_C = 3.8$ V$/20$ mA $= 190$ Ω.

The base current is 20 mA$/\beta = 20$ mA$/50 = 40$ μA. To ensure saturation, choose $I_B = 80$ μA. Then $R_B = V_{RB}/I_B = (2$ V $- 0.6$ V$)80$ μA $= 17.5$ kΩ.

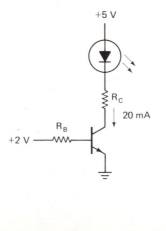

Figure 3-26 Transistor LED driver for Example 3-16.

Bipolar Two carrier types. The bipolar transistor requires both positively-charged holes and negatively-charged electrons.

Injection The process of passing electrons (*npn*) or holes (*pnp*) into the base of the transistor due to a forward bias on the base-emitter junction.

Cutoff This is a condition for the transistor in which no currents other than leakage currents are flowing.

Saturation This is a condition for the transistor in which excess base current floods the base with carriers and the collector is unable to absorb them all. It is identified by a low collector-emitter voltage and a forward-biased base-collector junction.

Active This is the normal biasing condition for the transistor to amplify. The base-emitter junction is forward-biased and the collector-base junction is reverse-biased.

Bias This refers to the dc voltages and currents applied to the transistor to make it operate at a specific Q point.

Q point The specific dc operating point of the transistor circuit, that is, I_B, I_C, V_{BE} and V_{CE}.

Amplification In an amplifier, the output signal may be greater than the input signal, resulting in amplification or gain. Amplifiers may have voltage, current, and power gain.

Signal Swing This is the amount of peak-peak signal variation. At the output of an amplifier, the signal swing is limited to some maximum value by the power supply voltage.

Distortion Ideally, the output of an amplifier is identical to the input waveform except for amplitude and phase. If this is not true, the output is said to be distorted. By overdriving an amplifier, the output waveform may actually become flat at the peaks due to saturation or cutoff of the amplifier stage. This is referred to as flat topping or clipping.

Logic Inverter A type of transistor circuit in which the transistor is operated in only one of two states: saturation or cutoff.

Family of Curves A two-dimensional graph may display only two quantities. By adding a third parameter, a series or family of graphs can be drawn for various values of this third quantity. This is the case for the transistor common-emitter characteristic curves. I_C is plotted against V_{CE} for various values of I_B.

————————————— QUESTIONS AND PROBLEMS —————————————

3-1 Indicate the normal biasing of the base-emitter and base-collector junctions for amplification. Does this apply for *pnp* as well as *npn* transistors?

3-2 List two considerations that are made in the manufacture of a bipolar transistor to ensure that a majority of the injected carriers reach the collector.

3-3   Draw a diagram similar to Fig. 3-3 for a *pnp* transistor. Be sure the polarities of the V_{CC} and V_{BB} batteries are correct.

3-4   Explain how electrons can flow across the collector-base junction in a typical *npn* transistor even though this junction may be reverse-biased.

3-5   If the base current is held constant, what will happen to the collector current as the collector-emitter voltage increases?

3-6   Why is the I_C versus V_{CE} set of common-emitter characteristic curves referred to as a *family* of curves?

3-7   Refer to the curves in Fig. 3-5. At approximately what voltage does the transistor breakdown?

3-8   Refer to Fig. 3-27 and classify the following operating points as saturated, cutoff, or active.
  (a) $I_C = 3$ mA; $V_{CE} = 15$ V
  (b) $I_C = 3$ mA; $I_B = 35\ \mu$A
  (c) $I_B = 10\ \mu$A; $V_{CE} = 5$ V
  (d) $I_B = 0$ A; $V_{CE} = 18$ V
  (e) $I_B = 30\ \mu$A; $V_{CE} = 0.2$ V

3-9   In Problem 3-8, supply the missing current or voltage by referring to Fig. 3-27. For example, in (a), supply I_B; in (b), supply V_{CE}.

3-10 Using Fig. 3-27, assume $I_B = 30\ \mu$A and calculate β at the following V_{CE} values:
  (a) 0.5 V
  (b) 1.0 V
  (c) 5.0 V
  (d) 15 V

3-11 When the transistor is saturated, is β a valid parameter? Explain why or why not.

3-12 A certain *npn* transistor operates at the following Q point: $I_B = 50\ \mu$A, $V_{BE} = 0.63$ V, $I_C = 6$ mA, and $V_{CE} = 5$ V. Calculate α and β.

3-13 Derive an equation for β in terms of α.

3-14 Refer to the data sheet for the 2N2221A. Find
  (a) β minimum at $I_C = 10$ mA
  (b) Minimum emitter-base breakdown voltage
  (c) Minimum collector-base breakdown voltage
  (d) $V_{CE(sat)}$ maximum at $I_C = 500$ mA
  (e) Absolute maximum for I_C
  (f) I_C test condition for the BV_{CEO} test
  (g) Test temperature for (f)

3-15 Why are the various breakdown voltages for the transistor specified as minimum values instead of maximum values?

3-16 Repeat Example 3-7 for $\beta = 200$.
    Note: Refer to Fig. 3-4 and assume $R_B = 27$ kΩ, $R_C = 7.5$ kΩ, $V_{BB} = 1.0$ V, and $V_{CC} = 20$ V. Use the curves in Fig. 3-27 for Problems 3-17 through 3-23.

3-17 Draw the base and collector load lines for this circuit. Find the two operating points.

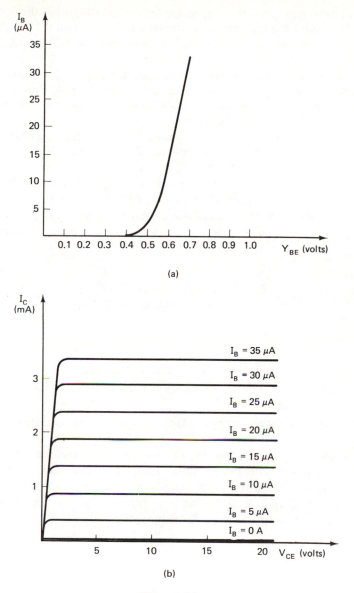

Figure 3-27

3-18 Assume the base voltage varies as a 0.1-V peak-peak sine wave and draw waveforms for v_{be}, v_{ce}, i_b and i_c. Be sure to include the dc levels.

3-19 Calculate the voltage gain of this circuit.

3-20 Calculate the ac and dc β at the Q point.

3-21 Calculate the power dissipation at the Q point.

3-22 How large an input signal can be applied before clipping of the output wave-form occurs? Will this clipping be due to saturation or cutoff?

3-23 If R_C is changed to a 10-kΩ resistor, determine the new operating point of the circuit.

3-24 A certain transistor amplifier is to be biased at V_{CE} = 10 V. If no heat sink is used, maximum power dissipation is limited to 750 mW. However, if a heat sink is used, this increases to 3 W. Calculate the maximum collector current at V_{CE} = 10 V for these two cases.

3-25 Design a logic inverter that will produce 0 V or 4 V across a 470-Ω load resistor. Assume the input voltage switches from 0 V to 2 V and a single 12-V power supply is available. Use a 2N2222A transistor.

3-26 Design a *pnp* transistor LED driver. Assume β_{min} = 75, V_{CC} = -12 V, and V_i = 0 V or -5 V. Bias the LED at I = 20 mA, V = 1.6 V.

3-27 Refer to Fig. 3-16. If R_B should open, what will happen to I_B, I_C, and V_{CE}?

3-28 What is the minimum value of β that will cause saturation of the amplifier in Fig. 3-16?

LABORATORY ASSIGNMENT 3:
TRANSISTOR CHARACTERISTICS

Objectives

1. To study the transistor diode characteristics.
2. To gain experience in identifying the various operating regions of the transistor.
3. To observe voltage amplification in a transistor amplifier.

Introduction In this laboratory assignment, various dc properties of the transistor will be studied, including the base-emitter diode characteristics, saturation and cutoff regions of operation, and the current gains α and β. A voltage amplifier will be assembled and a method of measuring its gain will be learned.

Components Required

1 2N2221A *npn* transistor
1 50-kΩ pot
1 1-MΩ pot
2 1-kΩ resistors
1 10-μF 15-V capacitor

Caution: To avoid damaging the transistor *set the power supply current limit to 20 mA.*

Part I: *Base Emitter Diode*

STEP 1 Set up the circuit shown in Fig. 3-28 and adjust the pot to midrange. Slowly advance V until 10 μA of current registers.

STEP 2 Without changing R or V, momentarily measure the base to ground (emitter) voltage.

STEP 3 Repeat the previous procedure to obtain I_B = 100 μA, 1 mA, and 10 mA, varying R and V as necessary.

STEP 4 Graph your results on semilog paper with I on the vertical (log scale) and V on the horizontal.

STEP 5 Reverse the ammeter and power supply connections and slowly raise V to obtain I_B = –1 μA, –5 μA, –10 μA, and –15 μA. Measure the base-emitter voltage for each current.

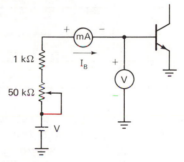

Figure 3-28 Test circuit to measure the IV characteristics of the base-emitter diode. The voltmeter should only be in the circuit for the length of measurement.

Question 1 Why is it necessary to adjust the current in the previous steps with the voltmeter disconnected?

Question 2 What parameter of the transistor is being measured in step 5? Refer to the transistor data sheet.

Part II: *Transistor Operation*

STEP 1 Set up the circuit shown in Fig. 3-29 and adjust the 1-MΩ pot to maximum resistance *before* applying power. Be sure the polarities of the meters are as shown.

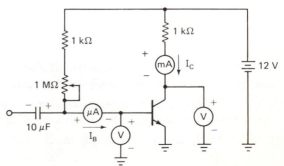

Figure 3-29 Test circuit for Part II

STEP 2 Apply power and adjust the pot until $I_B = 15\,\mu\text{A}$. Record I_C, V_B, and V_{CE} in Table 3-2. Calculate I_E, α, β, and V_{CB}.

STEP 3 Repeat step 2 for the other values of I_B in the table.

TABLE 3-2

I_B (μA)	15	30	50	70	100	1000
I_C						
$I_E = I_C + I_B$						
$\beta = I_C/I_B$						
V_C						
V_B						
$V_{CB} = V_C - V_B$						
$\alpha = I_C/I_E$						

NOTE A transistor in the active region generally has several volts from collector to emitter, and the collector-base junction is reverse-biased (V_{CB} is negative for an *npn* type). In saturation, V_{CE} falls to a low value, the collector base becomes forward-biased (V_{CB} is positive for an *npn* type), and β becomes a small number (actually β is meaningless in saturation).

A transistor in saturation has excessive base drive or more injection than the collector can absorb without raising V_{CE}.

STEP 4 Assuming $V_{CE} = 0$ V, *calculate* the collector current in saturation, $I_{C(\text{sat})}$. How does this value compare to the I_C measured when $I_B = 1000\,\mu\text{A}$?

Question 3 If I_B were increased to 2 mA, what do you think the value of β would be?

Part III: *Amplification*

STEP 1 Adjust the base resistor of Part II until $V_{CE} = 6$ V. This will bias the amplifier in the middle of the active region.

STEP 2 Apply a 50 mV peak-peak sine wave at $f = 1$ kHz to the base of the transistor through the 10-μF coupling capacitor.

STEP 3 With the oscilloscope in the *dc* position, record the waveforms at the base and collector terminals. Be sure to observe their relative phase relationships.

STEP 4   Calculate the ac voltage gain as v_{ce}/v_{be}.

STEP 5   Observe the effect of overdriving the amplifier by increasing the amplitude of the input signal. Eventually, the output will appear as a square wave. Record the output waveform and label the saturation and cutoff dc levels.

STEP 6   Reduce the input signal to a normal value (50-mV peak-peak). Now observe the effect of varying the dc bias point by adjusting the 1-MΩ pot. Which bias point (V_{CE}) yields optimum signal swing?

FOUR

MEASURING AMPLIFIER PERFORMANCE

A *transducer* is a device that converts energy from one form to another. For example, a microphone converts sound pressure to an electrical current that is representative of the amplitude and frequency of the applied sound pressure. Although the microphone's output is a current, it is of insufficient strength to directly drive a loudspeaker. An amplifier is therefore connected between the microphone and speaker, as shown in Fig. 4-1. The characteristics of the amplifier may be such that the sound level at the speaker may actually exceed the original sound level applied to the microphone (a P.A. system, for example). Amplifiers, therefore, are basically needed to increase the amplitude of a transducer's output.

Other examples of transducers are *phonograph cartridges*, which convert the mechanical vibrations of a phonograph needle to a varying voltage; radio and television *antennas*, which receive or radiate electromagnetic energy; and *photoelectric devices*, which convert light energy to an electrical signal.

Most transducers have one property in common. The electrical signal produced is quite small. Usually in the millivolt or microvolt range. In order for this signal to be of use so that it may drive a speaker, turn on an indicator light, or activate a relay, its amplitude must be increased.

In this chapter we discuss a general amplifier. Electrically, this amplifier consists of *passive* components (resistors, capacitors) and *active* components (bipolar transistors, field-effect transistors, vacuum tubes). For the purpose of this chapter, the particular components involved will not be important. Instead, we will treat the amplifier as a *black box* and discuss general amplifier terminology that applies to all amplifiers regardless of the active devices used and even the complexity of the particular design.

Figure 4-1 The input transducer (microphone) produces an output of insufficient strength to drive the output transducer (speaker). An amplifier is used to increase this signal's amplitude.

The definitions provided and equations developed will apply for the simple fixed-bias common-emitter amplifier of Chapter 3 as well as your 100-watt stereo amplifier at home.

One assumption that we will be making in the following sections is that the amplifier has been properly *dc biased* to provide amplification. As we saw in the previous chapter, the dc operating point of an amplifier stage has a great deal to do with its resultant output signal variations. A poorly chosen Q point may result in circuit operation in the saturation or cutoff regions.

For this reason, all results obtained in this chapter assume a proper dc bias point. We will only be concerned with *ac* quantities. We assume an ac signal is applied to the input and an amplified ac signal is generated at the amplifier's output. No dc power supplies are shown, although their presence is certainly assumed.

4.1 BASIC DEFINITIONS

Let's assume we have some sort of amplifier sitting before us. It may use vacuum tubes or transistors as the active devices but this will be of no consequence. Instead, we represent this amplifier with the *black box* shown in Fig. 4-2.

The amplifier must have an *input* and an *output* as represented by the two sets of terminals shown in the figure. At the input terminals, there is an input voltage V_i and an input current I_i. We will use rms quantities but peak or peak-peak values are just as valid. At the output terminals, the output voltage is represented by V_o and the output current by I_o. Note that $I_o = 0$ A if there is no load resistor for this current to flow into.

We can make a number of observations about this black box amplifier:

1. Although V_o, V_i, I_i, and I_o have polarity indications (+ and −, or arrowheads), this does not really represent polarity because the polarity of an ac signal is always changing. Instead, these indicators represent the relative *phase* between two signals. If $V_o = -2$ V, then V_o has the opposite phase to V_i or is $180°$ out of phase with respect to V_i.

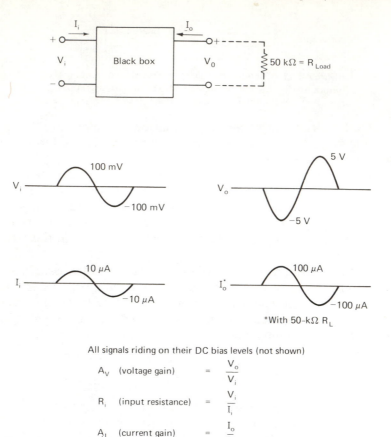

All signals riding on their DC bias levels (not shown)

A_V (voltage gain) $= \dfrac{V_o}{V_i}$

R_i (input resistance) $= \dfrac{V_i}{I_i}$

A_I (current gain) $= \dfrac{I_o}{I_i}$

G (power gain) $= A_V \times A_I$

Figure 4-2   A black-box amplifier.

2. As stated earlier, whether V_i is rms or peak does not matter. However, when comparing any of the four quantities (V_o, V_i, I_o, and I_i), we must be consistent. Trouble could arise when dividing rms volts by peak-peak volts! Also recall that to express the power level of an ac signal, rms *must* be used.

3. None of the conventions chosen for the amplifier depend in any way on the type of amplifier, the number of stages or the particular dc bias point.

Every amplifier can be characterized as a *voltage* amplifier, *current* amplifier, *power* amplifier, or some combination of the three. Let's define these three types of amplifiers and apply our definitions to the waveforms shown in Fig. 4-2.

Voltage Amplifier

When the output voltage of an amplifier exceeds its input voltage, that amplifier is said to have *voltage gain*. The actual amount of gain is expressed as

$$A_V \text{ (voltage gain)} = V_o/V_i \qquad (4\text{-}1)$$

The amplifier in Fig. 4-2 has a voltage gain equal to -5 V/100 mV = -5000 mV/100 mV = -50. Note that peak values for the voltages were used but rms quantities would yield the same result. This is because the 0.707 conversion factor would be common to both numerator and denominator in Eq. 4-1.

The voltage gain is -50 because V_o has opposite phase to V_i, as shown by their waveforms in Fig. 4-2.

Current Amplifier

As you might imagine, a *current amplifier* has a definition similar to the voltage amplifier:

$$A_I \text{ (current gain)} = I_o/I_i \qquad (4\text{-}2)$$

The amplifier in Fig. 4-2 has a current gain of 0 unless a load resistor is connected to the output terminals. In this case, and referring to Fig. 4-2, the current gain becomes 100 μA/10 μA = $+10$. The $+10$ indicates no phase reversal.

Power Amplifier

A *power amplifier* amplifies the input power and generates a signal with higher power level across the output load resistances. The power gain can be defined as

$$G \text{ (power gain)} = P_o/P_i \qquad (4\text{-}3)$$

Again referring to Fig. 4-2, the input power is found as (100 mV $\times$ 0.707) $\times$ (10 μA $\times$ 0.707) = 0.5 μW. The output power is found as (5 V $\times$ 0.707) $\times$ (100 μA $\times$ 0.707) = 250 μW. The power gain is therefore 250 μW/0.5 μW = 500.

Example 4-1 ——————————————————————————————

Show that Eq. 4-3 can also be written as $G = A_V \times A_I$

Solution   Substituting in the quantities for P_o and P_i

$$G = \frac{P_o}{P_i} = \frac{V_o \times I_o}{V_i \times I_i} = \frac{V_o}{V_i} \times \frac{I_o}{I_i}$$

and

$$G = A_V \times A_I \qquad\qquad\qquad (4\text{-}4)$$

Using our previous calculations for A_V and A_I, $G = |-50 \times 10| = +500$. Although $A_V \times A_I = -500$, the absolute value signs in Eq. 4-4 say to choose the positive magnitude only. The power gain is simply a number and carries no phase information with it.

4.2   DECIBELS

Anyone interested in the audio field has come across the term *decibel*. Taken literally, decibel means one tenth of a *bel*. The bel is a term named after Alexander Graham Bell, the inventor of the telephone. The concept of the decibel can be thought of as attempting to develop a relationship between the way our ears *hear* a sound and the corresponding power behind that sound.

For example, if you were to listen to a loud speaker driven by 10 W of power, how much louder would this sound appear to be if the power level was increased to 100 W? You might be surprised to learn that to most people the sound would only seem about *twice* as loud. This is because our ears respond *logarithmically* to sounds. This is illustrated graphically in Fig. 4-3. *Curve f* corresponds to the way our ears hear sound. Note that the power increases much more rapidly than the corresponding sound level. Also note that the curve *log f* results in a linear (straight line) relationship between power and sound level. This is reminiscent of the IV curve for the diode. If we express the power level in decibels by computing the log function, the mathematical relationship between power level and sound level becomes linear.

Power Gain in Decibels

One decibel (abbreviated *dB*) corresponds to the smallest change in a sound level that most people can detect. Notice the word *change* in this definition. A given decibel rating by itself has no meaning. To say that an amplifier is delivering 10 dB of power is incorrect and *meaningless*. Decibels always represent a ratio of numbers. If we had said the power output of the amplifier had increased by 20 dB from a 1-W

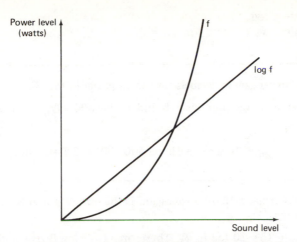

Figure 4-3 Our ears hear sound in a nonlinear (graph *f*) manner. The power level must increase much more than twice to increase the sound level by two. By computing the log of *f*, this relationship can be linearized.

reference level, then our statement would have meaning. Always remember that a reference level for comparison must be defined in order to use decibels correctly.

To convert the power gain of an amplifier to decibels, the following formula is used.

$$G_{(dB)} = 10 \log_{10} (P_o/P_i) \tag{4-5}$$

This formula gives the power gain in dB for an amplifier's output power with respect to its input power. Usually $\log_{10}$ is abbreviated log.

Those of you unfamiliar with logarithms may feel a bit uneasy about the $\log_{10}$ term in Eq. 4-5. Mathematically, the logarithm of a number is simply a mathematical function, just like $\sin X$ or X^3. In this case, it is $\log X$. If you have your calculator handy, try this sequence: 100 log. Your display should indicate 2. Try again using 1000. You should see 3. From this we can conclude that $\log X$ equals a number such that when this number is used as a power of 10, the result equals X. For example, log 100 = 2 and therefore 10 raised to the second power must be 100.

A number of other rules about logarithms also apply.

1. $\log (A \times B) = \log A + \log B$
   Example: $\log(10 \times 100) = \log 10 + \log 100 = 1 + 2 = 3$
2. $\log A/B = \log A - \log B$
   Example: $\log 1000/10 = \log 1000 - \log 10 = 3 - 1 = 2$

3. $\log A^n = n \log A$
   Example: $\log 10^2 = 2 \log 10 = 2 \times 1 = 2$.

Example 4-2 ───

Determine the power gain in decibels for the amplifier in Fig. 4-2.

Solution   Because G was previously found to be 500, $G_{(dB)}$ can be found using Eq. 4-5.

$$G_{(dB)} = 10 \log G = 10 \log 500 = 10 \times 2.7 = 27 \text{ dB}.$$

Example 4-3 ───

A certain amplifier has a 15-dB power gain. If the output power is 75 W, determine the input power.

Solution $G_{(dB)} = 15 = 10 \log P_o/P_i$. Therefore, $1.5 = \log P_o/P_i = \log X$, where $X = P_o/P_i$. X can be solved using the INV log function on your calculator. Press the following sequence: 1.5 INV log. Your display should indicate 31.6. The equation $1.5 = \log X$ is really saying $X = 10^{1.5}$ and X can be alternately found using the Y^X key on your calculator. Press the sequence: 10 Y^X 1.5 = . You should again see 31.6. Because $X = P_o/P_i = 75$ W/$P_i = 31.6$, $P_i = 75$ W/31.6 = 2.37 W. As a check, recalculate the power gain in dB.

$$G_{(dB)} = 10 \log 75/2.37 = 10 \log 31.6 = 15 \text{ dB}$$

Voltage and Current Gain in Decibels

Because it can be difficult to measure the input or output power of an amplifier in the laboratory, it is common practice to also express the *voltage* and *current* gains of an amplifier in decibels. Be sure to note, however, that decibels of voltage or current gain are *not* the same as decibels of power gain.

The corresponding formulas for A_V and A_I are

$$A_V \text{ (dB)} = 20 \log V_o/V_i \qquad (4\text{-}6)$$

$$A_I \text{(dB)} = 20 \log I_o/I_i \qquad (4\text{-}7)$$

Example 4-4 ───

Express the voltage and current gains for the amplifier in Fig. 4-2 in dB.

Solution $A_{V(dB)} = 20 \log 5$ V/0.1 V = $20 \log 50 = 34$ dB. Note that for the purpose of calculating the decibel gain, A_V is assumed positive. The log of a negative number is not allowed and is undefined. The current gain is similarly found.

$$A_{I(dB)} = 20 \log 100 \ \mu A/10 \ \mu A = 20 \text{ dB}.$$

As this last example illustrates, it is convenient to express power, voltage, and current gains in decibels. In this way, if an amplifier provides 20 dB of voltage gain, we interpret this to mean V_o is ten times as large as V_i. It is also important to remember that just because the voltage gain is 20 dB does not mean the power gain or current gain will also be 20 dB. In fact, because

$$G = \frac{P_o}{P_i} = \frac{(V_o^2/R_L)}{(V_i^2/R_i)} = \frac{V_o^2}{V_i^2} \times \frac{R_i}{R_L}$$

and

$$G_{(dB)} = 10 \log \frac{P_o}{P_i} = 10 \log \left(\frac{V_o^2}{V_i^2} \times \frac{R_i}{R_L} \right)$$

$$= 20 \log \frac{V_o}{V_i} + 10 \log \frac{R_i}{R_L}$$

We can see that $G_{(dB)} = A_{V(dB)}$ only for the special case of the load resistor equalling the amplifier input resistance, in which case

$$10 \log \frac{R_i}{R_L} = 0$$

The key point to remember is that we can express voltage, current, and power gain all in decibels, but the results obtained for each case are totally independent of the others.

The 3-dB Point

Often the bandwidth of an amplifier or filter is measured between the -3-dB points. This means the output voltage (or power level) of the circuit has fallen by 3 dB from its typical or midband value.

Example 4-5 ——————————————————————————————

What factor does -3 dB represent for (a) voltage gain? and (b) power gain?

Solution (a) For voltage gain: -3 dB $= 20 \log X$ or $\log X = -3/20 = -0.15$ and $X = 10^{-0.15}$ (or -0.15 INV log) $= 0.707$. Therefore, the output voltage will have fallen to about 71% of its typical value when A_V falls by 3 dB. (b) For power gain: -3 dB $= 10 \log X$ or $\log X = -0.3$ and $X = 0.5$. This means the power has fallen to 50% of its typical value when G falls by 3 dB.

Example 4-6 ————————————————————————————————

An amplifier with 1 W of input power has 9 dB of power gain. Calculate P_o.

Solution We could solve this straightforward as 9 dB = 10 log P_o/P_i; 0.9 = log P_o/P_i or P_o/P_i = $10^{0.9}$ (or 0.9 INV log) = 7.9. Therefore, P_o = 7.9 W. However, notice that +3 dB also represents a power *increase* of two (because –3 dB means a *halving* of power, +3 dB means a *doubling* of power). In 9 dB, there are three doublings of power or a factor of 2^3 and a power increase by a factor of 8. Therefore, P_o is approximately 8 W.

Example 4-7 ————————————————————————————————

The output voltage of a signal generator is adjusted to 1-V peak-peak. If this signal generator includes a coarse output adjustment control (often called an attenuator), graduated in 20-dB steps, what will the 1-V output become if –20 dB of attenuation is switched in?

Solution –20 dB = 20 log V_o/V_i; therefore –1 = log V_o/V_i or V_o/V_i = 10^{-1} (or –1 INV log) = 0.1 and V_o = 0.1 × V_i = 0.1 V peak-peak.

Sometimes the face of an analog voltmeter will also be graduated in decibels. Two methods are commonly used here. In one system, all voltage measurements are referenced to 1 mW across 600 Ω (the characteristic resistance of typical audio transmission cables). This corresponds to 0.775 V (the square root of 1 mW × 600 Ω) and is labelled 0 dBm on the meter face. On such a meter, a voltage reading of 0.245 V will be shown as –10 dBm (20 log 0.245/0.775). An important point to remember in using this meter is that the *decibel readings* are only accurate as long as the voltage measurements are taken across 600 Ω. The *voltage readings* however, will be accurate regardless of the load resistor's value.

In another system, all voltage measurements are referenced to 1 V independent of the load resistor. In this case, 1 V corresponds to 0 dBV. On this meter, 0.245 V would indicate –12.2 dBV (20 log 0.245/1).

4.3   THE AMPLIFIER EQUIVALENT CIRCUIT

It is convenient to draw an *equivalent circuit* or *model* for the various amplifiers we discuss. This model should be a simple representation of what the amplifier appears to be electrically to the real world.

These are only two "windows" into the amplifier, one at the input terminals and the other at the output terminals. Basically, a signal applied at the input terminals appears amplified (and possibly phase-shifted) at the output terminals. The input of the amplifier appears to be a simple resistor, R_i, called the *input resistance*. The output appears to be a signal source in series with a resistance. The amplitude of this signal source should equal the amplitude of the amplified input signal and the resistance will represent the amplifier's *output resistance, R_o*.

Figure 4-4 illustrates this amplifer equivalent circuit. A signal applied to the input terminals "sees" R_i and appears at the output terminals with amplitude kV_i. k represents the *open circuit* (no load resistor) voltage gain of the amplifier. If you recall Thevenin's Theorem, then you will recognize the output as a Thevenin equivalent.

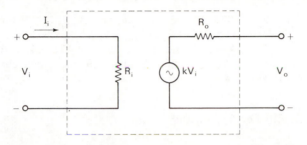

Figure 4-4 The amplifier equivalent circuit with input resistance R_i, and open circuit voltage gain k.

The following examples will demonstrate the usefulness of this equivalent circuit.

Example 4-8

An audio amplifier has the following properties: $R_i = 100\ \Omega$, $R_o = 8\ \Omega$, and $k = 50$. Assume this amplifier is driven by a signal generator whose open circuit output voltage is 0.1 V rms and whose characteristic resistance is 50 Ω. Calculate the power gain in dB to an 8-Ω load.

Solution The equivalent circuit is shown in Fig. 4-5. Note that the signal generator can be represented by a circuit similar to the output of an amplifier. Also note that due to the voltage drop across R_G, V_i of the amplifier will *not* equal E_G. This is why it is poor practice to set the output of a signal generator *before* connecting to the circuit under test. The test circuit will load the generator down, decreasing the preset generator output voltage.

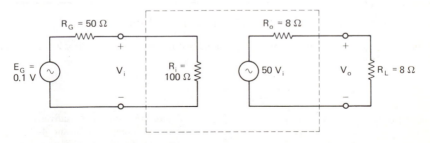

Figure 4-5 Amplifier equivalent circuit for Example 4-8.

V_i is calculated as $V_i = R_i/(R_i + R_G) \times E_G = 100/150 \times 0.1$ V $= 0.067$ V. The input power is then $V_i^2/R_i = (0.067)^2/100 = 44.9$ μW.

The output voltage divides between R_o and R_L; $V_o = (kV_i) \times R_L/(R_o + R_L)$ $= 50(0.067) \times (8/16) = 1.675$ V. P_o is then $V_o^2/R_L = (1.675)^2/8 = 0.35$ W.

Then $G_{(dB)} = 10 \log 0.35$ W$/44.9$ μW $= 38.9$ dB.

Example 4-9

Calculate the voltage and current gains in dB for Example 4-8.

Solution $A_V = V_o/V_i = 1.675$ V$/0.067$ V $= 25$, and $A_{V(dB)} = 20 \log 25 = 28$ dB. $I_i = V_i/R_i = 0.067$ V$/100$ $\Omega = 670$ μA, and $I_o = V_o/R_o = 1.675$ V$/8$ $\Omega = 0.21$ A. $A_I = I_o/I_i = 0.21$ A$/670$ μA $= 313$, and $A_{I(dB)} = 20 \log 313 = 49.9$ dB. Note that $G_{(dB)} = 10 \log (A_V \times A_I) = 10 \log (25 \times 313) = 38.9$ dB, which checks with the result obtained in Example 4-8.

General Results For A_V, A_I, and G

Example 4-10

Determine a general equation for the voltage gain of an amplifier in terms of its open circuit voltage gain k, output resistance R_o, and load resistance R_L.

Solution   Figure 4-6 illustrates a general amplifier connected to a load resistor R_L. The output voltage is found as a voltage divider between R_L and R_o. $V_o = R_L/(R_L + R_o) \times kV_i$. When this equation is divided by V_i, an equation for the voltage gain results.

$$A_V = \frac{V_o}{V_i} = k \times \frac{R_L}{R_L + R_o} \tag{4-8}$$

This equation predicts that the maximum voltage gain for any amplifier will occur under *open circuit* conditions ($R_L = \infty$) and will have a value of k. But this is only to be expected because k was defined as the open circuit voltage gain in the first place! The point, however, is that it is not possible to achieve a voltage gain greater than k.

Example 4-11

Repeat Example 4-10 for the current gain of a general amplifier.

Solution   Again referring to Fig. 4-6, the output current is found as $kV_i/(R_o + R_L)$.

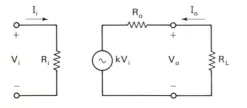

Figure 4-6 Amplifier equivalent circuit used in Examples 4-10 and 4-11 to predict the absolute maximum voltage and current gains possible.

The input current is V_i/R_i. The current gain is

$$A_I = \frac{kV_i/(R_o + R_L)}{V_i/R_i} = \frac{k/(R_o + R_L)}{1/R_i} = k \times \frac{R_i}{R_o + R_L} \tag{4-9}$$

This equation predicts a maximum value of $k \times R_i/R_o$ for the current gain when R_L is a *short circuit*. When R_L is an open circuit, the current gain is 0.

Figure 4-7 graphs A_V and A_I versus load resistance. Note that the optimum load for a voltage amplifier is an *open circuit*, but a *short circuit* for a current amplifier.

Example 4-12

Determine a general equation for the maximum power gain of an amplifier.

Solution Maximum power gain occurs when $R_L = R_o$, called the *matched-load* condition. Under this condition, $V_o = kV_i/2$ (because $R_L = R_o$), and $G = P_o/P_i =$

$$\frac{V_o^2/R_L}{V_i^2/R_i} = \frac{(kV_i/2)^2/R_L}{V_i^2/R_i}$$

or

$$G(\text{matched load}) = \frac{k^2}{4} \times \frac{R_i}{R_L} \tag{4-10}$$

Equation 4-10 tells us the power gain for a matched-load condition. This is the very best the amplifier can ever do. In a real amplifier,

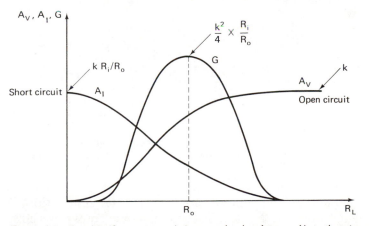

Figure 4-7 Graphs of A_V, A_I, and G versus load resistance. Note that A_V is maximum for an open circuit, A_I is maximum for a short circuit, and G is maximum when $R_L = R_o$.

the actual power gain and output power may be less than ideal for several reasons.

1. It is not always practical to achieve a matched-load condition. In this case, Eq. 4-10 does not apply.
2. In a real amplifier, the dc bias point may restrict the output signal swing to small values. Because the job of the power amplifier is to deliver power to a given load resistor, this may be more significant than the power gain itself. A large power gain is of little value if the output signal swing is limited (see Section 5.4).
3. The actual output power obtained from a given amplifier is affected by several factors besides the load resistor. These include heat sinking of the power devices, class of amplification (A or B), dc bias point, and coupling techniques.

Of course a majority of amplifiers are not intended to be used as power amplifiers at all. Voltage amplifiers, for example, work best into open circuits while current amplifiers work best when driving short circuits. These properties are summarized in Fig. 4-7.

Example 4-13 ————————————————————————————————

Show that the output circuit of an amplifier may also be represented as a current source and parallel resistance.

Solution   Although all of the amplifiers shown up to this point have used voltage sources in the output model, these could be converted to current sources using Norton's Theorem. The current source is kV_i/R_o and the resistor is R_o. Figure 4-8 shows the resulting circuit.

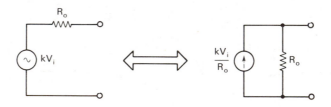

Figure 4-8   The output stage of the amplifier equivalent circuit can be converted to a current source and parallel resistor using Norton's Theorem.

Depending on the type of amplifier and its application, it may be convenient to use the Norton or current source equivalent circuit when trying to analyze a problem.

Example 4-14 ————————————————————————————————

List the desirable characteristics for R_i and R_o for ideal current and voltage amplifiers.

Solution (a) Refer to Fig. 4-9a for the ideal voltage amplifier. Because E_G voltage divides with R_G and R_i, it would be desirable for R_i to be very *large* compared to R_G to achieve maximum input voltage. Similarly, on the output, kV_i voltage divides with R_o and R_L. Here it is desirable for R_o to be very *small* compared to R_L.

(b) Refer to Fig. 4-9b for the ideal current amplifier, R_i should be *small* compared to R_G to achieve maximum input current while R_o should be *large* compared to R_L to achieve maximum output current.

You might think that the ideal amplifier is a useless concept because it does not exist in the real world. However, it does provide a reference against which real amplifiers can be compared. And, as we shall see, the modern operational amplifier closely approximates an ideal voltage amplifier and as such has numerous circuit applications (see Chapters 11 and 12).

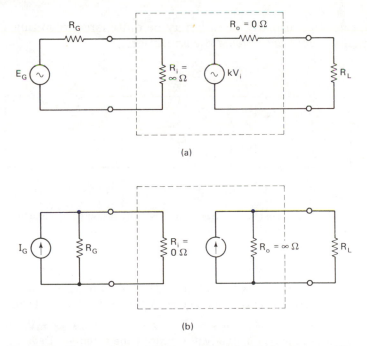

(a)

(b)

Figure 4-9 Characteristics of ideal amplifiers. (a) ideal voltage amplifier. (b) ideal current amplifier.

Once the equivalent circuit of an amplifier has been determined, a number of important characteristics about that amplifier may be readily determined as shown in the previous section. Its gain with various load resistors can be predicted and the maximum power gain possible can also be calculated.

But how do we determine R_i, R_o, and k in the first place? It is the purpose of this section to detail laboratory techniques that will allow you to measure these parameters. As with the amplifiers themselves, the methods used are independent of the type of amplifier.

Open Circuit Voltage Gain

Refer to Fig. 4-10. The open circuit voltage gain is easily measured by disconnecting any load resistance, applying an ac input signal and measuring V_o.

$$k = V_o/V_i \text{ (open circuit)} \qquad (4\text{-}11)$$

Notice that in many cases k may be quite large, requiring V_i to be very small to avoid *overdriving* the amplifier.

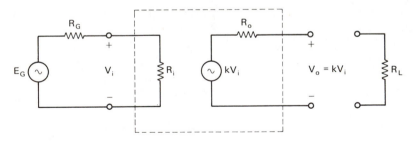

Figure 4-10 The open circuit voltage gain k is easily measured by disconnecting the load resistor. Then $k = V_o/V_i$.

Example 4-15 ──────────────────────────────────

A certain voltage amplifier has $k = 250$. If V_o is limited to 12 V peak-peak by the power supply, what is the maximum input signal that will not cause distortion?

Solution Because $k = V_o/V_i$, then $V_i = V_o/k = 12 \text{ V}/250 = 48 \text{ mV}$. Any input signal larger than 48 mV peak-peak will overdrive the amplifier. Depending on the signal generator being used, it may be necessary to construct a voltage divider to *attenuate* the signal generator's output to the necessary small level.

Input Resistance

Referring to Fig. 4-11, the input resistance may be found by placing a pot in series with the amplifier input. Monitoring E_G' and V_i on a dual channel oscilloscope, adjust the pot until $V_i = E_G'/2$. The value of the pot must now equal R_i.

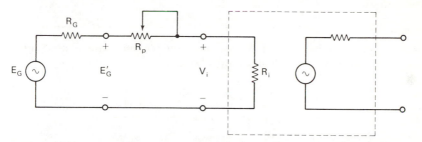

Figure 4-11 Measuring the input resistance of an amplifier. R_p is adjusted until $V_i = E_G'/2$. Then $R_i = R_p$.

Note that the amplifier input resistance is an *ac* quantity and cannot be measured with a conventional dc ohmeter. For this reason, you would also discover that R_i would change if the dc bias changes.

Output Resistance

Figure 4-12 illustrates a method for measuring R_o. Note that this is similar to R_i. The following steps should be performed:

1. Adjust V_i to obtain a typical output signal under open circuit conditions.
2. Place a pot across the amplifier output and adjust until V_o is one half that obtained in (1). The pot should now equal R_o.

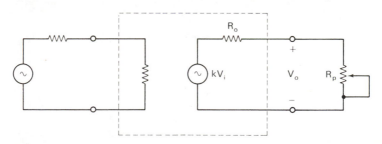

Figure 4-12 Test circuit for measuring R_o. R_p is adjusted until $V_o = kV_i/2$. Then, $R_p = R_o$.

Some amplifiers will go into distortion when attempting this method. This is because the output of the amplifier may not be able to deliver the necessary current when its load resistance is matched. This may be cured by reducing the input (and therefore the output) signal. If the distortion persists, a resistor somewhat larger than R_o may be placed across the output and from the resulting V_o, R_o may be determined.

Example 4-16

A certain amplifier has $V_o = 1$ V peak-peak under open circuit conditions and 0.8 V peak-peak with a 1-kΩ load resistor. Determine R_o.

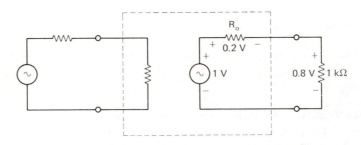

Figure 4-13 Alternate method for determining R_o as explained in Example 4-16.

Solution Refer to Fig 4-13. Because the open circuit voltage is 1 V peak-peak, this is the value of the internal generator kV_i. With the 1-kΩ load in place, the output current is 0.8 V/1 kΩ = 0.8 mA. From this we can determine that R_o = (1 V – 0.8 V)/0.8 mA = 0.2 V/0.8 mA = 250 Ω.

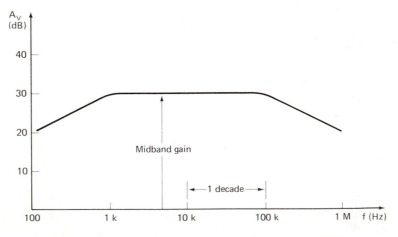

Figure 4-14 Typical frequency response of a voltage amplifier.

4.5 FREQUENCY RESPONSE

Every amplifier has a band of frequencies over which its amplification factor is relatively constant. The gain in this region is referred to as the *midband gain*. The particular frequency response of an amplifier is usually shown by graphing the amplification factor on the vertical with a linear scale while frequency is displayed on the horizontal using a logarithmic scale. This is shown in Fig. 4-14 for a hypothetical amplifier.

The voltage gain is plotted in decibels from 100 Hz to 1 MHz. Note that the logarithmic frequency scale is not linear. However, for each horizontal increment (1 k to 10 k, 10 k to 100 k, etc.), the frequency increases by a factor of ten. This factor of ten is often referred to as a *decade*.

Example 4-17 ───

Frequency increases from 1 kHz to 1 MHz. How many decades is this?

Solution 1 MHz/1kHz = $10^6/10^3 = 10^3$, or three factors of 10. This corresponds to three decades.

Often, the midband gain of an amplifier is selected as a reference level against which the gain at all other frequencies is compared. This is shown in Fig. 4-15. Note that the midband gain now corresponds to 0 dB. Actually, Fig. 4-15 is the same as Fig. 4-14 except that the vertical scale has been shifted so that the 30-dB midband gain now corresponds to 0 dB.

Bandwidth

An amplifier's bandwidth is measured between the *–3-dB* points of its frequency response. These points correspond to those frequencies where the current or voltage gain is 70.7% or the power gain 50% of the midband value.

Example 4-18 ───

Indicate a laboratory technique to measure the bandwidth of a voltage amplifier.

Solution When the voltage gain has fallen by 3 dB from its midband value, the upper and lower bandwidth frequencies may be found.

1. Monitor the input of the amplifier with one channel of the oscilloscope.
2. On the second channel, monitor V_o.
3. Adjust V_i such that $V_o = 1$ V in the midband range of frequencies.
4. Being sure that V_i remains constant, adjust the signal generator frequency until $V_o = 0.7$ V (70.7% of 1 V).

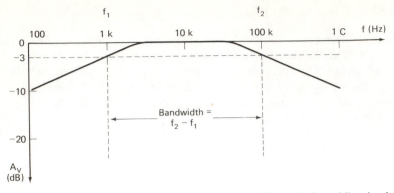

Figure 4-15 Frequency response of a voltage amplifier with the midband gain referenced to 0 dB. The bandwidth is measured between the two frequencies where the response is down by 3 dB.

5. Record the two frequencies where this occurs and compute the bandwidth as $f_2 - f_1$. This is illustrated in Fig. 4-15.

Shunt and Coupling Capacitors

Usually amplifiers are capacitively coupled. This is done because the dc bias levels within the various amplifier stages would be changed if direct connections were made. *Coupling* capacitors allow the ac signal to be passed while blocking any dc voltages. A typical situation for a two-stage amplifier is shown in Fig. 4-16a.

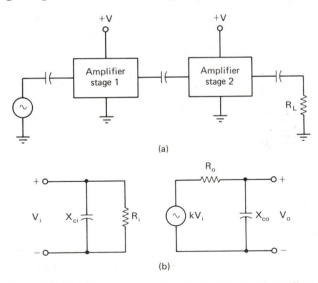

Figure 4-16 (a) Two-stage amplifier illustrating use of coupling capacitors. (b) the effect of shunt capacitors within the active devices is modelled as X_{ci} and X_{co}.

Because the capacitive reactance, X_c, becomes quite large at low frequencies, these coupling capacitors will have the effect of a high pass filter, blocking the low frequencies while passing the higher frequencies. This is shown in Fig. 4-17.

Inside the amplifier itself, the active devices (vacuum tubes, transistors) have small *shunt* or *parallel* capacitances. The effect of these capacitances could be modelled as shown in Fig. 4-16b. The low frequencies will be passed but the high frequencies are shunted to ground. This is characteristic of a low-pass filter, as shown in Fig. 4-18.

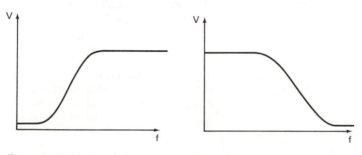

Figure 4-17 Characteristic of high-pass response produced by the coupling capacitors in a typical amplifier.

Figure 4-18 Characteristic low-pass response produced by internal shunt capacitances within the amplifier.

Putting Figs. 4-17 and 4-18 together, the typical frequency response curves shown in Figs. 4-14 and 4-15 results. The particular RC combinations involved will determine the actual upper and lower break points of the response.

Example 4-19

Show that the frequency reponse of the RC circuit in Fig. 4-19 falls off at 20 dB/decade.

Solution We can use a voltage divider to find V_o in terms of V_i remembering that ohms of X_c add to ohms of resistance by the Pythagorean method ($Z = \sqrt{R^2 + X^2}$).

$$V_o = \frac{X_c}{\sqrt{X_c^2 + R^2}} \times V_i = \sqrt{\frac{X_c^2}{X_c^2 + R^2}} \times V_i$$

or

$$\frac{V_o}{V_i} = \frac{1}{\sqrt{1 + R^2/X_c^2}} \tag{4-12}$$

Because $V_o/V_i = A_V$, we may gain more insight into Eq. 4-12 by computing some values for various values of X_c:

When

1. $X_c = R(f = f_o)$; then $A_V = 1/\sqrt{2} = 0.707$, or -3 dB
2. $X_c = R/10(f = 10f_o)$; then $A_V = 1/\sqrt{101} = 0.0995$, or -20 dB
3. $X_c = R/100(f = 100f_o)$; then $A_V = 1/\sqrt{10,001} = 0.0099$, or -40 dB

Note that as X_c decreases by a factor of ten, A_V decreases by 20 dB. Because $X_c = 1/(2\pi f_c)$, a factor of ten decrease in X_c corresponds to an increase in frequency by a factor of ten. The output of the circuit in Fig. 4-19 will therefore have a break frequency at f_o when $X_c = R$ and the response is 3 dB down. From then on, every *increase* in frequency by a factor of ten will result in a *decrease* in the output voltage by 20 dB.

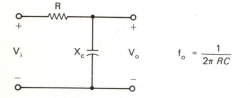

$$f_o = \frac{1}{2\pi RC}$$

Figure 4-19 Circuit used in Example 4-19 to demonstrate the effect of shunt capacitances. The output voltage falls at a rate of 20 dB/decade once the break frequency is reached.

Figure 4-19 models the internal shunt capacitances of the active devices within an amplifier and explains the high-frequency rolloff seen in Figs. 4-15 and 4-18.

When R and C are interchanged, the capacitor now functions as a coupling capacitor and causes the high pass effect explained previously. In this case, as the frequency *increases* by a factor of 10, the response *increases* by 20 dB up to the break frequency. This explains the low-frequency rolloff in Figs. 4-15 and 4-17.

Example 4-20 ————————————————————————

Refer to Fig. 4-19 and plot the phase of V_o versus frequency.

Solution If we assume V_o is driving a large impedance, then the phase diagram of Fig. 4-20a results. The desired phase angle is between V_i and V_c (V_o). Using the

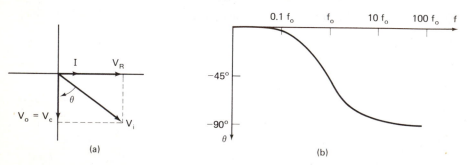

Figure 4-20 (a) Phase diagram for Fig. 4-19. (b) phase plot developed in Example 4-20. The phase angle approaches 90° as the frequency exceeds f_o.

tangent relationship, we see that

$$\Theta = \text{Arc tan} \frac{V_R}{V_c} = \text{Arc tan} \frac{R}{X_c}$$

Proceeding as in Example 4-19 we may calculate values for Θ at various frequencies related to f_o, the frequency where $X_c = R$.

f	X_c	R/X_c	Θ
$0.01f_o$	$100R$	0.01	0.6°
$0.1f_o$	$10R$	0.1	5.7°
f_o	R	1	45°
$10f_o$	$0.1R$	10	84°
$100f_o$	$0.01R$	100	89°

This data is graphed in Fig. 4-20b.

We can now appreciate the complete effect of coupling and shunt capacitances on an amplifier's frequency response. At high frequencies, the *shunt* capacitances will cause the *magnitude* of the voltage gain to begin to roll off such that the response is down by 3 dB at the break frequency (f_o) and falls by 20 dB/decade thereafter. As this is occuring, the *phase* of the output voltage relative to the input will also be changing such that V_o will *lag* V_i by 45° at the break frequency and approach 90° of phase shift as the frequency exceeds $10f_o$.

The same effect occurs due to coupling capacitors but at the opposite end of the frequency spectrum. In this case, as the frequency *decreases*, the gains *decrease* by 20 dB/decade and the phase angle again approaches 90°.

Because many amplifiers introduce 180° of phase shift due to their design, the preceeding analysis predicts a total of 270° of phase shift as the capacitors become active.

Graphs such as those in Figs. 4-19 and 4-20b are commonly referred to as *Bode plots* of magnitude and phase.

4.6 DISTORTION

Ideally, an amplifier can be thought of as a *gain block*. An input signal enters the amplifier and an exact duplicate of this signal appears at the output but increased in amplitude by the gain of the block.

When an amplifier distorts this input signal, the output waveshape is no longer an exact duplicate of the input. For example, overdriving an amplifier with too large an input sine wave signal can cause the output signal to become a square wave. This is a rather severe form of distortion!

Although we may think of distortion as affecting the waveshape of the output signal (*time domain*), we may also think of it in terms of the *frequency domain*. An amplifier with no distortion will amplify an input sine wave and reproduce an output only at the sine wave frequency. However, when distortion is present, many other frequency components will also be present in this output signal. This is because any waveshape can be made up of a sum of many sine waves of various frequencies and amplitudes. For this reason, a square wave has many frequency components in it while a pure sine wave has only a single frequency component. This can be observed using a *spectrum analyzer*.

A spectrum analyzer is similar to an oscilloscope but instead of displaying the time domain of a given signal, it displays the frequency domain. Figure 4-21 illustrates the display of a spectrum analyzer for a pure sine wave and a square wave of the same frequency. The numerous harmonics present in the square wave signal are readily apparent, while the sine wave can be seen to have only one frequency component.

Two types of distortion are common. These are *harmonic* distortion and *intermodulation* distortion. Harmonic distortion produces frequencies harmonically related to the input signal. For example, a 1-kHz sine wave might be harmonically distorted to produce frequency components at 2 kHz, 3 kHz, 4 kHz, and so on. Intermodulation distor-

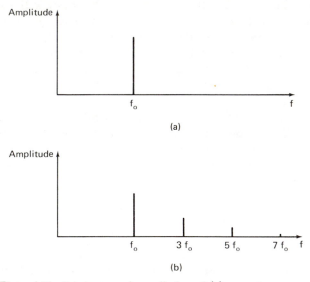

Figure 4-21 Spectrum analyzer display of (a) pure sine wave; (b) square wave of the same frequency.

tion occurs when frequency components other than the harmonic frequencies are generated in the output. This type of distortion is particularly undesirable as the human ear can detect these frequencies and the resulting sound is generally unpleasant.

––––––––––––––––––––––––––––––– KEY TERMS –––––––––––––––––––––––––––

Transducer Electromechanical device that converts energy from one form to another. Often the output of a transducer is an electrical signal that must be amplified by an amplifier.

Gain The gain of an amplifier is expressed as the ratio (not the difference) of output signal to input signal. As such, it may compare voltage, current, or power.

Decibel Decibels are used to express the ratio of two numbers. A sound level increase of 1 dB is just detectable by the human ear.

Amplifier Equivalent Circuit A simple linear model that can be used to represent any amplifier. Its components are the amplifier input and output resistance and the open circuit voltage gain factor.

Matched Load When the output resistance of an amplifier equals the load resistance, maximum power will be transferred to that load resistance. This is called the matched-load condition.

Frequency Response A graph of an amplifier's gain versus frequency is called a frequency response curve. For most amplifiers there is a midband region of frequencies where the gain of the amplifier is relatively constant and upper- and lower-frequency regions where the gain falls off from this midband value in multiples of 20 dB/decade.

Distortion An amplifier distorts its input signal when the output waveshape is not the same as the input waveshape or includes frequency components not found in the input. In practice, all amplifiers will introduce some degree of distortion.

––––––––––––––––––––––––––– QUESTIONS AND PROBLEMS –––––––––––––––––––––––

4-1 Explain why a loudspeaker can be considered a transducer.

4-2 A 1-V rms sine wave is applied to the input of an amplifier. 5 V peak-peak is measured across a 1-kΩ load. Calculate A_V. Is this the same as the open circuit voltage gain?

4-3 An attenuation network yields an output of 1 μW with an input of 1 mW. What is the decibel loss of the network?

4-4 The noise level of a particular tape recording is expressed as 30 dB below the signal level. If the signal power is 10 mW, what is the noise power?

4-5 One decibel is considered to be the smallest sound level change that most people can detect. What power increase does this represent? Express as a percent.

4-6   The 741 operational amplifier has a voltage gain of 200,000. How many dB is this?

4-7   The voltage gain of the 741 rolls off at 20 dB/decade beginning at 10 Hz. If the gain at 10 Hz is 100 dB, how high a frequency may this amplifier be operated at and still have at least a voltage gain of 100?

4-8   A certain audio amplifier is rated as "flat" (±3dB) from 10 Hz to 20 kHz. What does this mean?

4-9   A 25-W input signal is attenuated by 9 dB. What is the output power?

4-10 Draw the equivalent circuit for an amplifier with input resistance of 3 kΩ, output resistance of 100 Ω, and open circuit voltage gain of 20.

4-11 For the amplifier in Problem 4-10, determine the maximum possible power, voltage, and current gains and express as dB. State the conditions for each.

4-12 A 50-Ω signal generator with 100 mV open circuit output voltage produces 70 mV when connected to an amplifier. This amplifier has an open circuit voltage gain of 100 and produces 4 V across a 1-kΩ load. Draw the equivalent circuit for this problem and determine R_i and R_o for the amplifier.

4-13 A certain amplifier has R_i = 1 kΩ, V_i = 100 mV, k = 50, and R_o = 10 kΩ. Calculate the power and voltage gains in dB for R_L = 10 Ω, 100 Ω, 1 kΩ, 10 kΩ, 100 kΩ, and 1 MΩ. Plot your results on graph paper using a semilog scale for resistance on the horizontal and a linear scale for G on the vertical. Explain the significance of the intersection of the two curves.

4-14 Show that the output resistance of an amplifier can be calculated as $R_L(kV_i/V_o - 1)$.

4-15 Why is it preferable to use a logarithmic frequency scale instead of a linear scale when graphing frequency response curves?

4-16 Repeat Example 4-19 for the coupling capacitor case shown in Fig. 4-22.

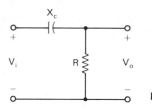

Figure 4-22

4-17 A P.A. system has 1 W of output power. In order for your voice to sound ten times as loud, what power level is needed?

4-18 In Example 4-19, if the resistor is 1 kΩ and the capacitor is 0.01 μF, at what frequency is the output down by 3 dB? *Hint:* X_c = R. What is the phase angle at this frequency?

4-19 An ac voltmeter is used to measure the voltage across a 600-Ω resistor. If the meter indicates 1 V, how many dBm is this?

4-20 In Problem 4-19, if the meter should indicate –3 dBm, what voltage is indicated across the 600-Ω resistor?

Objectives

1. Experience with an integrated circuit.
2. Practice in measuring and calculating amplifier performance data.

Introduction   In this laboratory assignment, an integrated amplifier (op-amp) will be assembled and then used as a test amplifier to measure

1. Input resistance
2. Open circuit voltage gain
3. Output signal swing
4. Output resistance
5. Frequency response.

Components

1 741 op-amp (8-pin DIP)
1 100-Ω pot
1 1-kΩ pot
1 6.8-kΩ resistor
1 100-Ω resistor

Part I: *Signal Generator Output Resistance*

NOTE: The signal generator may be modeled as a signal source in series with a resistor as shown in Fig. 4-23.

STEP 1   Adjust the signal generator output voltage under *open circuit* conditions for 0.5 V peak-peak, at $f = 1$ kHz.

STEP 2   Connect a 100-Ω pot across the output of the generator and adjust until the voltage across the pot is 0.25 V peak-peak ($E_G/2$). Measure R_p; $R_p = R_G$.

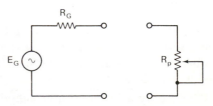

Figure 4-23   Equivalent circuit for a signal generator. R_p is used to determine the value of R_g.

Question 1   Refer to the service manual for the signal generator and determine its output resistance. Compare to your measurement.

Part II: *Open Circuit Voltage Gain and Signal Swing*

NOTE: The test amplifier has an equivalent circuit illustrated in Fig. 4-4. The open circuit voltage gain can be measured using the method outlined in Section 4-4 and illustrated in Fig. 4-10.

STEP 1   Assemble the test amplifier shown in Fig. 4-24 on your breadboard. Pay careful attention to the IC orientation for pin 1 and the ±12-V connections.

STEP 2   Apply a 0.1 V peak-peak sine wave to the input of the amplifier at $f = 1$ kHz.

STEP 3   Measure V_o and calculate k under *open circuit* conditions.

STEP 4   Increase the input signal amplitude until the output voltage begins to clip. Record the maximum peak-peak output voltage (signal swing) before distortion.

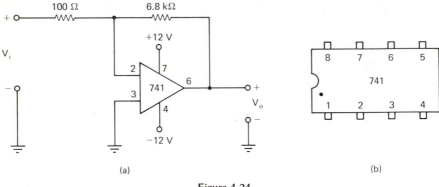

Figure 4-24

Question 2   What limits the absolute maximum output voltage for this circuit?

Part III: *Output Resistance*

NOTE: The output stage of the test amplifier is modeled as a signal source and series resistor as shown in Fig. 4-4. The value of R_o can be measured using the method outlined in Section 4-4 and illustrated in Fig. 4-12.

STEP 1   Apply a 10-mV input signal at $f = 1$ kHz to the amplifier and measure the *open circuit* output voltage.

STEP 2 Connect a 100-Ω pot across the output and adjust to obtain one half the output voltage measured in step 1. Measure R_p; $R_p = R_o$. If the output waveform becomes distorted, refer to Example 4-16 for an alternate method.

Part IV: *Input Resistance*

NOTE: The equivalent circuit for the input stage of the test amplifier is shown in Fig. 4-11. R_i can be measured following the method outlined in Section 4-4 and illustrated in Fig. 4-11.

STEP 1 Adjust E_G' to 0.5 V peak-peak at $f = 1$ kHz.

STEP 2 Adjust R_p until $V_i = 0.25$ V peak-peak. Measure R_p; $R_i = R_p$.

(NOTE: Be certain that E_G' does not change as the pot is varied. Use two channels of the oscilloscope to observe this.)

Part V: *Frequency Response*

NOTE: As discussed in Section 4-5, all amplifiers have a limited frequency range over which they provide constant gain. Eventually a frequency is reached where the gain begins to fall ("roll off") and ultimately reaches unity or less.

STEP 1 Disconnect all loads and test pots.

STEP 2 Adjust V_i to 0.1 V peak-peak and f to 100 Hz.

STEP 3 Record V_o and the voltage gain (V_o/V_i). Be certain V_o is not distorted and, if it is, reduce V_i.

STEP 4 Adjust f to 1 kHz, being certain V_i does not change, and repeat step 3. Continue recording V_o and the amplifier voltage gain at decade intervals to 1 MHz.

STEP 5 Graph your data using semilog paper, f on the log scale (horizontal) and k on the linear scale (vertical).

STEP 6 From your graph, determine the -3-dB frequency for the test amplifier. Note that this amplifier is *dc coupled* and therefore does not have the characteristic *low-frequency* rolloff associated with a coupling capacitor.

FIVE

AC AND DC EQUIVALENT CIRCUITS

In this chapter, we will learn to analyze a transistor amplifier by considering its dc and ac *equivalent* circuits. The actual performance of the amplifier will then be the sum (by superposition) of these two circuits.

When analyzing the transistor's dc bias (the dc equivalent circuit), the key points to remember are that $I_C = \beta I_B$ and that $V_{BE} = 0.6$ V for a silicon transistor. On the other hand, the ac equivalent circuit will center around a "model" for the transistor. This model will *simulate* the transistor as a linear circuit element made up of resistors and a current source. With this circuit, we can use Ohm's Law to calculate the important ac properties of the amplifier as discussed in the preceding chapter. You must always remember, however, that the ac equivalent circuit is only valid provided the transistor is biased in the *active region* (base-emitter junction forward-biased and collector-base junction reverse-biased).

5.1 A SMALL SIGNAL DIODE MODEL

Equivalent Circuits

Before proceeding to a transistor amplifier, let us consider a simpler diode circuit. Figure 5-1a is the schematic diagram of an amplitude-limiting circuit, similar to that found in FM receivers used to remove amplitude variations of the incoming signal. We might analyze this circuit by considering it to be *two* separate circuits in one. The dc bias is supplied by E and a dc current flows through the choke (inductor), resistor, and diode. Capacitor C blocks this dc from flowing into the ac source. A second circuit exists for ac. In this circuit, ac flows through C

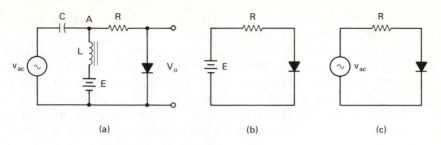

Figure 5-1 (a) Circuit used to develop the small signal model for the diode. (b) the dc equivalent circuit. (c) the ac equivalent circuit.

(X_c is assumed small), the resistor, and the diode. The choke blocks this current due to its X_L.

Only in the resistor and diode are the ac and dc components mixed together. In Fig. 5-1b and c, the dc and ac *equivalent* circuits are drawn. You should be able to see that these circuits just include those portions of the main circuit in which dc or ac only is flowing.

The dc source, E, biases the diode at some Q or operating point. This may be represented graphically as shown in Fig. 5-2. As the value

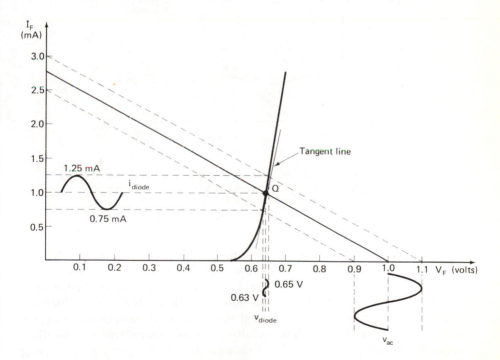

Figure 5-2 Load-line representation of the circuit in Fig. 5-1(a).

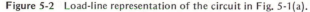

of E is changed, the load line moves parallel to the one shown and the Q point moves up or down the diode curve.

If we assume a dc bias of 1.0 V and an ac source voltage of 0.1 V peak, then by superposition the signal at point A in the circuit will be a 0.1 V peak sine wave "riding on" a 1.0-V dc level. This is shown in Fig. 5-3a. As far as the load line is concerned, there now appear to be *many* load lines just as if someone was varying the value of E from 0.9 V to 1.1 V. This is represented by the dashed load lines in Fig. 5-2.

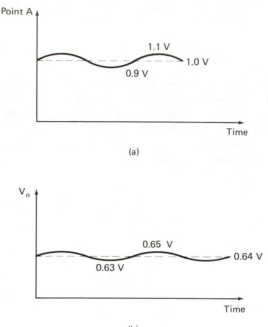

(a)

(b)

Figure 5-3 Combined ac and dc waveforms for the circuit in Fig. 5-1(a). In (a) the input ac signal is riding on the 1.0-v dc source level. In (b) a corresponding ac signal appears across the diode but riding on the Q point dc level.

Now the crucial point: the voltage and current of the diode must follow the load line. When point A swings upward to 1.1 V, the Q point must move up the diode curve to $V_F = 0.65$ V. Similarily, when point A falls to 0.9 V, the diode voltage falls to 0.63 V. Because the load line moves up and down in a sinusoidal manner, these variations across the diode are also sinusoidal. The resulting voltage waveform for the diode is shown in Fig. 5-3b. Note that this waveform is a sine wave riding on a dc level. In fact, the dc level is the Q point value for the diode when no ac is present.

Example 5-1 ———————————————————————————————————

Explain how a sine wave can appear across a diode. Doesn't this violate the concept that a diode can conduct in one direction only?

Solution Although a sine wave is observed across the diode, this sine wave is actually a form of *pulsating* dc. Referring to Fig. 5-3b, the diode voltage never goes negative and therefore the current through the diode is always in the same direction and the diode is always forward-biased. The *amplitude* of this current is varying and this is why a sine wave is observed. (*Note:* Your oscilloscope must be in the dc position to observe this combined ac–dc signal).

It is interesting to see if the value of the dc source voltage will affect the sine wave across the diode. For example, what happens if E is adjusted to 0 V? The Q point of the circuit must now move to the origin. Now the ac signal will cause the diode to conduct for the positive half-cycle but will reverse-bias the diode for the negative half-cycle. The circuit actually becomes a *half-wave rectifier*! A sine wave will only appear across the diode provided its Q point is somewhere on the vertical part of its IV curve.

Example 5-2 ———————————————————————————————————

Determine if the circuit shown in Fig. 5-1a functions as an amplifier or attenuator.

Solution The gain of the circuit is V_o/V_i and for this circuit 0.01 V peak/0.1 V peak = 0.1. Because the gain is less then 1, the circuit functions as an attenuator.

This is a general result to be expected with a diode. Diodes are *passive* components and cannot provide amplification. This result is particularly clear when the ac model of the diode is understood.

An AC Model for the Diode

The key word here is *ac*. The model must represent the IV characteristics of the diode for ac signals *only*. Referring to Fig. 5-2, the diode operating point varies up and down the characteristic curve as the ac source varies in amplitude. If the ac signal is not too large (small-signal), this variation about the Q point on the diode curve is nearly a *straight* line. Another way of saying this is that for small signals, the diode has a *linear* current-voltage relationship about the Q point. In Fig. 5-2, this is illustrated by the line-drawn *tangent* to the Q point.

Any time a component exhibits a linear current-voltage relationship, a resistor is implied. For small signal variations about the Q point, the diode appears to be a resistor to the ac signal source. The value of this resistance can be found by finding the inverse slope of the tangent line drawn through the Q point.

Example 5-3

Determine the ac resistance of the diode whose IV curve is shown in Fig. 5-2.

Solution   The ac resistance can be calculated as $\Delta V/\Delta I$ at the Q point. Figure 5-4 shows a triangle constructed out of the tangent line drawn in Fig. 5-2. From this we compute

$$r_{diode} = \frac{\Delta V}{\Delta I} = \frac{0.02 \text{ V}}{0.5 \text{ mA}} = 40 \text{ } \Omega$$

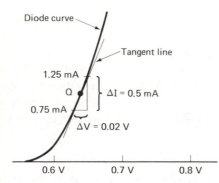

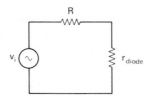

Figure 5-4 Triangle method for determining the ac resistance at the Q point. $r_{diode} = \Delta V/\Delta I$.

Because the diode ac equivalent circuit is just a resistor, a voltage divider results between this resistor and the current limiting resistor R. This is shown in Fig 5-5 where the ac equivalent circuit for the diode has been redrawn with a resistor in place of the diode. Obviously, no voltage gain is possible with this circuit no matter what the value of r_{diode}. This is what is meant by the diode being a *passive* component.

It is important to remember that the diode ac resistance is valid only for ac and only at the specific Q point used. At a new Q point, its value will be different.

Figure 5-5 The ac equivalent circuit with the diode replaced by its small-signal equivalent resistance.

The diode also has a *dc resistance*. This is just the ratio of dc voltage to dc current at the Q point. For our example, this is 0.64 V/ 1 mA = 640 Ω. This means that the diode could be replaced by a 640-Ω resistor and there would be no change to the *dc* current and voltages in the circuit. The 640 Ω is only valid at this one Q point and a new value would result if the Q point was changed. The dc resistance of the diode is not used too often but you should know the difference between it and the ac resistance. Sometimes the ac and dc resistances are referred to as the *dynamic* and *static* diode resistances, respectively.

It can be shown from calculus that the ac resistance of a silicon diode at room temperature can be approximated with the formula

$$r_{\text{diode}} = \frac{0.025 \text{ V}}{I_Q} \tag{5-1}$$

where I_Q is the dc current at the Q point. This is much more convenient than attempting to find the slope of a tangent line drawn through the Q point. Note that the larger the current, the lower the ac resistance. This corresponds to moving further up the diode curve where the characteristic is more vertical and the resistance is indeed less.

Example 5-4 ———————————————————————————————————

Calculate the ac resistance of the diode in Example 5-3 using Eq. 5-1.

Solution   At the Q point, $I = 1$ mA. Then, $r_{\text{diode}} = 0.025$ V/1 mA = 25 Ω. This compares to 40 Ω calculated from the graph in Example 5-3. Allowing for small errors in the graphical method, the correlation is reasonable.

5.2   A BIPOLAR TRANSISTOR MODEL

The main purpose for developing an ac model for any component is that we may then use conventional circuit analysis to *calculate* that circuit's performance. As we have just seen, estimating the ac voltage dropped across a diode is a simple matter once its ac resistance is known. Similarly, predicting the performance of a transistor amplifier is greatly simplified once its ac model is known. In fact, the *amplifier equivalent circuit* discussed in Chapter 4 can be derived directly once the transistor has been replaced by its ac model.

Deriving the Model

Let us assume we have a common-emitter transistor amplifier biased at some Q point in the *active region* and use the characteristic transistor curves to determine an ac model at this Q point. We will ignore the external biasing resistors and look for the moment at the transistor only. In Fig. 5-6, the *input* and *output* of such a circuit are defined.

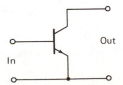

Figure 5-6 Defining the input and output of a common-emitter amplifier.

Starting with the *input* set of terminals, we see a current–voltage relationship such as that shown in Fig. 5-7a. This is the familiar (at least I hope it's familiar!) diode curve. When an ac signal is applied to the base of the transistor, the Q point will move up and down this charac-

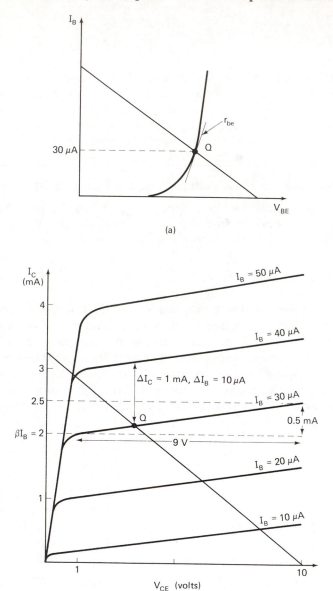

(a)

(b)

Figure 5-7 Characteristic curves used to derive the transistor's ac model.

teristic curve, and if this input can be considered small-signal, the ac equivalent circuit for the input will be a resistor. This is similar to the small-signal diode discussed previously. In fact, we can use the result of the preceding section to determine its value. Because this resistor appears between base and emitter it will be called r_{be}.

$$r_{be} = \frac{0.025 \text{ V}}{I_{BQ}} \qquad (5\text{-}2)$$

where I_{BQ} is the *dc* base current at the *Q* point.

When looking into the output set of terminals, the IV characteristics of Fig. 5-7b are seen. If the curves in the active region were perfectly flat, they could be modeled by a constant current source independent of V_{CE}. This type of source is called a *dependent* source because its value *depends* on I_B. Because our model is to apply for ac, we are really concerned about changes in I_B causing changes in I_C. For this reason, we will use the ac β and label the current source βi_b.

$$\beta_{ac} = \frac{\Delta I_C}{\Delta I_B} \qquad (5\text{-}3)$$

Most transistors do not have perfectly flat output curves and, as a result, if V_{CE} increases, the collector current will also increase as shown in Fig. 5-7b. An increasing voltage causing an increasing current is representative of Ohm's Law and this means our model should have a resistance component to model this effect. We will call this resistor the *transistor output resistance*, r_o. It is found as the inverse slope of the characteristic output curve at the *Q* point.

$$r_o = \frac{\Delta V_{CE}}{\Delta I_C} \qquad (5\text{-}4)$$

The completed transistor model is shown in Fig. 5-8. This model reproduces the input and output characteristics of the transistor: a small signal resistance between base and emitter and a collector current made up of βi_b plus v_{ce}/r_o. Although the model does not resemble the

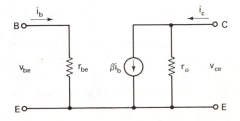

Figure 5-8 Equivalent circuit used for a bipolar transistor biased in the active region.

original transistor physically, it does include components we can work with (i.e., Ohm's Law is applicable).

Again, a critical point to remember is that the model applies for ac only. In addition, it assumes the transistor to be biased in the active region.

Example 5-5

Using the characteristic curves shown in Fig. 5-7, determine the values of r_{be}, β_{ac}, and r_o at the indicated Q points.

Solution   Referring to the $I_B V_{BE}$ set of curves, r_{be} may be calculated.

$$r_{be} = \frac{0.025 \text{ V}}{I_{BQ}} = \frac{0.025 \text{ V}}{30 \text{ } \mu A} = 833 \text{ } \Omega$$

Referring to the $I_C V_{CE}$ set of curves, the ac value of β can be estimated.

$$\beta_{ac} = \frac{\Delta I_C}{\Delta I_B} = \frac{1 \text{ mA}}{10 \text{ } \mu A} = 100$$

The value of r_o is found by finding the inverse slope of the characteristic curves at the Q point. Following the curve $I_B = 30 \text{ } \mu A$, I_C increases from 2 mA at $V_{CE} = 1$ V to 2.5 mA at $V_{CE} = 10$ V. From this, r_o is found as

$$r_o = \frac{\Delta V_{CE}}{\Delta I_C} = \frac{(10 \text{ V} - 1 \text{ V})}{2.5 \text{ mA} - 2 \text{ mA}} = 18 \text{ k}\Omega$$

The characteristic curves in Fig. 5-7 can be readily obtained using a *curve tracer*. This means that the parameters for a given transistor can be measured if a curve tracer is available. Of course, the manufacturer's data sheets can always be consulted for approximate figures. In this case, we need to know something about the h-parameters. This is discussed in the next subsection.

More Complex Models

If you are thinking that our model is the only one possible for the transistor, you should be aware that several others are also used. One such model is the hybrid or *h-parameter* model, shown in Fig. 5-9. This model applies the principals of Thevenin's and Norton's theorems, which state that any circuit can be reduced to a series resistance and voltage source, or parallel resistance and current source.

The h-parameter model is very similar to the model presented in the previous section, except for the voltage source $h_{re} v_{ce}$. In most cases,

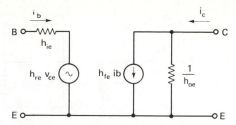

Figure 5-9   The *h*-parameter model.

however, h_{re} is quite small and the two models are then essentially the same. The h-parameter model has an advantage in that many transistor data sheets specify the various h-parameters. Table 5-1 compares parameters for the two models presented.

TABLE 5-1

Characteristic	h-Parameter	Small-signal Model
Input resistance	h_{ie}	r_{be}
Output resistance	$1/h_{oe}$	r_o
Current gain	h_{fe}	β_{ac}
Voltage feedback ratio	h_{re}	assumed 0

Using the h-parameters presents several problems, however. Refer back to the 2N2218A–2N2222A transistor data sheets in Fig. 3-11. Note that the h-parameters are included under the small-signal characteristics. Also note the wide variations from minimum to maximum for all parameters. The value of h_{fe}, for example, varies from a low of 75 to a high of 375 for the 2N2222A!

The point of this should be clear. Even though the h-parameters are given on typical transistor data sheets, a large variation from one transistor to another can be expected, as the data sheets indicate. This problem is further complicated by the fact that *conversion* formulas must be used to convert *common-emitter* h-parameters to *common-base* or *common-collector* parameters.

Of course, if the model parameters are accurate, very good correlation between calculations and the real world can be obtained. The problem is that the model itself must by necessity become more complex to account for *all* the various properties of the transistor. A model suitable for *computer simulation* is shown in Fig. 5-10. Note the inclusion of several capacitors and inductors. Our simple model has assumed operation in the *midband* range of frequencies where the X_c and X_L values are negligible. If the frequency response of an amplifier is to be calculated, these quantities must be taken into consideration.

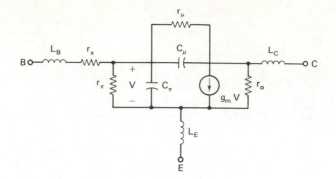

Figure 5-10 A more complex model for the transistor must include junction capacitances and lead inductances.

A good lesson to learn from this is not to expect too much from any model you use. The equations we develop using the small-signal model help us to see which components to change or adjust to meet a certain amplifier's requirements, and in this respect our model is extremely helpful. If high accuracy is required, a computer simulation must be done using a more complex model for the transistor.

5.3   APPLYING THE SMALL-SIGNAL MODEL

We are now prepared to apply our small-signal model to the simple fixed-bias transistor amplifier shown in Fig. 5-11. Our objective is to analyze this circuit and obtain equations for voltage and current gain and input and output resistance.

Step 1: The DC Equivalent Circuit

The dc equivalent circuit is obtained by drawing only those components with dc currents in them. In Fig. 5-11, the two capacitors, C_1

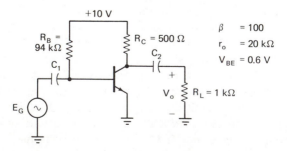

Figure 5-11 Fixed-bias common-emitter amplifier to be analyzed using the small-signal model.

and C_2, couple the ac signal into and out of the amplifier while blocking the dc current from entering the signal generator or load resistor. For this reason, the signal generator and load resistor are not included in the *dc equivalent circuit* shown in Fig. 5-12.

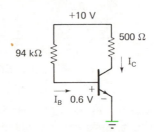

Figure 5-12 DC equivalent circuit for Fig. 5-11.

The dc voltages and currents for this circuit may be calculated as done previously in Chapter 3 and repeated in the following example.

Example 5-6 ─────────────────────────────────

Calculate I_B, I_C, and V_{CE} for the circuit shown in Fig. 5-12. Determine if the amplifier is biased in the active region.

Solution The base current is found using Eq. 3-4: $I_B = (V_{CC} - 0.6 \text{ V})/R_B = (10 \text{ V} - 0.6 \text{ V})/94 \text{ k}\Omega = 100 \ \mu\text{A}$. Now I_C is found as: $I_C = \beta I_B = 100 \times 100 \ \mu\text{A} = 10 \text{ mA}$, and $V_{CE} = V_{CC} - I_C R_C = 10 \text{ V} - 10 \text{ mA}(500 \ \Omega) = 5 \text{ V}$.

The transistor is indeed biased in the *active region* because the base-emitter junction is forward-biased while the base-collector junction has 0.6 V – 5 V = –4.4 V of reverse bias.

This completely solves the dc equivalent circuit.

Step 2: The AC Equivalent Circuit

As might be expected, the ac equivalent circuit includes all components with ac currents in them. We will assume all capacitors to have been chosen such that their reactances are negligible at the midband frequency of calculation and therefore replace them with *short circuits* in the ac equivalent circuit.

A slightly tricky point to understand is that the supply voltage, V_{CC}, will appear as *ac ground*. One way to see this is to recall that most power supplies include a large *filter capacitor* across their output that will have negligible reactance at our frequency of calculation. An even more fundamental explanation, however, is that any point in the circuit that maintains a constant voltage (i.e., dc) will appear to be 0 V to ac.

For these reasons, a circuit may have a *dc ground* and an *ac ground*. The dc supply voltage will always be considered an ac ground. The resulting *ac equivalent circuit* is shown in Fig. 5-13. Study this circuit carefully to be sure you understand how it is derived from Fig. 5-11.

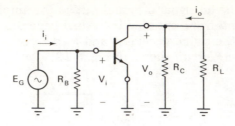

Figure 5-13 AC equivalent circuit for Fig. 5-11.

Even with the ac equivalent circuit, we are unable to analyze the amplifier until the transistor is replaced with its model. When this is done, the resulting circuit will only have components that obey *Ohm's Law* and can be analyzed accordingly.

In Fig. 5-14 the transistor has been replaced by its *small-signal* equivalent. The base, collector, and emitter terminals are shown with capital letters. When comparing Fig. 5-14 to the original amplifier circuit in Fig. 5-11, there does not seem to be much similarity. However, with the model in place, we can now calculate several properties for this amplifier.

$$R_B = 94 \text{ k}\Omega$$
$$r_{be} = 250 \ \Omega$$
$$r_o = 20 \text{ k}\Omega$$
$$R_C = 500 \ \Omega$$
$$R_L = 1 \text{ k}\Omega$$

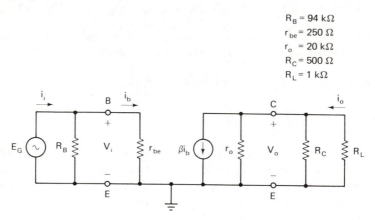

Figure 5-14 The ac equivalent circuit with the transistor replaced by its small-signal equivalent model. The base, collector, and emitter terminals are labeled.

Voltage Gain

From Chapter 4 we know that $A_V = V_o/V_i$. Applying this definition to the circuit shown in Fig. 5-14, we can calculate the *input voltage* as $i_b \times r_{be}$. For the output circuit, note that r_o, R_C, and R_L are all in parallel. If these three resistors are called R_L', then the output

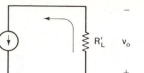

Figure 5-15 The output stage of Fig. 5-14 is redrawn, replacing the parallel combination of r_O, R_C, and R_L with R_L'. The ac output voltage is $-\beta i_b R_L'$.

stage can be redrawn as shown in Fig. 5-15. The ac *output voltage* is then $-\beta i_b R_L'$, where the minus sign indicates a 180° phase shift with respect to the input. The voltage gain is

$$A_V = \frac{V_o}{V_i} = \frac{-\beta i_b R_L'}{i_b r_{be}} = \frac{-\beta R_L'}{r_{be}} \tag{5-5}$$

Substituting numbers, $r_{be} = 0.025V/I_B = 0.025V/100\ \mu A = 250\ \Omega$; and $A_V = -100(328\ \Omega)/(250\ \Omega) = -131$, where R_L' is calculated to be 328 Ω (1 kΩ || 500 Ω || 20 kΩ).

NOTE: (||) signifies "in parallel with."

Current Gain

The current gain was defined in Chapter 4 to be the ratio of *output current* to *input current*. The easy way to determine this for Fig. 5-14 is to pretend R_B, r_o, and R_C are all infinitely large. For this special case, the input current is the same as i_b and the dependent source current, βi_b, all flows into R_L. In this case, the current gain is β. Of course, R_B, r_o, and R_C are *not* infinite and, as a result, some of the input current is lost to R_B and some of the βi_b current is lost due to r_o and R_C. Because the current gain is at *most* β, the formula for current gain is β multiplied by current divider equations for the input and output stages.

$$A_I = \beta \times \frac{R_B}{R_B + r_{be}} \times \frac{r_o \parallel R_C}{r_o \parallel R_C + R_L} \tag{5-6}$$

Again, substituting numbers

$$A_I = 100 \times \frac{94\ k\Omega}{94\ k\Omega + 0.25\ k\Omega} \times \frac{0.49\ k\Omega}{0.49\ k\Omega + 1\ k\Omega}$$

$$= 100 \times 0.997 \times 0.329 = 32.8.$$

Note that most of the current gain is lost due to the small R_C resistor compared to the load resistor.

Input Resistance

The input resistance is that "seen" by the *source* E_G. Referring to Fig. 5-14, this is simply R_B in parallel with r_{be}.

$$R_i = R_B \parallel r_{be} \tag{5-7}$$

In this case, $R_i = 94$ kΩ $\parallel$ 250 Ω = 249 Ω. Because r_{be} is usually in the kΩ range and R_B is usually a factor of ten greater than this, the input resistance is usually determined by r_{be}.

Output Resistance

The output resistance of the amplifier is defined to be the resistance "seen" by the *load resistor*. From Fig. 5-14, this is r_o in parallel with R_C.

$$R_o = r_o \parallel R_C \tag{5-8}$$

And for our circuit, R_o = 20 kΩ $\parallel$ 0.5 kΩ = 488 Ω.

Example 5-7

Draw the amplifier equivalent circuit of the fixed-bias common-emitter amplifier in Fig. 5-11. Determine the maximum possible power gain in dB.

Solution The previous sections have worked out all the parameters needed for the model. However, the voltage gain factor k must be calculated for *open circuit* conditions. We can use Eq. 5-5 to recalculate the voltage gain for an open circuit condition if R_L' is replaced with r_o in parallel with R_C only.

$$A_V \text{ (open circuit)} = k = -\frac{\beta(r_o \parallel R_C)}{r_{be}} = -\frac{100\,(488)}{250} = -195$$

The resulting amplifier equivalent circuit is shown in Fig. 5-16. The maximum possible power gain occurs for a *matched-load* condition and requires $R_L = R_o$ = 488 Ω. Using Eq. 4-10,

$$G = \frac{k^2}{4} \times \frac{R_i}{R_L} = \frac{(195)^2}{4} \times \frac{249}{488} = 4851$$

Converting G to decibels, 10 log G = 37 dB.

We have now completely analyzed the simple common-emitter amplifier in Fig. 5-11 and determined its dc bias and amplifier equivalent

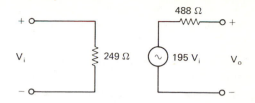

Figure 5-16 The amplifier equivalent circuit for Fig. 5-11.

circuit. Using the techniques presented in the last chapter it would now be possible to construct this amplifier in the lab and *measure* these same ac quantities. Figure 5-17 summarizes the equations for this circuit.

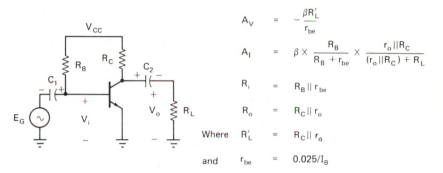

$$A_V = -\frac{\beta R'_L}{r_{be}}$$

$$A_I = \beta \times \frac{R_B}{R_B + r_{be}} \times \frac{r_o \| R_C}{(r_o \| R_C) + R_L}$$

$$R_i = R_B \| r_{be}$$

$$R_o = R_C \| r_o$$

Where $\quad R'_L = R_C \| r_o$

and $\quad r_{be} = 0.025/I_B$

Figure 5-17 The fixed-bias common-emitter amplifier and a summary of its important ac equations.

Coupling Capacitors

The two coupling capacitors used in the fixed-bias circuit (C_1 and C_2 in Fig. 5-17) have the job of *blocking* the dc current from the ac signal source and load resistor while *coupling* the ac signal into the amplifier and the amplified ac output signal to the load resistor. The values of these two capacitors are set by the *lowest* frequency of operation and by the resistance that the ac signal must be coupled into. Figure 5-18 illustrates these two cases for C_1 and C_2. In both cases, X_c should be *small* compared to the resistance. Because the amplifier may be called upon to work over a range of frequencies, C should be calculated at the *lowest* frequency of operation where X_c will be largest. In this way, X_c can only get smaller for higher operating frequencies. Accordingly, a common rule of thumb is to *choose X_c to be one tenth of R at the lowest frequency of operation.*

$$C \text{ (coupling)} = \frac{1}{2\pi f X_c} = \frac{1}{2\pi f (R/10)} \tag{5-9}$$

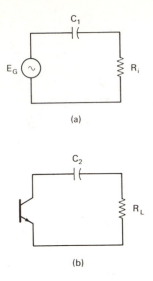

(a)

(b)

Figure 5-18 Coupling capacitors. (a) C_1 couples the ac signal into the amplifier. (b) C_2 couples the output ac signal to the load resistor.

Example 5-8 ——————————————————————————————————————

Determine values for C_1 and C_2 in Fig. 5-11. Assume $100\ \text{Hz} \leqslant f \leqslant 20\ \text{kHz}$.

Solution   Using Eq. 5-9 with $R_i = 249\ \Omega$ and $R_L = 1\ \text{k}\Omega$,

$$C_1 = \frac{1}{(2\,\pi \times 100 \times 24.9)} \cong 64\ \mu\text{F}$$

$$C_2 = \frac{1}{(2\,\pi \times 100 \times 100)} \cong 15.9\ \mu\text{F}$$

Because capacitors larger than $1\ \mu\text{F}$ are usually electrolytic, care must be taken to observe the voltage polarity markings when inserting these components into the circuit. As shown in Fig. 5-17, the positive ends of the capacitors should be connected to the amplifier.

5.4 GENERAL EFFECTS OF LOAD RESISTANCE

In Chapter 4, we saw that the addition of a load resistor reduces the voltage gain of an amplifier. As can be seen in Fig. 5-19, the amplified input signal *voltage divides* between R_o and R_L, causing this reduction in gain.

The voltage gain of the common-emitter amplifier is $-\beta R_L'/r_{be}$, which is directly dependent on R_L'. Because R_L' is the parallel combination of r_o, R_C, and R_L, A_V will have a maximum value when there is no load resistor (*open circuit*). In this case, A_V will be the open circuit voltage gain k.

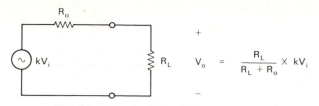

Figure 5-19 The output voltage of an amplifier is reduced due to the load resistor.

This reduction in voltage gain due to the load resistance is a general result that can be expected for all amplifiers. For example, an amplifier with an open circuit voltage gain of 100 and output resistance of 1 kΩ will only have a gain of nine when driving a 100-Ω load.

The opposite comments are true for the current gain of an amplifier. In Chapter 4, the general tendencies for A_V and A_I were covered and Fig 4-7 graphed these two characteristics against load resistance.

Signal Swing

There is one aspect of an amplifier's performance that its amplifier equivalent circuit does not cover. If a certain amplifier has an open circuit voltage gain of 100 and 1 V is applied to the input, will the output be 100 V? The answer is most likely no. The amplifier equivalent circuit predicts that the output signal will be a *linear* copy of the input signal but it is *not* possible to exceed the limits established by the power supply. If you try to do so, the amplifier will enter *saturation* and *cutoff* and the output signal will resemble a square wave rather than a sine wave.

Example 5-9

A common-emitter amplifier has an open circuit voltage gain of 100 and a 10-V power supply. What is the maximum input signal that will not overdrive the amplifier?

Solution If the bias point is properly chosen ($V_{CE} = 5$ V), the absolute maximum output signal swing will be 10 V peak-peak. Because $A_V = V_o/V_i$, V_i can be calculated: $V_i = V_o/A_V = 10$ V$/100 = 0.1$ V peak-peak. Due to the nonlinearity of the transistor's output curves in the active region, the input signal will probably have to be even less than this amount to prevent distortion.

The previous example assumed an open circuit condition and resulted in a maximum of 10 V peak-peak for the output signal swing. This is illustrated on the collector curves in Fig. 5-20. The Q point is selected at $V_{CE} = 5$ V and, following the *dc load line*, the signal swing is limited by saturation and cut-off to approximately 10 V peak-peak.

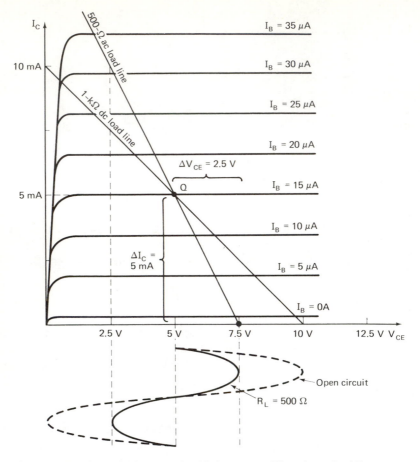

Figure 5-20 When a load resistor is added to an amplifier, the ac load line must be used and the voltage gain and signal swing are reduced.

What happens when a load resistor is added? We already know that the voltage gain will fall as previously discussed, but what happens to the signal swing? Can we still achieve 10 V peak-peak? What we must remember is that the collector "sees" a different resistance for ac than it does for dc. This means that an *ac load line* must be drawn on the characteristic curves when R_L is present.

Chapter 3 covered a method for determining the dc load line, and the ac load line is found in a similar way. When the ac input signal is *zero*, the operating point must be at the dc Q point. This is one point on the ac load line. In Fig. 5-20, R_L' is assumed to be 500 Ω. When the current *changes* by 5 mA, the collector voltage will have to change by 5 mA $\times$ 500 Ω = 2.5 V. This results in a second point for the ac load line at I_C = 0 A, and V_{CE} = 7.5 V (or a third point at 10 mA and 2.5 V). The ac load line is drawn between these two points.

Examining this load line, we can see that the signal swing is going to be less than that for the open circuit case. For the 500-Ω load line shown in Fig. 5-20, the maximum signal swing will be 2.5 V peak or 5 V peak-peak limited by the cutoff condition. In general, the lower the value of R_L' , the more *vertical* the ac load line and the *smaller* the signal swing possible. In summary, the effect of adding a load resistor to an amplifier will be to *reduce* the voltage gain and signal swing when compared to the open circuit case.

By examining the ac signal swing predicted in Fig. 5-20, we can also see that it is *not necessary* to choose a Q point centered between V_{CC} and 0 V. This is because the ac load line will restrict our signal swing to less than this full amount. In fact, the whole premise of the small-signal model is exactly that. The signal variations must be *small* about the dc operating point, generally, a few volts only. Attempts to obtain larger output voltages will only result in distorted output signals. For this reason, the Q point should be selected to cause the transistor to operate in its most *linear* region.

If it is desired to *maximize* the output signal swing, Fig. 5-20 indicates that this can be accomplished by (1) increasing the load resistance (eventually the ac and dc load lines will coincide) or (2) bias the transistor at a larger collector current (this will increase ΔI_C in Fig. 5-20). If true large signal operation is desired, other types of amplifiers should be considered (class B). This is discussed in Chapter 7.

Example 5-10 ——————————————————————————————

Design a fixed-bias common-emitter amplifier using the characteristic curves in Fig. 5-21a. Assume V_{CC} = 12 V, R_L = 4.7 kΩ, and r_o = 20 kΩ. Calculate A_V, A_I, R_i, and R_o, and estimate the maximum output signal swing.

Solution We must first select a Q point. Because V_{CC} = 12 V, we will choose 6 V for V_{CE} and bias the transistor in the middle of its linear region. There are many values of I_B with V_{CE} = 6 V and the Q point shown in Fig. 5-21 is chosen arbitrarily to ensure operation in the active region. (*Note:* If the power dissipation of the transistor was known, this could influence the Q point selection.).

Now that I_B is known, the value of R_B can be calculated. $R_B = (V_{CC} - V_{BE})/I_B$ = 11.4 V/15 μA = 760 kΩ. The collector resistor is found in a similar manner: $R_C = (V_{CC} - V_{CE})/I_C$ = 6 V/2 mA = 3 kΩ. The final circuit is shown in Fig. 5-21b.

The *ac performance* of the circuit can now be calculated: r_{be} = 0.025 V/15 μA = 1.7 kΩ. The ac β is found in the vicinity of the Q point using Eq. 5-3.

$$\beta_{ac} = \frac{\Delta I_C}{\Delta I_B} \cong \frac{0.6 \text{ mA}}{5 \text{ }\mu A} = 120.$$

Now A_V, A_I, R_i, and R_o can be calculated.

$$A_V = \frac{-\beta R_L'}{r_{be}} = \frac{-120 \text{ } (3 \text{ k}\Omega \parallel 4.7 \text{ k}\Omega \parallel 20 \text{ k}\Omega)}{1.7 \text{ k}\Omega} = -120$$

$$A_I = \beta \times \frac{R_B}{R_B + r_{be}} \times \frac{r_o \parallel R_C}{r_o \parallel R_C + R_L} = 120 \times \frac{760}{760 + 1.7} \times \frac{2.6}{2.6 + 4.7} = 42.6$$

$$R_i = R_B \parallel r_{be} = 760 \text{ k}\Omega \parallel 1.7 \text{ k}\Omega = 1.7 \text{ k}\Omega$$

$$R_o = R_C \parallel r_o = 2.6 \text{ k}\Omega$$

The signal swing can be predicted by drawing the ac load line on the collector curves with $R_L' = 1.7$ kΩ (3 k$\Omega \parallel$ 20 k$\Omega \parallel$ 4.7 kΩ). This is shown in Fig. 5-21a and

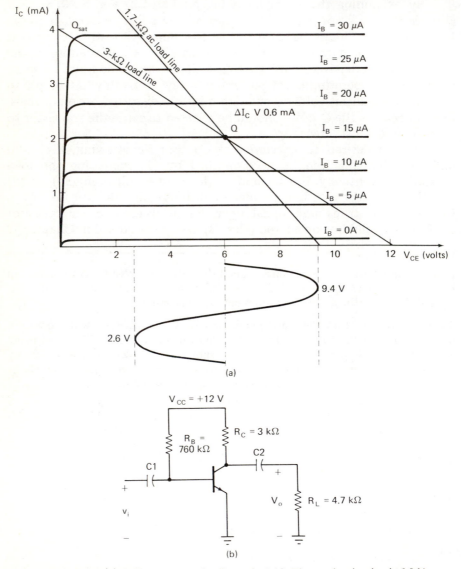

Figure 5-21 (a) Collector curves for Example 5-10. The ac signal swing is 6.8-V peak-peak.

results in approximately a 6.8-V peak-peak output signal. This means the maximum input signal that will not cause distortion is approximately 6.8 V/120 = 56 mV peak-peak.

5.5 BIAS STABILITY

The performance of any amplifier will depend on its dc operating point. The selection of this point not only affects the dc currents and voltages in the circuit but also ac quantities such as r_{be}, β, and the output signal swing. *Bias stability* refers to a circuit's ability to remain at a chosen output Q point as other parameters in the circuit change.

β Variations

Probably the most common variable in a transistor circuit is β. The following example will illustrate the bias *instability* of the fixed-bias amplifier we have been discussing as its β changes.

Example 5-11 ───

Recalculate the dc operating point of the circuit designed in Example 5-10 if β rises to 300.

Solution The base current does not depend on β and therefore remains the same (I_B = 15 μA). I_C will increase, however. $I_C = \beta I_B = 300 \times 15\ \mu A = 4.5$ mA and $V_{CE} = V_{CC} - I_C R_C = 12 - (4.5\ \text{mA})(3\ \text{k}\Omega) = -1.5$ V. But wait a minute. Can V_{CE} be a negative voltage? With only a positive power supply, the answer is no. What has happened to our circuit is that the collector current has increased until the operating point is no longer in the active region. In effect, the family of I_B curves have shifted up such that the Q point has moved to the point labeled Q_{sat} in Fig. 5-21. Once this occurs, the transistor saturates with I_C = 4 mA and V_{CE} = 0 V.

A question we might now logically ask ourselves is, Is it reasonable for β to vary from 120 to 300? Unfortunately, the answer to this question is yes. As we saw previously in this chapter, h_{fe} (β) varied from 75 to 375 for the 2N2222A.

Because the base current of the fixed-bias circuit is relatively constant, depending only on V_{BE}, V_{CC}, and R_B (none of which can be expected to change), the collector current is going to vary *directly* with β. This in turn means V_{CE} will vary *inversely* with β going towards or into saturation for high β transistors.

Temperature Variations

The situation worsens when we consider *temperature*. When the transistor is biased in the active region, its base-collector junction is reverse-biased. This means that a small leakage current flows across this junction at room temperature due to *minority carriers*. However, as temperature increases, this current will increase due to the greater number of thermally-created free electrons and holes, as discussed in Chapter 1. An approximate rule of thumb for silicon is to expect this current to *double* each time the temperature increases by $10°C$.

The total collector current is then made up of the normal transistor current βI_B plus a component due to this leakage. The actual equation is

$$I_C = \beta I_B + (\beta + 1) I_{CBO} \qquad\qquad (5\text{-}10)$$

where I_{CBO} is the reverse leakage current of the collector-base junction with the emitter open (refer to Chapter 3, Table 3-1). Note that the I_{CBO} current gets amplified by $\beta + 1$ due to transistor action and increases the effect of the leakage current by a factor of 100 or more, typically.

In addition to leakage current variations, the *base-emitter voltage* of the transistor can *also* be expected to vary with temperature. For silicon, this is approximately a *decrease* of 2 mV for each $1°C$ *increase* in temperature. In fact, this property is taken advantage of in some electronic temperature sensors.

Finally, the value of β is by no means a constant either. It does tend to be constant over a *range* of temperatures but increases at higher temperatures and decreases for very low temperatures. β also varies with the collector current, tending to have a stable midband value but decreasing if the collector current exceeds or falls below this midband region.

Example 5-12

Recalculate the dc operating point of the circuit in Example 5-10 if I_{CBO} = 10nA at $25°C$ and the temperature increases to $125°C$. Assume β remains constant at 120.

Solution As the temperature increases from $25°C$ to $125°C$, the base emitter voltage *decreases* by 200mV (2mV per degree). The base current becomes I_B = (12 V – 0.4 V)/760 kΩ = 15.3 μA. As can be seen, this is an insignificant change. The I_{CBO} leakage is not so insignificant however. With the stated temperature increase there are ten $10°C$ rises in temperature [(125 – 25)/10 = 10]. This means there are ten doublings of I_{CBO} as the temperature rises to $125°C$. Therefore, I_{CBO} = $2^{10} \times 10$ nA = 10.2 μA. But remember the transistor amplifies its own leakage current so that I_C = 120 (15.3 μA) + 121(10.2 μA) = 3.3 mA. V_{CE} is calculated as V_{CC} – $I_C R_C$ = 12 – (3.3 mA)(3 kΩ) = 2.1 V. Again, the transistor is being driven towards saturation.

A common phenomenon for the transistor is that the increased leakage current causes *heating* of the transistor, which in turn causes more leakage, which in turn causes further heating, and so on. This is called *thermal runaway* and can drive the transistor into saturation.

The point of these last two examples is to demonstrate that transistor parameters can be expected to vary a great deal from part to part (even among the same part numbers!) and with temperature. They also demonstrate that the fixed-bias circuit is a *very poor one.* In Example 5-10, we designed an amplifier with a voltage gain of 120. However, we have seen one case in which the circuit would not function at all due to a β variation within manufacturer stated specifications and another case in which the bias point has shifted by 50% due to a 100°C increase in temperature.

In the next chapter we study ways to design bias stable circuits by employing *negative feedback* to counteract increases or decreases in I_C due to β or temperature variations. A design goal you might be thinking about is to build a circuit independent of β. If this could be done, most any transistor capable of handling the bias current could be used in the design.

─────────────── KEY TERMS ───────────────

DC Equivalent Circuit This refers to that portion of the transistor circuit in which dc currents are flowing. Capacitors are commonly used to restrict the dc to certain portions of the circuit only.

AC Equivalent Circuit This refers to that portion of the transistor circuit in which ac currents are flowing. An inductor can be used to block the ac while allowing the dc to pass.

Small-Signal Model Using the ac-equivalent circuit, the transistor or diode is replaced with a small-signal model that reproduces the current-voltage characteristics of the device for small-signal variations about its dc Q point. This model consists of dependent sources and resistances.

r_{be} This is the ac resistance of the base-emitter diode as seen looking into the base terminal of the transistor. It is calculated as $0.025 \text{ V}/I_{BQ}$. Generally, r_{be} is on the order of 1–5 kΩ.

Coupling Capacitors These capacitors are used to pass the ac signal into the amplifier and to pass the amplified signal to the load resistance. They are necessary to block the dc currents and voltages of the amplifier from being upset by the signal source or load.

Signal Swing This refers to the amplitude of the output ac signal from the amplifier. It is limited to a maximum value of V_{CC} peak-peak, provided the Q point is centered between 0 V and V_{CC}. With the addition of a load resistor, the signal swing is generally reduced to less than this amount.

AC Load Line This is similar to the dc load line but represents only the ac resistance seen by the collector of the transistor. It must pass through the dc Q point and intersect the horizontal axis at $(I_{CQ} \times R_L') + V_{CEQ}$.

Bias Stability This refers to the stability of I_C and V_{CE} in a transistor circuit, as other parameters such as β or temperature are caused to vary. The fixed-bias common-emitter amplifier has very poor bias stability.

5-1 Why does the diode appear to be a resistance for ac small signals?

5-2 Why is the diode model inaccurate for a large-signal variation?

5-3 Explain why a diode cannot function as an amplifier.

5-4 Sketch a 0.2-V peak-peak sine wave riding on a 1-V dc level.

5-5 Referring to Figs. 5-1 and 5-2, determine the ac voltage across the diode if the input signal source is 0.3 V peak-peak.

5-6 As the dc base current of a transistor increases, its ac base-emitter resistance decreases. Explain why.

5-7 If an ohmmeter is connected across the base-emitter terminals of a transistor, will it indicate the r_{be} resistance of the transistor? Explain.

5-8 A load resistor is connected to the output of an amplifier. What two effects can be expected to this amplifier's performance?

5-9 Explain the following statement: The Q point of a small-signal amplifier is chosen as $\frac{1}{2} V_{CC}$ for maximum linearity, not maximum output signal swing.

5-10 A certain amplifier has an open circuit voltage gain of 200 and output resistance of 100 Ω. What is the voltage gain for a 500-Ω load resistance?

5-11 A transistor is biased at the indicated Q point in Fig. 5-20. Sketch the ac small-signal model for this transistor and determine the values of all model parameters.

5-12 Design a fixed-bias common-emitter amplifier circuit to operate at the Q point shown in Fig. 5-20 with $V_{CC} = 12.5$ V.

5-13 Calculate A_V, R_i, and R_o for the circuit in Problem 5-12. Sketch the amplifier ac equivalent circuit.

5-14 Determine the maximum possible input signal that will not overdrive the amplifier in Problem 5-12.

5-15 Calculate the voltage gain and maximum signal swing of the amplifier in Problem 5-12 if a 1-kΩ load resistor is used.

5-16 Sketch the dc and ac equivalent circuits of the *pnp* amplifier shown in Fig. 5-22.

5-17 Determine the dc operating point of the amplifier in Fig. 5-22.

5-18 Determine the amplifier equivalent circuit for the amplifier in Problem 5-16.

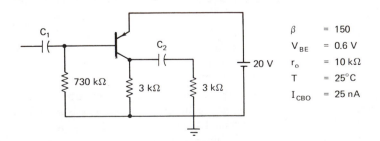

Figure 5-22

5-19 Refer to Fig. 5-21a. Estimate the maximum peak-peak output signal swing if the ac load line is 1 kΩ.

5-20 Determine values for C_1 and C_2 in Fig. 5-22. Assume an operating frequency from 500 Hz to 30 kHz.

5-21 Repeat Problem 5-20 for the circuit in Fig. 5-21b.

5-22 If the ambient temperature of the circuit in Fig. 5-22 increases to 75°C, re-calculate the collector current and V_{CE} due to changes in I_{CBO} and V_{BE}.

5-23 A particular *npn* transistor has $V_{BE} = 0.6$ V at 25°C. Calculate the approximate value of V_{BE} at 125°C and –55°C.

LABORATORY ASSIGNMENT 5:
THE FIXED-BIAS COMMON-EMITTER AMPLIFIER

Objectives

1. Graphical design of a common-emitter amplifier using a curve tracer.

2. Observation of bias stability.

3. Practice in measuring amplifier ac characteristics.

Introduction   In this laboratory assignment, a fixed-bias common-emitter amplifier will be designed using a load-line method and a curve tracer. The completed amplifier will then be tested for *bias stability* against changes in temperature and β. Finally, the voltage gain and input and output resistance will be measured. The basic circuit is shown in Fig. 5-23.

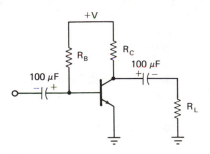

Figure 5-23

Components Required

1 *npn* transistor
2 100-μF capacitors
Miscellaneous ¼-W resistors
1 1-kΩ, 10-kΩ pot

Part I: *Design and Bias Stability*

STEP 1 Select an *npn* silicon transistor and use the curve tracer to display its characteristic common-emitter output curves. Use $I_B = 10 \mu A/$ step and adjust the vertical sensitivity (I_C) to obtain six I_B traces. Set the horizontal sensitivity to 2 V/division.

STEP 2 Using a sheet of graph paper and straight edge, copy the curves for the *active region*. Label each I_B trace appropriately.

STEP 3 Select a Q point in the middle of the active region with V_{CE} centered between 0 V and your choice of V_{CC}.

STEP 4 Use the value of I_B at the Q point (I_{BQ}) to calculate R_B.

$$R_B = (V_{CC} - V_{BE})/I_{BQ}$$

STEP 5 Similarily, calculate R_C at the Q point as

$$R_C = (V_{CC} - V_{CEQ})/I_{CQ}$$

STEP 6 Obtain resistor values within 10% of R_B and R_C and assemble the amplifier. Prepare a data table and record the dc *design* goals in this table.

STEP 7 Measure these same quantities and record in your table under the label Q_1. Repeat this step for two other transistors of the same part number but with *different* IV characteristics. Label these entries Q_2 and Q_3.

STEP 8 Replace Q_1 in the circuit. Hold the tip of a hot soldering iron against the transistor case for about five seconds and record V_C.

$$V_C \text{ (5 s of heat)} = \underline{\qquad}$$

Question 1 How do you account for the differences in bias voltages for transistors Q_2 and Q_3?

Question 2 As heat was applied to the transistor, does the collector current increase or decrease? Does the collector voltage measured in step 8 verify this?

Question 3 Comment on the bias stability of this simple amplifier circuit.

Part II: *AC Analysis*

STEP 1 Using the measured value of I_{BQ} for Q_1, calculate r_{be}.

$$r_{be} = \underline{\qquad}$$

STEP 2   Calculate the transistor output resistance, r_o, and β_{ac} using the characteristic curves obtained in Part I.

STEP 3   *Calculate A_V, R_i, and R_o for the amplifier.*

$$A_V = \underline{\hspace{2cm}}$$

$$R_i = \underline{\hspace{2cm}}$$

$$R_o = \underline{\hspace{2cm}}$$

STEP 4   Using $f = 10$ kHz, *measure* the quantities calculated in step 3 using the methods discussed in Chapter 4. Prepare a table comparing calculated and measured ac quantities. Determine the maximum undistorted output signal swing.

STEP 5   Draw the amplifier equivalent circuit.

STEP 6   Add a 1-kΩ load resistor to the amplifier and remeasure the voltage gain and signal swing.

$$A_V = \underline{\hspace{2cm}}$$

$$V_o \,(\text{max}) = \underline{\hspace{2cm}}$$

Question 4   What was the effect of adding the 1-kΩ load resistor? Sketch the corresponding ac load line on the characteristic curves and estimate the signal swing.

SIX

SOME REAL CIRCUITS

The fixed-bias transistor amplifier of the preceding chapter is a fairly simple circuit to analyze and understand. It is not used too often, however, due to its poor bias stability. In this chapter and the next we conclude our study of bipolar transistors with some more realistic and practical circuits. *Single-stage* amplifiers are covered in this chapter and *multistage* amplifiers and coupling methods are covered in the next.

The novice to transistor circuit analysis can get discouraged and confused by the many equations that can be developed for the various circuits. Here are two comments or suggestions concerning this:

1. This chapter uses a *universal voltage divider bias circuit*. By learning this one circuit, a majority of the others can be solved by choosing resistor values to be either zero or infinite ohms and then applying the universal circuit equations.

2. This comment is really in the form of a *challenge*. Most students have access to a microcomputer or time-shared computer system running the BASIC computer language. It would certainly be possible to write a BASIC program to solve all the various transistor circuits to be presented in this chapter. Doing this is an excellent way to learn the various circuits and, once completed, provides a powerful tool to be used in conjunction with your laboratory work. Circuits may be "designed" on the computer and later breadboarded in the laboratory. As new circuits are studied, they can be added to your existing program. An example of this will be provided later in this chapter.

As we cover the various circuits in this chapter, the following format is used: circuit name and schematic, dc bias, ac solution (using the small-signal model), comments, and examples.

Figure 6-1 illustrates a voltage divider bias common-emitter amplifier. Throughout this text this circuit is referred to as the *universal circuit* for reasons that will become clearer as we proceed. We can make a number of observations about this circuit:

1. Because the input is applied to the base and the output is taken at the collector, this circuit is considered a common-emitter amplifier. This is true even though the emitter is not at ground potential.

2. If I_B is small compared to I_1, R_1 and R_2 are essentially in *series* and the *base* voltage (not base-emitter voltage) is determined by the voltage divider relationship between these two resistors. Hence, the circuit's name.

3. The addition of resistor R_E causes a *negative feedback* affect to occur. This means that if I_C should increase (and therefore I_E also), the voltage drop across R_E will increase ($I_E \times R_E$). Because the base voltage is held constant by R_1 and R_2, the base-emitter voltage, $V_B - V_E$, must decrease. This in turn decreases I_C, acting to *stabilize* the circuit. The opposite will occur if I_C should tend to decrease.

4. The bias voltages and currents for this circuit are actually independent of β. This is true provided I_1 is much larger than I_B, placing R_1 and R_2 in series. Then, $V_B = R_2/(R_1 + R_2) \times V_{CC}$. $V_E = V_B - 0.6$ V, and $I_E = V_E/R_E$. For a reasonable β, $I_C \cong I_E$. Finally, $V_C = V_{CC} - I_C R_C$. The entire circuit has been solved without having to know the value of β!

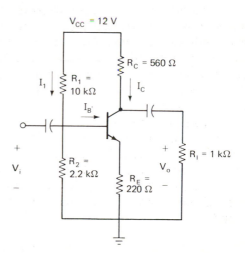

Figure 6-1   The universal voltage divider bias common-emitter amplifier.

In summary, the voltage divider bias circuit is a common-emitter amplifier employing *voltage divider* base bias and *negative feedback* for bias stability.

Example 6-1 ───

Use the method just outlined and solve the circuit in Fig. 6-1 for all dc-bias currents and voltages. Assume $\beta = 100$ and a silicon transistor. Verify the assumption that $I_1 \gg I_B$.

Solution   Although β is given to be 100, we will not need it to find the dc bias.

$$V_B = (2.2k/12.2k) \times 12 \text{ V} = 2.2 \text{ V}$$

$$V_E = V_B - 0.6 \text{ V} = 1.6 \text{ V}$$

$$I_E = 1.6 \text{ V}/220 \ \Omega = 7.3 \text{ mA}$$

$$I_C \cong I_E = 7.3 \text{ mA}$$

$$V_C = 12 \text{ V} - (7.3 \text{ mA})560 \ \Omega = 7.9 \text{ V}$$

$$V_{CE} = V_C - V_E = 6.3 \text{ V}$$

We have now solved the circuit for dc except for I_B. This can be found using β. $I_B = I_C/\beta = 7.3 \text{ mA}/100 = 73 \ \mu\text{A}$. This can be compared to $I_1 = V_{R_1}/10k = (12 - 2.2)/10k = 980 \ \mu\text{A}$. I_1 is 980/73 = 13 times as large as I_B. You should be able to see that the *higher* the β, the *smaller* the base current and the better the assumption that $I_1 \gg I_B$. This in turn means that R_1 and R_2 are in series, stabilizing the base voltage and the circuit.

The Thevenin Equivalent

The previous example illustrated a simple method for determining the dc bias of the voltage divider bias circuit. There is one problem with this method, however. If the value of β should fall, the base current will have to *increase* to support the same collector current. Eventually, the assumption of $I_1 \gg I_B$ will become invalid. This in turn will mean that R_1 and R_2 are no longer in series and the entire method will yield inaccurate results.

A more exact solution to the dc bias can be obtained by determining the *Thevenin equivalent* of the base circuit. This is illustrated in Fig. 6-2a. We break the circuit at the base terminal and determine the Thevenin equivalent seen by the base looking back into these terminals.

Figure 6-2b indicates the resulting circuit. R_{TH} is found by "killing" all sources and measuring the resistance between the two terminals a and b.

$$R_{TH} = R_1 \parallel R_2 \tag{6-1}$$

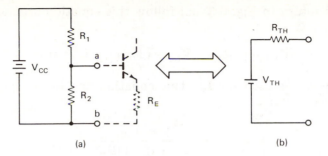

Figure 6-2   Applying Thevenin's Theorem, the equivalent circuit seen by the base of the transistor becomes a single voltage source and resistor.

V_{TH} is found as the open circuit voltage measured between the two terminals.

$$V_{TH} = \frac{R_2}{R_1 + R_2} \times V_{CC} \tag{6-2}$$

Now proceeding with the circuit solution, Kirchoff's Law can be used to write an equation for the new transistor circuit, shown in Fig. 6-3.

Before doing this, it is useful to note that all currents in the transistor can be written in terms of I_B and β. Check the following.

$$I_C = \beta I_B \tag{6-3}$$

and, because $I_C + I_B = I_E$,

$$I_E = \beta I_B + I_B = (\beta + 1)I_B \tag{6-4}$$

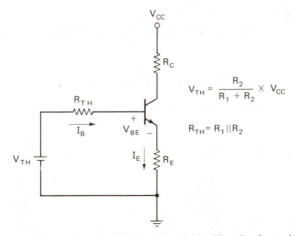

Figure 6-3   The simplified voltage divider bias circuit resulting from an application of Thevenin's Theorem to the base circuit.

Now return to Fig. 6-3 and follow this equation around the base loop.

$$V_{TH} - I_B R_{TH} - V_{BE} - (\beta + 1)I_B R_E = 0 \text{ V} \qquad (6\text{-}5)$$

where Eq. 6-4 is used for I_E. This equation can be solved for I_B by combining like terms.

$$I_B = \frac{V_{TH} - V_{BE}}{R_{TH} + (\beta + 1)R_E} \qquad (6\text{-}6)$$

This is the *key* equation for the voltage divider bias circuit. Once I_B is known, I_C and I_E can be found using Eqs. 6-3 and 6-4.

Example 6-2 ──

Use the Thevenin equivalent method to solve the circuit in Fig. 6-1. Compare the results to that obtained using the voltage divider method. Use $\beta = 100$.

Solution The first step is to find R_{TH} and V_{TH}. Referring to Eqs. 6-1 and 6-2, $R_{TH} = 10 \text{ k}\Omega \parallel 2.2 \text{ k}\Omega = 1.8 \text{ k}\Omega$, and $V_{TH} = 2.2/12.2 \times 12 \text{ V} = 2.2 \text{ V}$. Now, I_B is found from Eq. 6-6:

$$I_B = \frac{2.2 \text{ V} - 0.6 \text{ V}}{1.8 \text{ k}\Omega + (101)0.22 \text{ k}\Omega} = 66.6 \ \mu\text{A}.$$

$$I_C = \beta I_B = 6.7 \text{ mA}$$

and

$$I_E = (\beta + 1)I_B = 6.7 \text{ mA}$$

$$V_E = I_E R_E = (6.7 \text{ mA})0.22 \text{ k}\Omega = 1.5 \text{ V}$$

$$V_B = V_E + 0.6 \text{ V} = 2.1 \text{ V}$$

$$V_C = V_{CC} - I_C R_C = 12 \text{ V} - (6.7 \text{ mA})(0.56 \text{ k}\Omega) = 8.2 \text{ V}$$

Finally,

$$V_{CE} = V_C - V_E = 8.2 \text{ V} - 1.5 \text{ V} = 6.7 \text{ V}$$

The two methods are compared below.

	Voltage Divider Method (Example 6-1)	Thevenin Method (Example 6-2)
I_B	73 μA	66.6 μA
I_C	7.3 mA	6.7 mA
I_E	7.3 mA	6.7 mA
V_B	2.2 V	2.1 V
V_C	7.9 V	8.2 V
V_E	1.6 V	1.5 V
V_{CE}	6.3 V	6.7 V

Although the two methods do not give *exactly* the same results, the differences are quite small. The reason for the discrepancy is that R_1 and R_2 are *not* actually in series, as the voltage divider method assumes.

A logical question to ask at this point is, which method should I use? You might answer that question yourself if you consider that an equation is useless unless it can accurately predict what will happen with the physical circuit. This includes several new variables such as *resistor tolerances*, *power supply variations*, and *meter errors*. Anyone of these could "mask out" the difference between the two methods. For these reasons, I would use the *voltage divider method* because it is the simplest and yields only a small error compared to the longer but more accurate Thevenin method.

In many cases, we only desire a quick estimation of a dc voltage in the circuit to determine if the circuit is functioning properly. If the measured voltage is within 10 to 20% of our quick calculation, we would probably look elsewhere for the circuit problem.

A final word of caution is in order regarding the voltage divider method. If R_1 and R_2 cannot be considered in series, this method will give totally *erroneous* results. A good rule-of-thumb to follow is for I_1 to be at least *ten* times greater than I_B at the worst case value of β.

Bias Stability

In our initial discussion of the voltage divider bias circuit, we said that the addition of resistor R_E allowed negative feedback to stabilize the circuit. The following example will illustrate this stability.

Example 6-3 ─────────────────────────────────────

Use the Thevenin equivalent method to solve the circuit in Fig. 6-1 if β increases to 300.

Solution   The base current is found similar to Example 6-2. R_{TH} and V_{TH} have not changed.

$$I_B = \frac{2.2 \text{ V} - 0.6 \text{ V}}{1.8 \text{ k}\Omega + (301)0.22 \text{ k}\Omega} = 23.5 \ \mu\text{A}$$

Now $I_C = \beta I_B = 7.1$ mA, and $I_E = (\beta + 1)I_B = 7.1$ mA.

$$V_E = I_E R_E = 1.6 \text{ V}$$

$$V_B = V_E + 0.6 \text{ V} = 2.2 \text{ V}$$

$$V_C = 12 - (7.1 \text{ mA})(0.56 \text{ k}\Omega) = 8.0 \text{ V}$$

$$V_{CE} = 8.0 \text{ V} - 1.6 \text{ V} = 6.4 \text{ V}$$

The results for the two cases of β are shown below.

	$\beta = 100$	$\beta = 300$
I_B	66.6 μA	23.5 μA
I_C	6.7 mA	7.1 mA
I_E	6.7 mA	7.1 mA
V_B	2.1 V	2.2 V
V_C	8.2 V	8.0 V
V_E	1.5 V	1.6 V
V_{CE}	6.7 V	6.4 V

The stability of this circuit should be readily apparent. Only slight changes in any of the parameters occured. You can imagine what this increase in β would have done to the fixed-bias circuit of the preceding chapter!

Actually, one parameter did change quite a bit. That was I_B. In fact, I_B is about one third of its value compared to when β was 100. This is actually an indication of the negative feedback at work. As β tripled, I_B was reduced by about one third, which caused I_C (βI_B) to stay nearly constant!

The main problem with the fixed-bias circuit is that I_B is *constant*, independent of β. The voltage divider bias circuit has I_B a function of β:

$$I_B = (V_{TH} - V_{BE})/[R_{TH} + (\beta + 1)R_E]$$

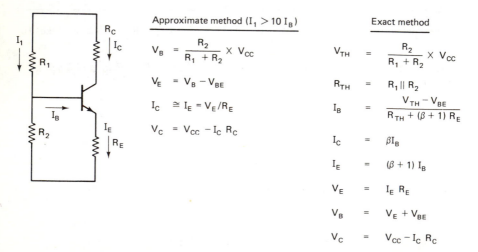

Figure 6-4   A summary of the important dc-bias equations for the universal circuit.

In fact, if $(\beta + 1)R_E \gg R_{TH}$, then

$$I_C = \beta I_B \cong \beta \, \frac{V_{TH} - V_{BE}}{(\beta + 1)R_E} \cong \frac{V_{TH} - V_{BE}}{R_E}$$

which depends only on *resistor values* and V_{CC}.

The important dc equations for the voltage divider bias circuit are summarized in Fig. 6-4.

Example 6-4

Calculate the dc bias for the circuit in Fig. 6-5.

Solution At first glance, this appears to be a totally new circuit. But is it? If we assume $R_2 = \infty$ and $R_E = 600 \; \Omega$ (we can ignore C_E for dc purposes), all of the Thevenin equations apply directly.

$$R_{TH} = R_1 \parallel R_2 = 86 \text{ k}\Omega$$

$$V_{TH} = R_2/(R_1 + R_2) \times 10 \text{ V} = \infty/(86 \text{ k}\Omega + \infty) \times 10 \text{ V} = 10 \text{ V}$$

The remainder of the problem can be worked out using the equations in Fig. 6-4. For practice, work out the answers and compare to the following:

$$I_B = 95 \; \mu\text{A} \qquad V_C = 5.3 \text{ V}$$

$$I_C = 1.9 \text{ mA} \qquad V_E = 1.2 \text{ V}$$

$$I_E = 2.0 \text{ mA} \qquad V_B = 1.8 \text{ V}$$

$$V_{CE} = 4.1 \text{ V}$$

This example illustrates why the circuit is called the *universal* voltage divider bias circuit. It is left to you as a homework problem

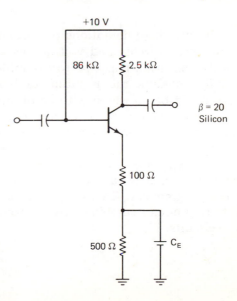

Figure 6-5 A nonstandard circuit to which the universal equations can still be applied.

to show that even the fixed-bias circuit of Chapter 5 can be solved using these universal equations. One caution to be observed is that the *Thevenin equivalent* equations must be used for these special circuit variations. The voltage divider technique only works if there is an R_1 and R_2 that can be considered in series.

Computer Simulation

Earlier it was mentioned that a computer could be programmed to solve the many transistor circuits. In this section, an example of such a program is given.

There are a number of computer circuit analysis programs available for large computers and minicomputers (ECAP, SPICE). The program presented here is *not* in this category. Our program will simply assume *linear* circuit operation and proceed to solve the Thevenin equations presented earlier in this chapter. In this respect, the computer is being used like a large programmable calculator but with alphanumeric capabilities.

A listing of the program written in BASIC is given in Fig. 6-6 along with a sample run. This program has several limitations:

1. Only dc solutions to a circuit are provided.
2. Only common-emitter universal bias circuits and variations there of are permitted.
3. Linear operation is assumed with β constant and independent of I_C.

Despite these limitations the program is capable of accurately predicting bias currents and voltages when fed accurate information. A typical application might be to assemble a circuit in the laboratory and carefully measure the resulting dc-bias conditions. The component values could be measured with an ohmmeter and the β of the transistor determined from a curve tracer. Armed with this information, the circuit analysis program is now run. Typical correlation is within 5%.

Although the program in Fig. 6-6 may be of limited value as given, it can be expanded to include *ac* solutions and other circuit types. Most computer programs have a tendency to grow as new features are added by the programmer. Eventually, the program becomes a powerful tool for testing circuit combinations without having to actually breadboard these circuits.

```
 10 PRINT "               *********************************"
 15 PRINT "               *   TRANSISTOR CIRCUIT SOLVER   *"
 17 PRINT "               *********************************"
 20 PRINT
 30 PRINT "      THIS PROGRAM WILL PROVIDE DC BIAS SOLUTIONS"
 40 PRINT " FOR COMMON EMITTER AMPLIFIERS USING THE UNIVERSAL"
 50 PRINT " VOLTAGE DIVIDER BIAS CIRCUIT.  I WILL PROMPT YOU"
 60 PRINT " FOR THE FOLLOWING INFORMATION:"
 70 PRINT "          (1) SUPPLY VOLTAGE"
 80 PRINT "          (2) RESISTOR VALUES IN K OHMS (100 OHMS = .1)"
 85 PRINT "          (3) BASE-EMITTER DIODE DROP"
 90 PRINT "          (4) BETA"
 92 PRINT
 95 PRINT "      AFTER TYPING THE PROPER NUMERIC KEY VALUE PUSH THE";
 97 PRINT " RETURN KEY"
100 PRINT
110 PRINT "REMEMBER:  MANY COMMON EMITTER CIRCUITS CAN BE SOLVED FROM"
120 PRINT "THE UNIVERSAL CIRCUIT BY CHOOSING CERTAIN RESISTORS TO BE"
130 PRINT "VERY LARGE OR O OHMS."
140 PRINT
200 PRINT "PUSH P AND THE RETURN KEY TO PROCEED"
210 INPUT A$
220 PRINT
230 PRINT
250 PRINT "PLEASE ENTER THE SUPPLY VOLTAGE: ";
260 INPUT V
270 PRINT "NOW I NEED TO KNOW THE DC BETA: ";
280 INPUT B
285 D=.6
290 PRINT "IS VBE=.6V OK WITH YOU (Y/N) ";
300 INPUT A$
310 IF A$="Y" THEN 350
320 PRINT "WHAT VALUE DO YOU WANT ";
330 INPUT D
350 PRINT "ENTER R1(TOP) AND R2(BOTTOM) THE BASE BIASING RESISTORS";
355 PRINT " IN K OHMS"
360 INPUT R1,R2
370 PRINT "NOW ENTER RC THE COLLECTOR RESISTOR AND RE THE EMITTER";
375 PRINT " RESISTOR"
380 INPUT R3,R4
390 PRINT
395 PRINT
400 PRINT "DC SOLUTIONS FOR YOUR COMMON EMITTER AMPLIFIER"
405 PRINT
410 PRINT "PARTS LIST:"
420 PRINT "R1 = ";R1;"K",TAB(15),"R2 = ";R2;"K"
430 PRINT "RC = ";R3;"K",TAB(15),"RE = ";R4;"K"
440 PRINT "VCC = ";V;"VOLTS"
450 PRINT "BETA = ";B
460 PRINT "VBE = ";D;"VOLTS"
470 PRINT
475 PRINT "PUSH P AND THE RETURN KEY TO PROCEED"
477 INPUT A$
480 PRINT "           DC BIAS"
490 V1=(R2/(R1+R2))*V
500 R5=(R1*R2)/(R1+R2)
510 I1=(V1-D)/(R5+((B+1)*R4))
520 I2=B*I1
```

Figure 6-6 BASIC computer program for solving the dc equations for the universal circuit. The program listing is given along with a sample run.

```
530 I3=I1+I2
540 V2=V-(I2*R3)
550 V3=I3*R4
560 V4=V2-V3
570 V5=V3+D
580 IF V5>=V2 THEN 750
590 IF V4>=(.98*V) THEN 780
600 PRINT
610 PRINT "RTH= ";R5;"K",TAB(15),"VTH= ";V1;"VOLTS"
620 PRINT "IB= ";I1;"MA",TAB(15),"IC= ";I2;"MA"
630 PRINT "IE= ";I3;"MA"
635 PRINT "VB= ";V5;"VOLTS",TAB(15),"VC= ";V2;"VOLTS"
640 PRINT "VE= ";V3;"VOLTS",TAB(15),"VCE= ";V4;"VOLTS"
650 PRINT
655 IF R4=0 THEN 730
660 I4=(V-V5)/R1
670 IF I4<5*I1 THEN 730
680 IF I4<=20*I1 THEN 710
690 B$="EXCELLENT"
700 GOTO 740
710 B$="GOOD"
720 GOTO 740
730 B$="POOR"
740 PRINT "BIAS STABILITY:------";B$
745 GOTO 800
750 PRINT
760 PRINT "W A R N I N G ....YOUR CIRCUIT IS SATURATED!"
770 GOTO 800
780 PRINT "W A R N I N G ....YOUR CIRCUIT IS CUTOFF!"
800 PRINT
805 PRINT "YOUR OPTION:"
810 PRINT "              (1) NEW CIRCUIT"
820 PRINT "              (2) NEW BETA"
830 PRINT "              (3) QUIT"
835 PRINT
840 PRINT "ENTER THE NUMBER OF YOUR CHOICE:";
850 INPUT N
855 IF N>=4 THEN 805
860 ON N GOTO 220,900,865
865 END
900 PRINT
910 PRINT "WHAT IS THE NEW BETA ";
920 INPUT B
930 GOTO 390

*


RUN
              ********************************
              *   TRANSISTOR CIRCUIT SOLVER  *
              ********************************

     THIS PROGRAM WILL PROVIDE DC BIAS SOLUTIONS
FOR COMMON EMITTER AMPLIFIERS USING THE UNIVERSAL
VOLTAGE DIVIDER BIAS CIRCUIT.   I WILL PROMPT YOU
FOR THE FOLLOWING INFORMATION:
          (1) SUPPLY VOLTAGE
          (2) RESISTOR VALUES IN K OHMS (100 OHMS = .1)
          (3) BASE-EMITTER DIODE DROP
          (4) BETA

  AFTER TYPING THE PROPER NUMERIC KEY VALUE PUSH THE RETURN KEY

REMEMBER:  MANY COMMON EMITTER CIRCUITS CAN BE SOLVED FROM
THE UNIVERSAL CIRCUIT BY CHOOSING CERTAIN RESISTORS TO BE
VERY LARGE OR 0 OHMS.
```

Figure 6-6 (cont.)

```
PUSH P AND THE RETURN KEY TO PROCEED
?P


PLEASE ENTER THE SUPPLY VOLTAGE: ?12
NOW I NEED TO KNOW THE DC BETA: ?125
IS VBE=.6V OK WITH YOU (Y/N) ?NO
WHAT VALUE DO YOU WANT ?.65
ENTER R1(TOP) AND R2(BOTTOM) THE BASE BIASING RESISTORS IN K OHMS
?68,12
NOW ENTER RC THE COLLECTOR RESISTOR AND RE THE EMITTER RESISTOR
?1.4,.35


DC SOLUTIONS FOR YOUR COMMON EMITTER AMPLIFIER

PARTS LIST:
R1 =  68 K        R2 =  12   K
RC =  1.4   K   RE =  .35  K
VCC =  12    VOLTS
BETA =  125
VBE =  .65  VOLTS

PUSH P AND THE RETURN KEY TO PROCEED
?P
                DC BIAS

RTH=  10.2   K    VTH=  1.8   VOLTS
IB=  .211786E-01  MA            IC=  2.64733   MA
IE=  2.6685 MA
VB=  1.58398   VOLTS           VC=  8.29374   VOLTS
VE=  .933977   VOLTS           VCE=  7.35977  VOLTS

BIAS STABILITY:-------GOOD

YOUR OPTION:
                (1) NEW CIRCUIT
                (2) NEW BETA
                (3) QUIT

ENTER THE NUMBER OF YOUR CHOICE:?3
*
```

Figure 6-6 (cont.)

6.2 SOLVING THE CIRCUIT FOR AC

The operation of any amplifier circuit is best understood by separately analyzing the dc and ac operation of that circuit. Up to this point we have studied the dc operation of the voltage divider bias circuit and found it to be quite stable compared to the fixed-bias type of amplifier.

Now let us consider the ac operation of this circuit. The basic amplifier is redrawn in Fig. 6-7a. Following the procedure outlined in Chapter 5, we will assume operation at a midband of frequencies where all capacitors are assumed to have negligible reactance. The resulting *ac* equivalent circuit is presented in Fig. 6-7b.

Finally, the transistor is replaced with its small-signal model, resulting in the circuit shown in Fig. 6-7c. Note that R_1 and R_2 are in

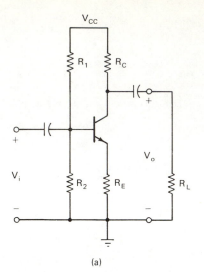

(a)

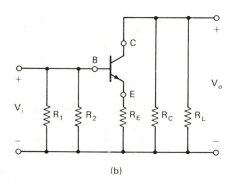

(b)

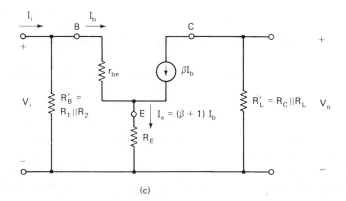

(c)

Figure 6-7 (a) Universal voltage divider bias circuit. (b) equivalent circuit for ac. (c) the small-signal model.

parallel for ac and are referred to as R_B'. Similarily, R_L and R_C are in parallel and referred to as R_L'.

The transistor output resistance, r_o, has been left off. This is an assumption that makes the mathematics simpler. Generally, r_o is large and does not affect the ac solutions to a significant degree.

With the exception of R_E, the circuit in Fig. 6-7c is similar to that of the fixed-bias amplifier. In fact, the output voltage is the same $(-\beta I_b R_L')$. The input voltage now includes $I_b r_{be}$ *plus* the drop across resistor R_E. This means that to obtain the *same* output voltage, a *larger* input voltage will be required. In other words, the voltage gain has been reduced. This is the price we must pay for the inclusion of R_E (and the resulting dc stability).

The exact ac equations are found as follows (refer to Fig. 6-7c).

$$V_i = I_b r_{be} + (\beta + 1)I_b R_E = I_b(r_{be} + (\beta + 1)R_E)$$

and

$$V_o = -\beta I_b R_L'$$

The voltage gain is

$$A_V = \frac{V_o}{V_i} = \frac{-\beta I_b R_L'}{I_b(r_{be} + (\beta + 1)R_E)} = \frac{-\beta R_L'}{r_{be} + (\beta + 1)R_E} \qquad (6\text{-}7)$$

Note that if $R_E = 0\ \Omega$, this equation is identical to that for the fixed-bias circuit. Equation 6-7 also indicates that the voltage gain will be reduced due to R_E because the denominator of this equation may now be quite large.

The input resistance is found as

$$R_i = V_i/I_i = R_B' \parallel \frac{I_b r_{be} + (\beta + 1)I_b R_E}{I_b} = R_B' \parallel (r_{be} + (\beta + 1)R_E) \qquad (6\text{-}8)$$

Again, note that if $R_E = 0\ \Omega$, the results are identical to the fixed-bias circuit. However, with R_E present, the input resistance may exceed r_{be}.

The current gain is found similar to the method used in Chapter 5. If R_B' and R_C were open circuits, the input current would be I_b and the output current would be βI_b. The current gain is β. In reality, R_B' and R_c' are not infinite and some current is lost due to the current dividers on the input and output.

$$A_I = \beta \times \frac{R_B'}{R_B' + r_{be} + (\beta + 1)R_E} \times \frac{R_C}{R_C + R_L} \qquad (6\text{-}9)$$

Finally, looking into the output terminals, R_o is the resistance seen by the load resistor R_L. In this case

$$R_o \cong R_c \qquad (6\text{-}10)$$

Let's work out an example to appreciate some typical numbers we might expect.

Example 6-5 ──

Determine the ac solution for the circuit in Fig. 6-1. This circuit was solved for dc in Example 6-1 and 6-2.

Solution As a first step, r_{be} must be found. Using I_B = 66.6 μA, found in Example 6-2, r_{be} = 0.025 V/66.6 μA = 375 Ω. Now the voltage gain can be found using Eq. 6-7.

$$A_V = \frac{-100(0.56 \text{ k}\Omega \parallel 1 \text{ k}\Omega)}{0.375 \text{ k}\Omega + 101(0.22 \text{ k}\Omega)} = -1.6$$

The input resistance is found using Eq. 6-8.

$$R_i = 10 \text{ k}\Omega \parallel 2.2 \text{ k}\Omega \parallel [0.375 \text{ k}\Omega + 101(0.22 \text{ k}\Omega)] = 1.7 \text{ k}\Omega$$

The current gain is

$$100 \times \frac{10\text{k} \parallel 2.2\text{k}}{(10\text{k} \parallel 2.2\text{k}) + 0.375\text{k} + 101(0.22 \text{ k})} \times \frac{0.56\text{k}}{0.56\text{k} + 1\text{k}} = 2.8$$

Finally, $R_o \cong R_C$ = 560 Ω.

As previously mentioned, the voltage gain is reduced. In fact, in this example it is very low. By varying component values it is possible to achieve gains of 50 to 75, but this must be done carefully so that the dc stability and input resistance are not compromised. In the next section we consider design techniques.

Although the results of this example may not clearly indicate it, the input resistance has been increased due to R_E. Without R_E, the input resistance is usually equal to r_{be} (375 Ω in this example). With the addition of R_E, the resistance looking into the base terminal alone is much larger. For this case, it is 22.6 kΩ (0.375 kΩ + 101(0.22 kΩ)). However, this is shunted by a 1.8-kΩ R_B' and the resulting input resistance is only 1.7 kΩ. A judicious change in the R_1 and R_2 component values could result in a higher input resistance.

Figure 6-8 illustrates the amplifier's equivalent circuit. Note that k

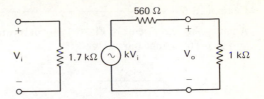

Figure 6-8 The amplifier equivalent circuit for Fig. 6-1.

(the open circuit voltage gain) is not indicated because the voltage gain calculated in the example included the 1-kΩ load resistor.

Figure 6-9 summarizes the important equations for the voltage divider bias circuit. Again, realize that these equations can be used for a number of variations of this basic circuit.

Voltage Divider Bias Circuit Summary

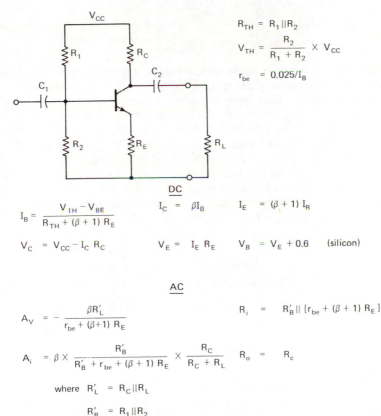

$$R_{TH} = R_1 \| R_2$$

$$V_{TH} = \frac{R_2}{R_1 + R_2} \times V_{CC}$$

$$r_{be} = 0.025/I_B$$

DC

$$I_B = \frac{V_{TH} - V_{BE}}{R_{TH} + (\beta + 1) R_E}$$

$$I_C = \beta I_B$$

$$I_E = (\beta + 1) I_B$$

$$V_C = V_{CC} - I_C R_C$$

$$V_E = I_E R_E$$

$$V_B = V_E + 0.6 \quad \text{(silicon)}$$

AC

$$A_V = -\frac{\beta R'_L}{r_{be} + (\beta + 1) R_E}$$

$$R_i = R'_B \| [r_{be} + (\beta + 1) R_E]$$

$$A_i = \beta \times \frac{R'_B}{R'_B + r_{be} + (\beta + 1) R_E} \times \frac{R_C}{R_C + R_L}$$

$$R_o = R_c$$

$$\text{where } R'_L = R_C \| R_L$$

$$R'_B = R_1 \| R_2$$

Figure 6-9 Important ac and dc equations for the voltage divider bias circuit.

Example 6-6

Analyze the *pnp* circuit in Fig. 6-10 by determining the dc bias voltages and currents and ac performance parameters.

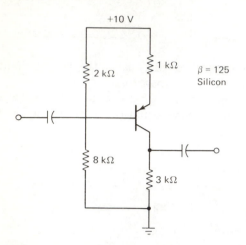

Figure 6-10   A *pnp* amplifier analyzed in Example 6-6.

Solution   Although this is a *pnp* circuit, this only means that the bias currents and voltages are of opposite polarity compared to a similar *npn* circuit. All of our equations still apply. Using the approximate method, we will assume the 8-kΩ and 2-kΩ resistors to be in *series* and calculate the base voltage as $V_B = 8/(2 + 8) \times 10$ V = 8 V. The emitter will be ~0.6 V *higher* than this or $V_E = 8.6$ V. The emitter and collector currents are found as V_{RE}/R_E, or (10 V - 8.6 V)/1 kΩ = 1.4 mA. V_C is calculated to be $I_C R_C = 1.4$ mA $\times$ 3 kΩ = 4.2 V. Finally, $V_{CE} = V_C - V_E = 4.2$ V - 8.6 V = -4.4 V. Note that this requires $V_{CB} = 4.2$ V - 8 V = -3.8 V. This means that the collector base junction is reverse-biased, as it should be for amplification.

   Proceeding to the ac solution, r_{be} must be found. This in turn requires I_B. Because I_C is known, $I_B = I_C/\beta = 1.4$ mA/125 = 11.2 μA. Then, $r_{be} = 0.025$ V/11.2 μA = 2.2 kΩ.

$$A_V = \frac{-125(3 \text{ k}\Omega)}{2.2 \text{ k}\Omega + 126(1 \text{ k}\Omega)} = -2.9$$

$$R_i = 2 \text{ k}\Omega \parallel 8 \text{ k}\Omega \parallel [2.2 \text{ k}\Omega + 126(1 \text{ k}\Omega)] = 1.6 \text{ k}\Omega$$

$$R_o = 3 \text{ k}\Omega$$

$$A_I = 0 \text{ (no load)}$$

6.3   DESIGN TECHNIQUES

When designing an amplifier circuit, what considerations are important? After our experience with the fixed-bias circuit, *dc stability* must certainly be near the top of our list! Others might include voltage and current gain, input and output resistances, and signal swing. The circuits we are considering now are considered *class A* amplifiers and are usually best for *voltage* amplification. Current amplification and the resulting power gain are usually accomplished using *class B* power amplifiers. With this in mind, our ac design goal might be an *ideal voltage amplifier*, that is, large input resistance and small output resistance.

Naturally a large voltage gain is desirable, but what about gain *stability*? If the voltage gain is 150 with one transistor but 100 with another, we may have problems using this amplifier in a practical system. Examining Eq. 6-7 for voltage gain, note that if $(\beta + 1)R_E$ is much larger than r_{be}, the equation becomes

$$A_V \cong \frac{-\beta R_L'}{(\beta + 1)R_E} \cong -R_L'/R_E$$

This is the ultimate in stability, depending only on resistor ratios. Unfortunately, this ratio must usually be small (<10) in order for the basic assumption $[(\beta + 1)R_E \gg r_{be}]$ to be true. Nevertheless, it may be worthwhile sacrificing some voltage gain in order to obtain more stable ac operation.

Table 6-1 summarizes the desirable properties for a voltage amplifier and their significance to the voltage divider bias circuit.

TABLE 6-1

Parameter	Desired Property	Significance on Circuit
DC bias	Stable	Use voltage-divider bias with R_E and β large, R_1 and R_2 small.
R_i	Large	β, R_1, R_2, and R_E large
R_o	Small	R_C small
A_V	Large	β, R_C large, and R_E small

A close study of this table quickly indicates that several of the parameters *conflict* with each other. A large R_E stabilizes the circuit for dc but kills the ac voltage gain. A small R_o is desirable for low output resistance but again reduces overall voltage gain. There really is no optimal combination and the best we can hope for is a *compromise* that meets a particular application. However, let us not be too discouraged. Transistor amplifier design is fast becoming a dying art as the 25¢ *operational amplifier* replaces more and more discrete transistor designs. The op-amp typically has a megohm input resistance, output resistance less than 100 Ω, and very accurate and repeatable voltage gains. More about this circuit later.

Considering the previous comments, let's assume a rather modest design goal. Our main concern will be *dc bias stability*. In this way, most any transistor will provide a stable operating point in the active region and the circuit should amplify. We will not be too concerned about input and output resistances and voltage gain. Once the circuit is working, we can fine tune it to approach the ac characteristics we might want.

We'll adopt three rules to ensure the dc stability and function of the amplifier:

$$(1) \quad V_{RE} = \frac{1}{10} \, V_{CC} \qquad (6\text{-}11)$$

As V_{RE} increases, the dc stability increases, but the voltage gain and signal swing decreases. $\frac{1}{10} \, V_{CC}$ is a compromise.

$$(2) \quad V_{CE} = \frac{1}{2} \, V_{CC} \qquad (6\text{-}12)$$

Because we are concerned with *small-signal* amplifiers, we don't expect large signal swings between saturation and cutoff. The bias point should be chosen to ensure optimum *linear* operation without distortion. Anywhere in the center one third of the characteristics should be satisfactory.

$$(3) \quad I_{R_1} \geqslant 10 \, I_B \qquad (6\text{-}13)$$

When this equation is true, resistors R_1 and R_2 are nearly in series and the base voltage is held constant by the $R_1 R_2$ voltage divider. This is crucial in establishing a stable bias point.

Example 6-7

Design a voltage divider bias *npn* common-emitter amplifier circuit using the previous rules. Assume $V_{CC} = 10$ V, $50 \leqslant \beta \leqslant 200$, and $I_C = 2$ mA. [The choice of collector current is somewhat arbitrary but determines the power dissipation of the transistor ($I_C \times V_{CE}$). Reference to data sheets will usually indicate maximum and typical I_C values.]

Solution

1. $V_E = \frac{1}{10} \, V_{CC} = 1$ V. If $I_E = I_C$, then $R_E = V_E / I_C = 1$ V/2 mA = 500 Ω
2. $V_{CE} = \frac{1}{2} \, V_{CC} = 5$ V. Then $V_{RC} = V_{CC} - V_{CE} - V_{RE} = 10$ V – 5 V – 1 V = 4 V. $R_C = V_{RC} / I_C = 4$ V/2 mA = 2 kΩ.
3. $V_B = V_E + 0.6$ V = 1.6 V

4. $I_B = I_C/\beta$. The *lowest* β should be used because this will require the *maximum* I_B. If I_{R_1} is ten times *this* value, the circuit can only improve for higher values of β. $I_B = 2$ mA/50 = 40 μA.

5. $I_{R_1} \geqslant 10I_B = 400$ μA. $R_1 = (V_{CC} - V_B)/I_{R_1} = 8.4$ V/0.4 mA = 21 kΩ. $R_2 = V_B/I_{R_1} = 1.6$ V/0.4 mA = 4 kΩ.

The final circuit is shown in Fig. 6-11.

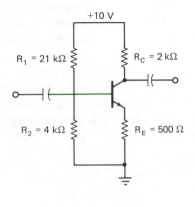

Figure 6-11 Common-emitter amplifier designed in Example 6-7.

Example 6-8

Determine the ac solution for the circuit in Fig. 6-11. Assume $\beta = 150$.

Solution r_{be} must first be found. $V_B = 4/25 \times 10$ V = 1.6 V, and $V_{RE} = 1.6$ V − 0.6 V = 1.0 V. Then $I_C = 1.0$ V/500 Ω = 2 mA. $I_B = 2$ mA/150 = 13.3 μA. This means $r_{be} = 0.025$ V/13.3 μA = 1.9 kΩ.

$$A_V = -150 \times 2 \text{ k}\Omega/[1.9 \text{ k}\Omega + 151(0.5 \text{ k}\Omega)] = -3.9$$

$$R_i = 21 \text{ k}\Omega \parallel 4 \text{ k}\Omega \parallel [1.9 \text{ k}\Omega + 151(0.5 \text{ k}\Omega)] = 3.2 \text{ k}\Omega$$

$$R_o = 2 \text{ k}\Omega$$

Example 6-9

Recalculate the voltage gain in Example 6-8 if β increases to 300.

Solution I_B now decreases to 2 mA/300 = 6.7 μA, and $r_{be} = 3.7$ kΩ. Then $A_V = -300 \times 2$ kΩ/[3.7 kΩ + 301(0.5 kΩ)] = −3.9.

This is an interesting result. Although β has doubled, the voltage gain has remained constant. The circuit is not only *dc stable* but also *ac stable*. It was previously mentioned that the voltage gain approaches R'_L/R_E if $(\beta + 1)R_E \gg r_{be}$. This indicates a very stable voltage gain dependent only on resistor values. For this example, $R'_L/R_E = 4$ and compares well with the 3.9 calculated.

The amplifier we have designed has been shown to have a stable ac gain. The dc operating point should also be stable for β and temperature variations (see Problem 6-10).

One serious drawback to this amplifier is its low voltage gain. This is primarily due to the emitter resistor. A technique for improving the voltage gain of a common-emitter amplifier involves *bypassing* the emitter resistor (or a portion of it) with a capacitor. This is shown schematically in Fig. 6-12a. The dc operation of the circuit is unaffected as this capacitor simply charges to the emitter dc voltage. However, if the value of the capacitor is properly chosen, it will appear to have a very low reactance in ohms compared to R_E and bypass all of the emitter *ac* signal to ground. This means that R_E is effectively 0 Ω for ac purposes. The effect of this capacitor is to *increase* the voltage gain dramatically.

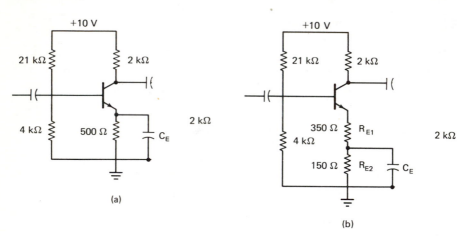

Figure 6-12 The voltage gain of an amplifier can be increased by bypassing the emitter resistor. In (a) 100% bypassing is shown, while partial bypassing is shown in (b).

Example 6-10 ─────────────────────────────────────

Calculate the ac voltage gain and input resistance for the circuit in Fig. 6-12a. Assume $\beta = 150$.

Solution In Example 6-8, r_{be} was found to be 1.9 kΩ. Using Eq. 6-7 for voltage gain with $R_E = 0$ Ω,

$$A_V = \frac{-150 \ (2 \ k\Omega)}{1.9 \ k\Omega} = -158$$

R_i is found using Eq. 6-8.

$$R_i = 21 \text{ k}\Omega \parallel 4 \text{ k}\Omega \parallel 1.9 \text{ k}\Omega = 1.2 \text{ k}\Omega$$

As was expected, the voltage gain increased considerably. This is because all of the input signal drops across the base-emitter junction rather than being shared with R_E. The same output signal can now be achieved with a much smaller input signal. It may not have been expected that R_i would decrease (3.2 kΩ to 1.2 kΩ). Unfortunately, emitter bypassing involves another *tradeoff situation*. As the voltage gain *increases*, the input resistance *decreases*. When all of R_E is bypassed, we see only r_{be} when looking into the base terminal. In fact, for the case of 100% bypassing, this circuit has an *ac* equivalent circuit identical to the *fixed-bias* circuit. This unfortunately means low input resistance and a voltage gain dependent on β.

A good compromise is to use *partial* emitter bypassing. This involves breaking the emitter resistor into *two* resistors and bypassing only one. This is illustrated in Fig. 6-12b. As an example, consider 30% of R_E to be bypassed. This means 70% is unbypassed or R_{E1} = 350 Ω and R_{E2} = 150 Ω. The equations for A_V and R_i can now be reworked substituting R_E = 350 Ω. Table 6-2 indicates several combinations of emitter bypassing and the resulting voltage gains and input resistances using the circuit in Fig. 6-12b.

TABLE 6-2

	No Bypass	30%	50%	80%	100%
R_{E1}	500 Ω	350 Ω	250 Ω	100 Ω	0 Ω
R_{E2}	0	150 Ω	250 Ω	400 Ω	500 Ω
A_V	-3.9	-5.5	-7.6	-17.8	-158
R_i	3.2 kΩ	3.2 kΩ	3.1 kΩ	2.8 kΩ	1.2 kΩ
R_C/R_{E1}	4	5.7	8	20	—

As more of R_E is bypassed, the trend of *increasing* voltage gain and *decreasing* input resistance is apparent. It is interesting to note that even with 80% of R_E bypassed, the gain is still relatively low compared to its 100% bypassed value.

Do not be misled by the data in Table 6-2, however. We cannot make a general statement such as "always use 80% (or 90%) bypassing." The specific results depend on other factors such as β, R_C, R_L, and r_{be}. The main point is simply that the gain will *increase* and R_i will *decrease* as more of R_E is bypassed. The exact amount of bypassing to use depends on the application.

The bottom line of Table 6-2 indicates the ratio of R_L' (R_C in this case) to R_{E_1}. This should approximate the voltage gain and is an indication of the ac gain *stability*, as discussed previously. Note that as more of R_E is bypassed, this ratio becomes increasingly *inaccurate*. This may be another consideration when determining the amount of emitter bypassing.

Example 6-11

Determine the value of the bypass capacitor in Fig. 6-12a assuming the amplifier is to be used from 100 Hz to 20,000 Hz.

Solution Because 100% of R_E is bypassed, X_C must effectively bypass 500 Ω. Following the method discussed in Chapter 5, X_C should be $^1/_{10}$ of 500 Ω at $f =$ 100 Hz (the *lowest* frequency of operation). Then $C = 1/(2 \pi f X_C) = 1/[2 \pi (100)(50)] = 32 \mu F$. A 33- or 47-$\mu F$ capacitor rated at 16 V or more should be used.

Signal Swing

An important aspect of any amplifier is its output signal swing. This is most easily visualized by sketching the ac and dc *load lines* on the characteristic curves.

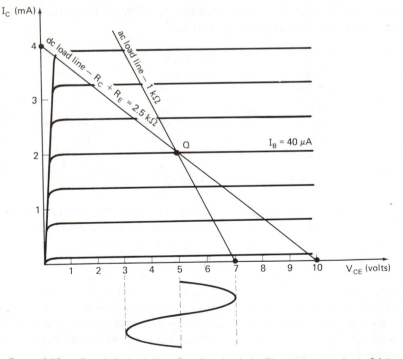

Figure 6-13 AC and dc load lines for the circuit in Fig. 6-11, assuming a 2-kΩ load resistor. The output signal swing is 4V peak-peak.

Consider the circuit we designed shown in Fig. 6-11. Assume a 2-kΩ load resistor has been added. The dc load line is determined by finding V_{CE} and I_C at cutoff and saturation.

$$V_{CE} \text{ (cutoff)} = 10 \text{ V} \qquad I_C = 0 \text{ A}$$

$$V_{CE} \text{ (sat)} = 0 \text{ V} \qquad I_C = 10 \text{ V}/2.5 \text{ k}\Omega = 4 \text{ mA}$$

Note that due to R_E, the saturation current is found as $V_{CC}/(R_C + R_E)$. These two points, along with the Q point found in Example 6-11, are shown in Fig. 6-13.

The ac load line must pass through the Q point and would be identical to the dc load line except for R_L. For ac only, the collector sees $R_C \parallel R_L$ or a 1-kΩ load. When I_C changes by 2 mA, the ac collector-emitter voltage changes by 2 mA × 1 kΩ = 2 V. A second point on the ac load line is therefore $I_C = 0$ A and $V_{CE} = 7$ V, ($\Delta I_C = 2$ mA, $\Delta V_{CE} = 2$ V). The maximum signal swing is seen to be 4 V peak-peak compared to 10 V peak-peak without R_L. As discussed in Chapter 5, it is inevitable that the addition of a load resistor will decrease both the voltage gain and output signal swing of the amplifier.

6.4 OTHER COMMON EMITTER CIRCUITS

Although most common-emitter amplifiers use a form of the universal voltage-divider bias circuit, other configurations do exist. As an example, consider the *collector self-bias circuit*.

Example 6-12 ──────────────────────────────────────

Solve the circuit shown in Fig. 6-14 for ac and dc and comment on its dc stability.

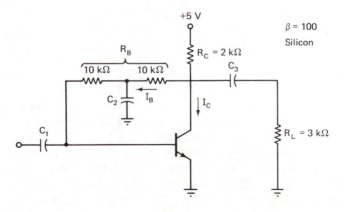

Figure 6-14   Collector self-bias circuit discussed in Example 6-12.

Solution This circuit gets its name from the fact that the base is self-biased by the collector instead of V_{CC}. This is accomplished by the two 10-kΩ resistors labelled R_B in the figure. The current through the 2-kΩ collector resistor is $I_B + I_C = I_B + \beta I_B = (\beta + 1)I_B$. A loop equation can be written to determine I_B. $V_{CC} - (\beta + 1)I_B R_C - I_B R_B - V_{BE} = 0$ V, and therefore

$$I_B = \frac{V_{CC} - V_{BE}}{R_B + (\beta + 1)R_C} \tag{6-14}$$

In this equation the two base resistors have been combined as a single base resistor R_B (for dc only). Plugging into this equation, $I_B = (5 \text{ V} - 0.6 \text{ V})/[20 \text{ k}\Omega + 101(2 \text{ k}\Omega)] = 19.8 \ \mu\text{A}$. Now $V_{CE} = V_C$ is found as $V_{CE} = V_{CC} - (\beta + 1)I_B R_C = 5 \text{ V} - 101 (19.8 \ \mu\text{A}) 2 \text{ k}\Omega = 1$ V. Finally, $V_B = 0.6$ V.

This circuit employs *negative voltage feedback* to stabilize the dc operating point. This can be seen two ways. If a high β transistor is used in the circuit, the collector current will tend to *increase*. This causes V_{CE} to decrease due to the increased drop across R_C. This in turn *decreases* the voltage fed back to the base, tending to turn the transistor OFF and decreasing the base and collector currents and, hence, stabilizing the circuit.

The stability can also be seen if Eq. 6-14 is multiplied by β. Then $\beta I_B = I_C = \beta(V_{CC} - V_{BE})/[R_B + (\beta + 1)R_C$. Now, if $(\beta + 1)R_C \gg R_B$, $I_C = (V_{CC} - V_{BE})/R_C$, independent of β. This result is similar to the result found for the voltage divider circuit where $(\beta + 1)R_E \gg R_{TH}$ for dc stability.

Turning to the ac solution, you should be able to see that C_2 causes the two base resistors to be in parallel with the input and output circuits respectively. First, we calculate r_{be}: $r_{be} = 0.025 \text{ V}/19.8 \ \mu\text{A} = 1.3 \text{ k}\Omega$. Then

$$A_V = \frac{-\beta R_L'}{r_{be}} = \frac{-101 \ (2 \text{ k}\Omega \parallel 3 \text{ k}\Omega \parallel 10 \text{ k}\Omega)}{1.3 \text{ k}\Omega} = -83$$

$$R_i = r_{be} \parallel 10 \text{ k}\Omega = 1.2 \text{ k}\Omega$$

$$R_o = R_C \parallel 10 \text{ k}\Omega = 1.7 \text{ k}\Omega$$

The Thermistor

The dc stability of an amplifier can also be enhanced through the use of *nonlinear* components. An example of this is illustrated in Fig. 6-15. This circuit is already familiar to us except for resistor R_T. This is a *thermistor*. A thermistor is a component whose resistance varies with temperature. In this application, a *negative* temperature coefficient is desired. As the temperature of the transistor (and thermistor) increases, the collector current tends to increase but R_T tends to decrease. This in turn causes I_T to increase and raise the emitter voltage. If the base is held constant by the $R_1 R_2$ voltage divider, the base-emitter voltage $(V_B - V_E)$ will decrease and tend to offset the increase in collector

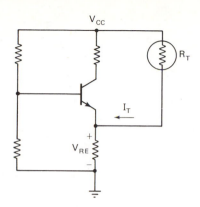

Figure 6-15 Use of a thermistor to stabilize the operating point of a common-
emitter amplifier.

current. For this circuit to be effective, the thermistor and transistor
must physically share the same heat sink or other mounting hardware.

6.5   THE EMITTER FOLLOWER

A second major type of circuit configuration involving bipolar transistors is called the *common-collector* or *emitter-follower* circuit. A typical
bias-stable emitter follower is shown in Fig. 6-16. This circuit should
look familiar because it uses voltage divider bias similar to the universal
circuit we have been studying. As illustrated in the figure, there are two
differences between this circuit and the universal common-emitter
circuit. First, the output signal is taken at the emitter and not the
collector. The output voltage "*follows*" the base voltage. Second, the
collector is tied directly to V_{CC}. There is no R_C resistor. This condition
is not absolutely necessary but is usually the case.

When analyzing the dc bias of the emitter follower, we can use the
same procedures as used for the voltage divider bias circuit. In fact,
everything that made that circuit stable $[I_1 \geqslant 10I_B , (\beta + 1)R_E \geqslant R_{TH}]$
also holds true for the emitter follower.

The ac equivalent circuit is shown in Fig. 6-17a. Note that the
collector is at ac ground and common to both input and output circuits
(hence the name common collector). The input voltage is divided
between r_{be} and the emitter resistor in parallel with R_L. This is shown
in Fig. 6-17b. The output voltage is the portion of this input voltage
dropped across the parallel combination of resistors R_E and R_L. The
significance of this is that V_o will always be *less* than V_i, although it is
common to assume nearly *unity* voltage gain for this circuit.

The input resistance may be large due to the β multiplication effect
of R_E (similar to the common-emitter amplifier with unbypassed R_E).

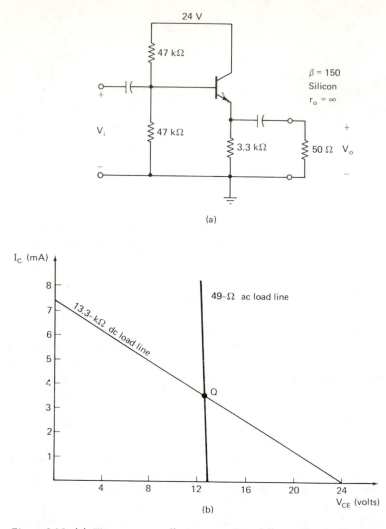

Figure 6-16 (a) The common-collector or emitter-follower circuit. The output voltage, taken at the emitter, follows the base voltage. (b) the dc and ac load lines used in Example 6-13. The output signal swing is approximately 0.4 V peak-peak.

Although the voltage gain is less than 1, the current gain is greater than 1 because the emitter (output) current is related to the base (input) current by a factor of $\beta + 1$.

Finally, although it is not obvious from the ac equivalent circuits, the output resistance may be quite small. This is shown mathematically in the next section.

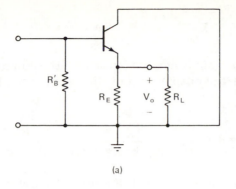

(a)

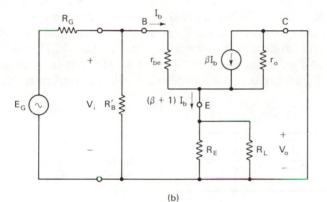

(b)

Figure 6-17 (a) The ac equivalent circuit for the emittet follower. (b) the small signal model.

The AC Equations

We begin by determining the exact equation for A_V.

$$A_V = \frac{V_o}{V_i} = \frac{(\beta + 1)I_b \times (R_E \parallel R_L \parallel r_o)}{I_b r_{be} + (\beta + 1)I_b \times (R_E \parallel R_L \parallel r_o)}$$

$$= \frac{(\beta + 1)R_L'}{r_{be} + (\beta + 1)R_L'} \tag{6-15}$$

where $R_L' = R_E \parallel R_L \parallel r_o$. Note that A_V must be less than 1 and that there is no phase reversal.

The *input resistance* will include R_B' in parallel with the resistance seen looking into the base terminal. This later term is found as V_i/I_b.

$$R_i = R_B' \parallel \frac{I_b r_{be} + (\beta + 1)I_b R_L'}{I_b} = R_B' \parallel (r_{be} + (\beta + 1)R_L') \qquad (6\text{-}16)$$

The *current gain* is nearly identical to the common-emitter equivalent except that the output current divides between R_E and R_L instead of R_C and R_L.

$$A_I = \beta \times \frac{R_B'}{R_B' + r_{be} + (\beta + 1)R_L'} \times \frac{R_E \parallel r_o}{R_E \parallel r_o + R_L} \qquad (6\text{-}17)$$

The *output resistance* of the emitter follower can be found with the aid of Fig. 6-18. The output resistance of any amplifier is found by removing R_L and looking into the output terminals with the input source "killed" or turned OFF. A source V_o applied to the output terminals will generate a current I_o such that $V_o/I_o = R_o$. This is illustrated in Fig. 6-18. The input source is OFF but its resistance is still present as R_G. This is in turn in parallel with R_B' and in series with r_{be}.

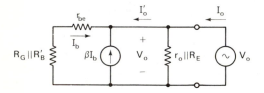

Figure 6-18 Circuit used to analyze the output resistance of the emitter-follower circuit.

Finally, the βI_b generator has been turned upside down. Carefully study Fig. 6-18 until you can see how it is obtained from Fig. 6-17. The output resistance is

$$R_o = r_o \parallel R_E \parallel \frac{V_o}{I_o'} = r_o \parallel R_E \parallel \frac{-I_b[(R_G \parallel R_B') + r_{be}]}{-(\beta + 1)I_b}$$

$$= r_o \parallel R_E \parallel \frac{(R_G \parallel R_B') + r_{be}}{\beta + 1} \qquad (6\text{-}18)$$

If β is relatively large, it may result in a *very low* output resistance. In fact, the emitter follower is sometimes referred to as a *buffer* because its output resistance can *match* a very small load resistance and step it up to become a much larger input resistance. Unfortunately, no voltage gain is possible.

Example 6-13 ————————————————————————

Analyze the circuit in Fig. 6-16 and determine the ac and dc solutions. Estimate the output signal swing.

Solution Using the voltage divider method:

$$V_B = \frac{47}{94} \times 24 \text{ V} = 12 \text{ V}$$

$$V_E = V_B - 0.6 \text{ V} = 11.4 \text{ V}$$

$$I_C = I_E = \frac{11.4 \text{ V}}{3.3 \text{ k}\Omega} = 3.5 \text{ mA}$$

$$I_B = \frac{I_C}{\beta} = 23 \text{ }\mu\text{A}$$

$$V_C = V_{CC} = 24 \text{ V}$$

$$V_{CE} = 24 - 11.4 \text{ V} = 12.6 \text{ V}$$

NOTE: $I_{R1} = 12 \text{ V}/47 \text{ k}\Omega = 255 \text{ }\mu\text{A}$, and $10I_B = 230 \text{ }\mu\text{A}$, justifying the previous method.

For ac we have:

$$r_{be} = \frac{0.025 \text{ V}}{23 \text{ }\mu\text{A}} = 1.1 \text{ k}\Omega$$

$$R_i = 47 \text{ k}\Omega \parallel 47 \text{ k}\Omega \parallel [1.1 \text{ k}\Omega + 151(3.3 \text{ k}\Omega \parallel 0.05 \text{ k}\Omega)] = 6.3 \text{ k}\Omega$$

$$A_V = \frac{151 (3.3 \text{ k}\Omega \parallel 0.05 \text{ k}\Omega)}{1.1 \text{ k}\Omega + 151(3.3 \text{ k}\Omega \parallel 0.05 \text{ k}\Omega)} = 0.87$$

$$R_o = \frac{(47 \text{ k}\Omega \parallel 47 \text{ k}\Omega) + 1.1 \text{ k}\Omega}{151} \parallel 3.3 \text{ k}\Omega = 155 \text{ }\Omega$$

$$A_I = 150 \times \frac{23.5 \text{ k}\Omega}{23.5 \text{ k}\Omega + 1.1 \text{ k}\Omega + 151(0.049 \text{ k}\Omega)} \times \frac{3.3 \text{ k}\Omega}{3.3 \text{ k}\Omega + 0.05 \text{ k}\Omega} = 109$$

The signal swing can be predicted following a procedure similar to that used for the common-emitter circuit. Figure 6-16b indicates the dc and ac *load lines* for this circuit. Note that *both* load lines must pass through the dc operating point (this corresponds to 0 input signal). For ac purposes, the current must follow the 49-Ω (3.3 kΩ $\parallel$ 50 Ω) ac load line. When I_C changes by 3.5 mA, V_{CE} will change by 3.5 mA $\times$ 49 Ω = 0.2 V. The maximum signal swing is therefore only 0.4 V peak-peak across the 50-Ω load.

Although this signal swing is quite small, consider the current needed to develop 5 V peak-peak across a 50-Ω load: I = 5 V/50 Ω = 100 mA peak-peak!

An important point to note in this example is that the 50-Ω load has been transformed into 6.3 kΩ at the emitter-follower input. This is 126 times its actual size! This is the *buffer* action we spoke of previously and is illustrated in Fig. 6-19.

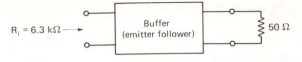

Figure 6-19 The emitter-follower can be thought of as a buffer, transforming the 50-Ω load resistance to 6.3 kΩ at the input terminals.

An unfortunate side effect to this circuit is that the voltage gain is less than 1. The circuit actually functions as a voltage attenuator. However, if you are thinking ahead a bit, you might be able to visualize a *two*-stage amplifier, the first stage of which develops voltage gain by driving an *emitter follower second-stage* acting as a buffer. Because voltage gain depends on R_L, an amplifier driving 6.3 kΩ will have much higher gain than it would attempting to drive 50 Ω directly. We talk more about this in the next chapter.

The Darlington Connection

The buffer action of the emitter follower can be enhanced through the use of the *Darlington* connection of two transistors. The improved emitter-follower circuit is shown in Fig. 6-20. Note that the base current of Q_1 is amplified by the β of Q_1 to become the base current of Q_2. This current is in turn amplified by the β of Q_2 to become the actual output current. This means $I_{C2} = \beta^2 I_{B1}$. The *effective* β of the Darlington pair transistors may be as high as 50–70,000!

The large β possible with the Darlington connection allows the emitter-follower circuit to be designed to have a very large input

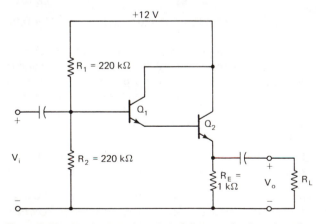

Figure 6-20 An improved emitter-follower circuit using the Darlington connection.

resistance. This is true for several reasons. First, the $(\beta + 1)R_L'$ term in Eq. 6-16 must be very large. Second, due to the larger β, I_B will be quite small ($I_B = I_C/\beta^2$) and r_{be} may become quite large ($r_{be} = 0.025$ V/I_B), contributing to the large input resistance. Finally, because I_B is small, the base biasing resistors, R_1 and R_2, can be made large and still satisfy the $I_{R_1} \geqslant 10I_B$ criteria for dc stability. Without the Darlington connection, the input resistance is often limited to a low value due to these two resistors. Now, with the Darlington connection, these resistors may be large and the input resistance will be approximately $R_1 \parallel R_2$.

As might be expected, the high effective β value increases the *current gain* compared to non-Darlington circuits. This high β also causes r_o to be *smaller* than the typical value for an emitter follower.

The higher β does *not* increase the voltage gain when used as a common-emitter amplifier however. This is because the effective r_{be} also increases due to the decrease in I_B ($A_V = -\beta R_L'/r_{be}$).

There are two practical considerations to remember when using the Darlington connection. Because the two transistors will run with significantly *different* collector currents, the β's of the two transistors will probably not be equal. Q_1 has a much lower I_C than Q_2. This may mean $\beta_1 < \beta_2$.

Another problem is that the *leakage current* of Q_1 is amplified by Q_2. This means that the composite transistor has a relatively high leakage current that could lead to dc stability problems.

In summary, the Darlington connection is most often used in the emitter-follower amplifier to increase the input resistance and current gain and decrease the output resistance. However, matching problems and leakage currents may make it undesirable in some applications. Often the two transistors are fabricated on a single piece of silicon, enhancing the matching between the two transistors. The composite is then packaged as a single transistor with three leads.

6.6 THE COMMON BASE

A third type of amplifier that found widespread use in the early days of transistors is the *common-base* circuit illustrated in Fig. 6-21. The base is common to the input and output circuits, with the input signal applied to the emitter and the output taken at the collector.

A major *disadvantage* of this configuration is that the input resistance is extremely low, typical values being 20 to 40 Ω. This is because looking into the base, we see a voltage $I_b \times r_{be}$, but an input current of $(\beta + 1)I_b$. R_i is then $\sim r_{be}/\beta$. This very low resistance makes it difficult to develop an ac signal across the input of the circuit because most sources will be loaded down by this low R_i.

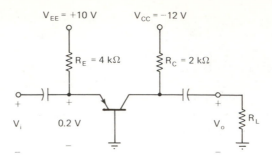

Figure 6-21 The common base amplifier.

Another disadvantage is that the current gain is less than 1. This must be true because the input and output currents are I_e and I_c respectively, and I_c must always be less than I_e.

The common base circuit is not without its attributes, however. Referring to Fig. 6-21, the emitter potential is ~0.2 V (for a germanium transistor) and $I_E = (10 - 0.2\ V)/4\ k\Omega = 2.5$ mA. If $I_C \cong I_E$, then V_C = -12 V + (2.5 mA)2 kΩ = -7 V. Note that β was not needed to determine the dc bias point. This circuit is *inherently* dc stable, deriving its bias stability from the emitter potential, which depends on V_{BE}, not β.

Actually, when we assumed $I_C \cong I_E$, we were assuming $\alpha = 1$. Because $\alpha = \beta/(\beta + 1)$, if $\beta = 10$, $\alpha = 10/11 = 0.91$, and there is a 9% error in the previous bias calculation. In the early 1950's, most transistors were germanium with relatively low β's (less than 50 not uncommon), and the common base proved to be a stable amplifier.

The voltage gain of the circuit is comparable to the common emitter but there is *no phase inversion*. Another advantage is that this gain tends to remain constant even at high frequencies (>10 MHz), where the gain of the common-emitter circuit begins to fall off.

———————————————— KEY TERMS ————————————————

Bias Stability This refers to variations in a transistor's collector current due to changes in circuit conditions such as temperature or transistor β. For a stable circuit, the collector current will change only slightly (< 5%).

Common-Emitter Amplifier This is a type of transistor amplifier in which the emitter is common to both the input and output circuits. It is characterized by relatively high voltage and current gains, low input resistance, and medium output resistance.

Negative Feedback This is a design technique that tends to stabilize the ac and dc properties of an amplifier but reduce its ac voltage gain. For a common-emitter amplifier, negative feedback is easily accomplished by adding an emitter resistor. In this circuit, as collector current tends to increase, the negative feedback causes the circuit to react by reducing the collector current, thereby stabilizing the circuit.

Bypassing This is a circuit technique in which a component is shunted by a capacitor. This causes the ac current to bypass the component. In the common-emitter amplifier, the emitter resistor is often bypassed by a capacitor to ground. This has the effect of maintaining the circuit's dc stability without reducing the voltage gain.

Emitter Follower (Common Collector) This type of transistor circuit acts as a buffer. Its voltage gain is less than unity but the current gain may be large. Because it may be designed to have a very low output resistance and a moderately large input resistance, it is often used for impedance matching or buffering.

Darlington Connection This refers to a compound transistor made up of two similar transistors such that the collector current of one supplies the base current of the other. The composite transistor may have a very large β (on the order of β^2). It is most often used in the emitter-follower configuration due to its very high input resistance.

Common Base Amplifier This is a transistor amplifier characterized by a very low input resistance, current gain less than 1, large voltage gain, and medium output resistance. It has excellent frequency response but, due to its low input resistance, is not used as often as the common-emitter amplifier.

――――――――――――――― QUESTIONS AND PROBLEMS ―――――――――――――――

6-1 Refer to the voltage divider bias circuit in Fig. 6-9. Assume V_{CC} = 12 V, R_1 = 20 kΩ, R_2 = 4.7 kΩ, R_C = 1.5 kΩ, R_E = 470 Ω, and R_L = 10 kΩ. If β = 100, use the *Thevenin* method to calculate the dc solution to this circuit.

6-2 Repeat Problem 6-1 if β = 300. Is the bias point stable to changes in β?

6-3 Repeat Problem 6-1 using the voltage divider method. Compare your results to those in Problem 6-1.

6-4 Determine the ac solution for the circuit in Problem 6-1 and draw the amplifier equivalent circuit. What is the approximate maximum output signal swing?

6-5 Show the algebra required to derive Eq. 6-6 from Eq. 6-5.

6-6 What two factors contribute to making the voltage divider bias circuit dc stable?

6-7 Refer to Fig. 5-17 and determine values for R_1, R_2, R_{TH}, V_{TH}, and R_E in Eq. 6-6. Compare the resulting equation to Eq. 3-4.

6-8 Referring to the circuit in Problem 6-1, prepare a table comparing R_i and A_V for the following cases of emitter resistance bypassing:
(a) 100%
(b) 80%
(c) 50%
(d) 0%

6-9 What factors should be considered when deciding on the amount of emitter resistance bypassing needed in a common-emitter amplifier?

6-10 Check the bias stability of the circuit designed in Example 6-7. Let $\beta = 200$ and $I_{CBO} = 10$ nA and use the Thevenin method to calculate V_B, V_C, V_E, and V_{CE}. Assume the junction temperature increases by $50°C$.

6-11 Determine the dc bias for the *pnp* amplifier in Fig. 6-22, $\beta = 100$.

6-12 What is the lowest value of β that will satisfy the dc bias stability criteria $(I_{R1} \geqslant 10 I_B)$ for the circuit in Fig. 6-22?

6-13 Calculate the collector voltage for cutoff and saturation for the circuit in Fig. 6-22.

6-14 Calculate the ac solution for the amplifier shown in Fig. 6-22, $R_L = 1$ kΩ.

6-15 Estimate the maximum peak-peak output signal swing for the *pnp* amplifier in Fig. 6-22.

6-16 Design a dc and ac stable common-emitter amplifier to meet the following specifications: $V_{CC} = +12$ V, $40 \leqslant \beta \leqslant 300$, silicon *npn* transistor, $A_V \geqslant 15$, and $R_i \geqslant 3$ kΩ.

6-17 Analyze the circuit shown in Fig. 6-23 and calculate the dc bias (*Hint:* Write a loop equation for the base-emitter circuit).

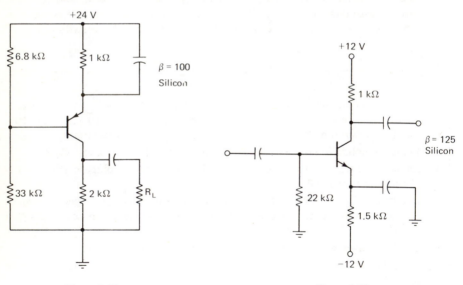

Figure 6-22   Figure 6-23

6-18 Determine the ac and dc solution for the collector self-bias circuit shown in Fig. 6-14 if the two R_B resistors are changed to 20 kΩ each and R_C is changed to 500 Ω.

6-19 Recalculate the dc solution found in Problem 6-18 if β doubles to 200.

6-20 An emitter-follower circuit is shown in Fig. 6-24. Determine the dc bias point. Calculate the ac solution and determine how much buffering of the load takes place; $\beta = 200$.

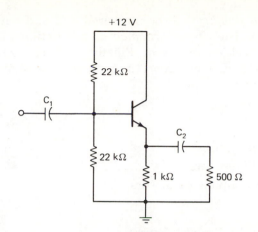

Figure 6-24

6-21 Assuming the amplifier in Problem 6-20 is to operate over the frequency range 100 Hz $\leqslant f \leqslant$ 10 kHz, determine values for the coupling capacitors, C_1 and C_2.

6-22 Estimate the maximum output signal swing possible across the 500-Ω load resistance for the emitter-follower circuit shown in Fig. 6-24.

6-23 Using the result obtained in Problem 6-22, determine the power gain in dB for this amplifier.

6-24 Compare and contrast the amplifier types listed considering the following performance factors: voltage and current gain, phase of input voltage relative to output, input resistance, and output resistance.
(a) Common emitter
(b) Emitter follower
(c) Darlington emitter follower
(d) Common base

6-25 Design a dc stable emitter-follower circuit assuming the following: $50 \leqslant \beta \leqslant 200$, silicon *pnp* transistor, V_{CC} = +10 V, $R_i \geqslant$ 10 kΩ, $A_V \geqslant 0.8$, and $R_o \leqslant$ 500 Ω. Assume R_L = 500 Ω and r_o = 20 kΩ.

_____   LABORATORY ASSIGNMENT 6:   _____
VOLTAGE DIVIDER BIAS TRANSISTOR AMPLIFIERS

Objectives

1. To learn to design a dc stable transistor amplifier circuit.
2. To perform a practical observation of the effects of emitter resistance bypassing.
3. To observe how the emitter follower is used as a buffer.

Introduction   In this laboratory assignment you will design a voltage divider bias common-emitter amplifier circuit using concepts presented in this chapter. Its dc *stability* will be measured as β and temperature are caused to change. The *ac performance* as a function of emitter-resistance bypassing will be observed. Finally, two emitter-follower circuits will be constructed in order to compare the standard and Darlington configurations.

Components Required

Various silicon *npn* transistors
Miscellaneous ¼-W resistors
Miscellaneous μF coupling capacitors

Part I: *Common-Emitter Amplifier Design*

NOTE: The circuit to be used is shown in Fig. 6-9. It provides a dc stable operating point via the voltage divider bias at the base terminal and negative feedback effect of the emitter resistor.

STEP 1   Use the curve tracer (if available) and select a Q point that will provide linear operation. You should select $V_{CEQ} = \frac{1}{2}V_{CC}$, and $I_C = 2$ mA – 5 mA.

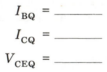

$$I_{BQ} = \underline{\qquad}$$

$$I_{CQ} = \underline{\qquad}$$

$$V_{CEQ} = \underline{\qquad}$$

STEP 2   Calculate the dc β at the Q point.

$$\beta_{dc} = \underline{\qquad}$$

STEP 3   For dc stability with reasonable signal swing, choose $V_E = \frac{1}{10} V_{CC}$.

$$V_E = \underline{\qquad}$$

STEP 4   Because $I_C \cong I_E$, and $V_E = I_E \times R_E$, then $R_E = V_E / I_C = \underline{\qquad}$
STEP 5   The voltage drop across R_C is $V_{CC} - V_{CEQ} - V_E = V_{RC}$. Calculate R_C as

$$R_C = V_{RC} / I_C = \underline{\qquad}$$

STEP 6   Calculate $V_B = V_E + 0.6\ V = \underline{\qquad}$

STEP 7 For dc stability, the base voltage should be determined by R_1 and R_2 in series. This requires $I_1 \geqslant 10I_B$ (i.e., I_B is negligible compared to I_1).

$$I_1 = 10I_B = \underline{\hspace{2cm}}$$

STEP 8 Now find $R_2 = V_B/I_1 = \underline{\hspace{2cm}}$, and $R_1 = (V_{CC} - V_B)/I_1 = \underline{\hspace{2cm}}$

Part II: *DC Measurements and Stability*

STEP 1 Obtain resistors with values as close as possible to your calculations and assemble the circuit. Compare calculated and measured results. Explain any discrepancies greater than 10%.

STEP 2 Select two or three other transistors (preferably with a higher β) and record V_B, V_C, and V_E in the test circuit. Comment on the bias stability as β changes.

STEP 3 Using your original transistor, heat the case of the transistor (a battery-powered soldering iron works well) for about five seconds. Again record V_B, V_C, and V_E. Comment on the bias stability as temperature changes.

Part III: *AC Performance*

STEP 1 Calculate values for coupling capacitors C_1 and C_2 at $f = 1$ kHz. Use $R_L = 1$ kΩ.

STEP 2 Using $f = 1$ kHz, *measure* A_V, R_i, R_o, and the output signal swing for the amplifier. Record these values under the heading "0% bypassed."

STEP 3 Repeat step 2 with 70% and 100% of R_E bypassed. *Note:* It will be necessary to *break* R_E into two resistors for the 70%-bypassed case.

STEP 4 Comment on the trade-offs involved when bypassing the emitter resistance.

STEP 5 Sketch the ac load line and predict the maximum output signal swing. Compare with your measurement.

Part IV: *The Emitter Follower*

NOTE: Two emitter-follower circuits are shown in Fig. 6-25a and b. Ideally, these circuits will buffer the load resistance to a much larger

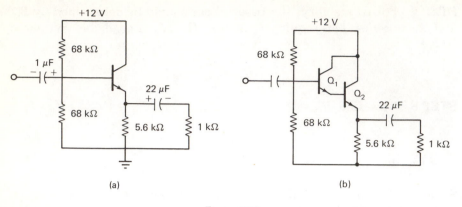

(a)   (b)

Figure 6-25

value at the input terminals while providing near unity voltage gain and no phase shift.

STEP 1   Assemble the circuit shown in Fig. 6-25a.

STEP 2   Measure the dc bias voltages and calculate $I_C = V_{RE}/R_E$.

STEP 3   Apply a 1-kHz input signal and measure A_V, R_i, R_o, and the output signal swing. Does the circuit function as a buffer? *Note:* The Darlington emitter follower in Fig. 6-25b improves upon the basic emitter-follower circuit due to the large effective β.

STEP 4   Repeat steps 1 to 3 for the Darlington amplifier in Fig. 6-25b. Use two silicon transistors with approximately the same IV characteristics for best results.

SEVEN

MULTISTAGE AMPLIFIERS AND COUPLING TECHNIQUES

In this final chapter on bipolar transistors, we discuss how the various single-stage transistor amplifiers we have been studying may be combined into *multistage* circuits to achieve the overall desirable properties of a particular amplifier.

Let's begin by using the *amplifier equivalent circuit* to help us understand multistage amplifiers.

7.1 MULTISTAGE AMPLIFIERS

Amplifier Equivalent Circuits

Figure 7-1a illustrates the amplifier equivalent circuit of a single-stage amplifier. What will happen if *two* of these stages are connected together? This is shown in Fig. 7-1b.

The output voltage of the first stage becomes the input voltage to the second. This voltage is now amplified by stage 2 to become the final output signal. Although this is a two-stage amplifier, we can draw an equivalent circuit for it similar to the equivalent we have drawn for the one-stage circuit. This is shown in Fig. 7-1c. The parameters for this new circuit can be found as follows:

1. Assume $V_{i1} = V_i = 1$ V. This may result in some unrealistically large output voltages, but our assumption is that the amplifier is *linear* and its gain does not depend on the magnitude of the input signal. The 1-V input could just as well be 0.01 V. Now, V_{o1} (or V_{i2}) can be found as $V_{o1} = 9$ kΩ/(1 kΩ + 9 kΩ) $\times$ 10 V = 9 V. This means the voltage gain of stage 1 is 9 (1 V in and 9 V out).

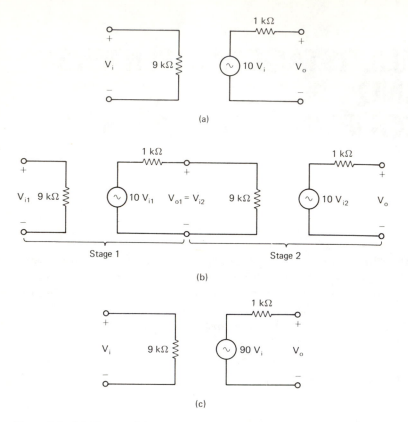

Figure 7-1 (a) The amplifier equivalent circuit for a single-stage amplifier. (b) two stages cascaded together. (c) the final equivalent circuit representing both stages cascaded.

2. The 9-V input signal to stage 2 is amplified by stage 2 to become 90 V at the output terminals. The overall open circuit voltage gain is then 90 V/1 V = 90. This is also found as

$$A_{VT} = A_{V1} \times A_{V2} \qquad (7\text{-}1)$$

For this case, $9 \times 10 = 90$. It is important to realize that Eq. 7-1 requires the *in-circuit* voltage gains and not the *open circuit* amplifier gains. In this example, the voltage gain of stage 1 is *loaded down* due to the input resistance of stage 2.

3. $R_i = 9$ kΩ. The amplifier equivalent input stage is identical to the stage 1 input circuit.

4. $R_o = 1$ kΩ. The amplifier equivalent output stage has the same resistance as the circuit of stage 2.

Be aware of one caution. In cases (3) and (4), errors may creep in depending on the circuit types. For example, the input resistance to some amplifiers *changes* as a load is added (e.g., the emitter follower) and similarily the output resistance of some amplifiers may change as the input source changes. In either case, the individual stage equivalent circuit must be corrected accordingly.

The benefits of cascading (putting in series) amplifier stages should be clear. The amplifier equivalent circuit in Fig. 7-1a might have represented an ac and dc stable common-emitter amplifier. The main drawback to this circuit is its low voltage gain. Now by combining stages, the overall gain has risen to 90, but without upsetting the basic stability of the circuit. This solves one of the main problems discussed in the previous chapter—how to obtain reasonable voltage gains without sacrificing circuit stability.

If two stages are good, then why not use three or four stages of amplification? After all, the gain in our example would approach 900 if another stage was added.

As is usually the case in electronics, be cautious when everything seems to be going well! As more stages of amplification are added, *stray capacitance* begins to increase due to the increased wiring complexity. This capacitance may act to couple a portion of the output signal back to the input stage. If this feedback signal is *in phase* with the original input, the amplifier may now be generating its own input signal. Actually the circuit has become an *oscillator* and not an amplifier. If the gain around the feedback loop is equal to 1, the oscillations are maintained and the circuit will be useless as an amplifier. Usually, the oscillations can be prevented by lowering the overall circuit gain, but this defeats the purpose of the multistages.

Most practical amplifiers employ only two or three voltage-gain stages and even then careful attention is given to tailoring the frequency response to fall off before a high frequency oscillation can begin.

Example 7-1 ───────────────────────────────────

Determine the voltage gain in dB for the three-stage amplifier shown in Fig. 7-2.

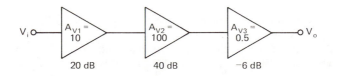

Figure 7-2 A three-stage amplifier. The overall voltage gain is
$10 \times 100 \times 0.5 = 500$, or $20 + 40 - 6 = 54$ dB.

Solution   The total gain may be found using Eq. 7-1 for three stages.

$$A_{VT} = 10 \times 100 \times 0.5 = 500$$

When converted to dB,

$$A_{VT} = 20 \log 500 = 54 \text{ dB}$$

We should also note that if the individual gains are expressed in dB, Eq. 7-1 becomes

$$A_{VT} \text{ (dB)} = A_{V_1} \text{ (dB)} + A_{V_2} \text{ (dB)} + A_{V_3} \text{ (dB)} \qquad (7\text{-}2)$$

For this example, $A_{VT} = 20 + 40 - 6 = 54$ dB.

7.2 CAPACITIVE COUPLING

Granted the advantages of a multistage amplifier, how do we go about connecting the individual stages together? Can we simply connect the collector of stage 1 to the base of stage 2? And if we do, won't this upset the *dc bias* of the two stages?

This method of coupling two amplifier stages together is called *direct coupling*, and although it can be done, special care must be taken. The dc bias of the two stages must be worked out together. Direct coupling is discussed in the next section.

A far simpler method of coupling is to use a *capacitor*. In this way, the ac signal is transferred from stage 1 to stage 2 while the dc voltages are blocked.

The following example will illustrate the analysis method for multistage amplifiers.

Example 7-2 ————————————————————————————

Determine the dc bias voltages, input and output resistance, and overall voltage gain for the two-stage capacitively-coupled amplifier shown in Fig. 7-3.

Solution   (a) DC. Beginning with stage 1 and assuming the voltage divider relationship to be true, $V_{B_1} = (10.7/66.7) \times 12 \text{ V} = 1.9$ V. Then, $V_{E_1} = 1.9 - 0.6 \text{ V} = 1.3$ V, and $I_{C_1} = 1.3 \text{ V}/0.5 \text{ k}\Omega = 2.6$ mA. $V_{C_1} = 12 \text{ V} - 2 \text{ k}\Omega \text{ (2.6 mA)} = 6.8$ V, and $V_{CE_1} = 5.5$ V.

Following a similar procedure for stage 2, $V_{B_2} = 6.5$ V, $V_{E_2} = 5.9$ V, $V_{C_2} = 12$ V, $V_{CE_2} = 6.1$ V, and $I_{C_2} = 6.7$ mA.

(b) AC. The input resistance of the overall amplifier will be the input resistance to stage 1. $R_i = 56 \text{ k}\Omega \parallel 10.7 \text{ k}\Omega \parallel [1.4 \text{ k}\Omega + 151(0.1 \text{ k}\Omega)] = 5.8 \text{ k}\Omega$, where $r_{be1} = 1.4$ kΩ.

The output resistance is that seen by the 8-Ω load. Because stage 2 is an emitter follower,

$$R_o = \frac{(2 \text{ k}\Omega \parallel 12 \text{ k}\Omega \parallel 14 \text{ k}\Omega) + 560 \text{ }\Omega}{151} \parallel 880 \text{ }\Omega = 14 \text{ }\Omega$$

where $r_{be2} = 560 \text{ }\Omega$.

Finally, the overall voltage gain is found. Note that the input resistance of stage 2 will act as the load resistance for stage 1.

$$R_{i2} = 12 \text{ k}\Omega \parallel 14 \text{ k}\Omega \parallel [560 \text{ }\Omega + 151(880 \text{ }\Omega \parallel 8 \text{ }\Omega)] = 1.4 \text{ k}\Omega$$

Then $A_{V1} = -R'_L/R_E = -(2 \text{ k}\Omega \parallel 1.4 \text{ k}\Omega)/0.1 \text{ k}\Omega = -8.2$. The gain of the emitter-follower stage is $151(8)/[151(8) + 560] = 0.68$. Finally, $A_{VT} = A_{V1} \times A_{V2} = -8.2 \times 0.68 = -5.6$.

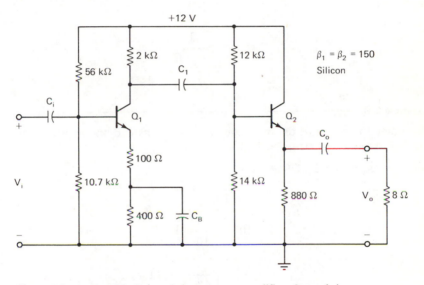

Figure 7-3 A capacitively-coupled two-stage amplifier. Stage 1 is a common emitter and Stage 2 is an emitter follower.

There are several comments that can be made about this circuit. Note that C_1 is the *coupling capacitor* between stages 1 and 2. This capacitor allows the ac signal at the Q_1 collector to pass to the Q_2 base. It also allows us to solve the dc bias conditions for both circuits completely independent of each other.

This particular amplifier is using a common-emitter first stage to develop voltage gain and an emitter follower acting as a buffer for the second stage. Although the gain is only 5.6, it would have been much lower if the 8 Ω had been driven directly by stage 1.

A key point is that multistage amplifiers are not always designed to achieve large voltage gains. Often, one stage may act as a buffer, allowing another stage to function normally. This is the case in the last example.

Determining the Value of C

The capacitive coupling technique used between amplifier stages is identical in concept to the capacitive coupling used between signal generator and amplifier input or load resistance and amplifier output as discussed in Section 5.3. As mentioned in that section, the value of X_C must be low compared to the resistance into which it is coupling — in this case, R_{i2}. Again adopting the $\frac{1}{10}R$ criteria, Eq. 5-9 for a two-stage amplifier becomes

$$C_C = \cfrac{1}{2\pi f \cfrac{R_{i2}}{10}} \tag{7-3}$$

Example 7-3

Determine the values of the three coupling capacitors and one bypass capacitor in Fig. 7-3. Assume $100 \text{ Hz} \leqslant f \leqslant 20 \text{ kHz}$.

Solution   Applying Eq. 5-9 and 7-3,

$$C_i = \cfrac{1}{2\pi\,(100)\left(\cfrac{5.8 \text{ k}\Omega}{10}\right)} = 2.7\ \mu\text{F}$$

$$C_1 = \cfrac{1}{2\pi\,(100)\left(\cfrac{1.4 \text{ k}\Omega}{10}\right)} = 11.4\ \mu\text{F}$$

$$C_o = \cfrac{1}{2\pi\,(100)\left(\cfrac{8\ \Omega}{10}\right)} = 1989\ \mu\text{F}$$

$$C_B = \cfrac{1}{2\pi\,(100)\left(\cfrac{400\ \Omega}{10}\right)} = 39.8\ \mu\text{F}$$

As expected, the output coupling capacitor must be quite large to couple the output signal to the small load resistance.

An important consideration of any coupling technique is the effect it will have on frequency response. This was discussed in general in

Section 4.6. As mentioned in that section, coupling capacitors cause a *high-pass* type of frequency response in which the gain rolls off at low frequencies, as shown in Fig. 4-17. This means that capacitively-coupled amplifiers *cannot* amplify a *dc* input voltage.

7.3 DIRECT COUPLING

A direct-coupled amplifier is one in which no coupling components are used. The output of one stage is *directly* connected to the input of the next. Advantages to this type of coupling include elimination of the large coupling capacitor and no low frequency gain roll-off.

A unique *disadvantage* to this type of coupling is that variations in the dc operating point are *amplified* by the following stage. This means temperature stability problems may be amplified due to the direct coupling.

Although successive common-emitter stages may be direct-coupled as shown in Fig. 7-4, this is infrequently done. As can be seen in this figure, the dc operating point of the collector must *rise* as each new stage is added. This is because V_{B2} must be at the same potential as V_{C1} but less than V_{C2}. In addition, if the collector voltage of Q_1 should change, Q_2 will amplify this change, causing an even larger change at the collector of Q_2.

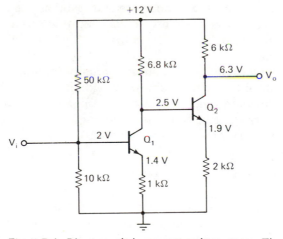

Figure 7-4 Direct-coupled common-emitter stages. The bias voltages increase from left to right for each additional stage.

An alternative to Fig. 7-4 is to use opposite type transistors, as shown in Fig. 7-5. Q_1 and Q_2 are still common-emitter amplifiers, but Q_2 has been replaced with a *pnp* transistor. Now the collector of Q_2 is actually at a *lower* potential than the Q_1 collector.

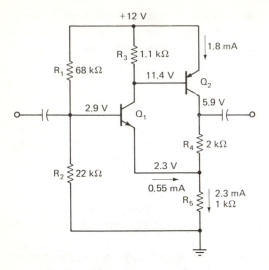

Figure 7-5 An improved direct-coupled amplifier. Both *npn* and *pnp* transistors are used and negative feedback is established via R_5.

This circuit also enjoys an improved dc stability when compared to the circuit in Fig. 7-4. This is due to Q_1 and Q_2 *sharing* resistor R_5. In essence, a negative feedback has been established. If, due to temperature, the collector currents of Q_1 and Q_2 should tend to increase, V_{R_5} will increase, causing Q_1 to conduct less (V_{BE_1} decreases), which in turn causes Q_2 to conduct less (Q_1 sinks the Q_2 base current). This means I_{C_1} and I_{C_2} will tend to *decrease* and stabilize the circuit.

Coupling capacitors are shown on the input and output of Fig. 7-5 and they prevent the operating point from being affected by the source or load resistances.

Some direct-coupled amplifiers are actually *intended* to amplify dc voltages. A dc amplifier with a gain of 10 will amplify 0.1 V dc to 1.0 V dc. The modern *operational amplifier* is an example of a dc amplifier. This amplifier uses a two-transistor direct-coupled circuit called the *differential* pair. This is covered in Chapter 10.

7.4   LARGE- AND SMALL-SIGNAL AMPLIFIERS

Figure 7-6 is a block diagram of a typical multistage amplifier. As was discussed in Chapter 4, an amplifier typically receives its input from a transducer. This signal is usually small and must be voltage-amplified without distorting the original waveshape. This amplification is usually accomplished using one or two small-signal amplifier stages.

We are by now very familiar with this type of circuit. Small-signal amplifiers are characterized by small variations about a central Q or operating point. By restricting the amplifier to a small-signal variation, linearity is ensured. In fact, the transistor model we have been using is based upon the assumption of small-signal operation.

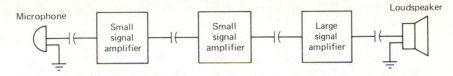

Figure 7-6 A typical multistage amplifier. Capacitive coupling is used between stages and the last stage is a power amplifier designed to drive the output transducer.

However, it eventually becomes necessary in any realistic amplifier to drive an output transducer, a speaker for example. This requires an amplifier stage capable of generating a reasonable amount of *power*. In fact, it is not unheard of for a stereo amplifier to produce 100 W or more per channel!

In Fig. 7-6, this particular stage is referred to as a *large-signal* amplifier. This amplifier must take the voltage-amplified transducer signal and increase its power level so that it may drive an output transducer.

A parameter of particular importance when discussing a power amplifier is *efficiency*. This is defined as

$$\eta = \text{efficiency} = \frac{P_{o(ac)}}{P_{i(dc)}} \tag{7-4}$$

This indicates how efficient the amplifier is in converting dc power to ac power. It should not be confused with the power *gain* of the amplifier, which is the ratio of ac output power to ac input power.

As Fig. 7-7 illustrates, the job of a power amplifier is to amplify the input signal and provide an output signal with a larger power level. It does this by drawing an average dc power from the power supply

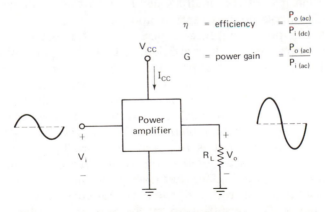

Figure 7-7 A typical power amplifier. The ratio of ac output power to dc input power is called the efficiency of the amplifier.

considered to be $P_{i(dc)}$ and generating an output ac power across the load called $P_{o(ac)}$.

Typical efficiencies for the amplifiers we have been discussing to this point are *less* than 25%. This means 1 W of dc power from the power supply is converted to 0.25 W of ac power in the load. If this seems rather low to you, it is! It is also the reason that most of the amplifier stages we have been studying are used as *small-signal* amplifiers only.

You might be wondering where those 0.75 W of power went in the previous example. As mentioned, 0.25 W got to the load. The remainder is lost in heating the transistor itself ($I_{CQ} \times V_{CEQ}$) and in $I^2 R$ heating of the collector resistor.

The next few sections illustrate several types of power amplifiers and their relative efficiencies. In addition, we discuss another method of interstage coupling that has the unique property of impedance transformation and matching.

7.5   CLASS A AMPLIFIERS

A *class A* amplifier is one in which the transistors are biased in the *active region* and the circuit draws current from the power supply at all times. This type of amplifier turns out to be very *inefficient* because power is consumed even when there is no input ac signal.

All of the amplifiers we have discussed in this text so far have been class A amplifiers. In fact, we have taken great pains to center our Q points in the active region. This results in *linear* operation for *small-signal* inputs and low distortion. This is the typical application of a class A amplifier. On occasion, the class A amplifier may be called upon to amplify *large signals*. In this case, the output will contain more distortion than it does for small signals (due to the nonlinearity of the characteristic curves near saturation and cutoff) but often less distortion than other power amplifier types, class B for example. The compromise, however, is very low efficiency. It may require 500 W of dc power to achieve 100 W of ac power out. This represents a large investment in power transistors, heat sinks, and power supplies.

Figure 7-8 illustrates the typical Q point for a class A amplifier. This might represent the operating conditions for a simple fixed-bias amplifier with capacitively-coupled load. We can see that with no input signal, the collector current is I_{CQ} and there are V_{CEQ} volts dropped across the collector-emitter of the transistor. This represents heat lost in the transistor despite the fact that there is no input signal.

As an ac signal is introduced, the collector current will vary equally above and below I_{CQ}, resulting in an average or dc collector current

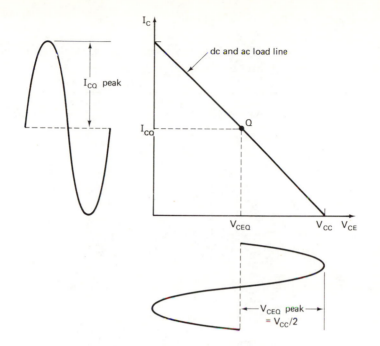

Figure 7-8 Class A bias point. The maximum as output power is $I_{CQ}V_{CC}/4$.

equal to I_{CQ}. This means the dc power drawn from the power supply will be $I_{CQ} \times V_{CC}$.

If the Q point is *centered* and the ac and dc load lines coincide, the very best the class A amplifier can do is to deliver an output sine wave whose *peak* current is I_{CQ} and whose *peak* voltage is V_{CEQ} ($V_{CC}/2$). This represents an ac output power of

$$I_{CQ}(0.707) \times \left(\frac{V_{CC}}{2}\right)(0.707) \text{ (rms is used for power)} = \frac{I_{CQ} \times V_{CC}}{4}$$

Applying Eq. 7-4 yields

$$\eta \text{ (class A)} = \frac{P_{o(ac)}}{P_{i(dc)}} = \frac{I_{CQ}\dfrac{V_{CC}}{4}}{I_{CQ}V_{CC}} = \frac{1}{4} = 25\%$$

The sad fact is that this is the *absolute best* the class A amplifier can do. Usually the efficiency will be much less than this.

There is one way this efficiency can be improved. This involves using a transformer coupling technique.

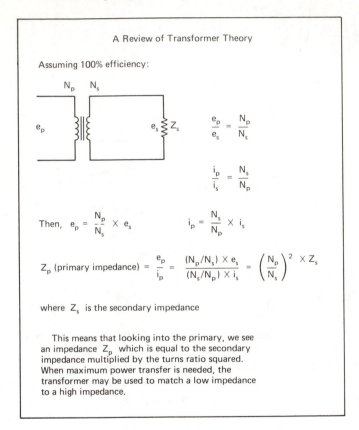

A Review of Transformer Theory

Assuming 100% efficiency:

$$\frac{e_p}{e_s} = \frac{N_p}{N_s}$$

$$\frac{i_p}{i_s} = \frac{N_s}{N_p}$$

Then, $e_p = \dfrac{N_p}{N_s} \times e_s$ $\qquad i_p = \dfrac{N_s}{N_p} \times i_s$

$$Z_p \text{ (primary impedance)} = \frac{e_p}{i_p} = \frac{(N_p/N_s) \times e_s}{(N_s/N_p) \times i_s} = \left(\frac{N_p}{N_s}\right)^2 \times Z_s$$

where Z_s is the secondary impedance

This means that looking into the primary, we see an impedance Z_p which is equal to the secondary impedance multiplied by the turns ratio squared. When maximum power transfer is needed, the transformer may be used to match a low impedance to a high impedance.

Figure 7-9a illustrates a transformer-coupled class A amplifier driving a low-resistance load. The efficiency of this circuit is better than the capacitively-coupled circuit due to the fact that no dc power is wasted in a collector resistor. There is actually a small amount of dc resistance in the primary of the transformer, but the power lost is small compared to that in the usually large collector resistor.

Example 7-4 ────────────────────────────────

Assuming the transformer in Fig. 7-9a is ideal, determine the power transferred to the 3.75-Ω load resistor and the dc power consumed by the circuit. Compute the efficiency.

Solution (a) The base current is calculated as $I_B = (12 \text{ V} - 0.6 \text{ V})/5.7 \text{ k}\Omega = 2 \text{ mA}$. The collector current is $2 \text{ mA} \times 100 = 0.2 \text{ A}$.

(b) If the transformer is ideal, it has no dc resistance and $V_C = V_{CE} = V_{CC} = 12$ V. The dc load line is 0 Ω and appears as a *vertical* line. This is shown on the characteristic curves in Fig. 7-9b, along with the circuit Q point.

(c) The 3.75-Ω load is reflected into the primary as $Z_p = 4^2 \times 3.75$ $\Omega = 60$ Ω.

(d) The ac load line is found using a previous method. If I_C decreases by 0.2 A, then V_{CE} will *increase* by 0.2 A $\times$ 60 $\Omega = 12$ V, to 24 V. Similarly, if I_C *increases* by 0.2 A, V_{CE} decreases by 12 V to 0 V. The ac load line is shown in Fig. 7-9b.

(e) Ignoring saturation and cutoff, the largest possible voltage swing across the primary is 12 V with a 0.2-A peak current. The ac output power is 12 V(0.707) $\times$ 0.2 A(0.707) = 1.2 W.

(f) The average collector current is 0.2 A, and the dc power from V_{CC} is 0.2 A $\times$ 12 V = 2.4 W.

(g) The efficiency is $P_{o(ac)}/P_{i(dc)}$ = 1.2 W/2.4 W = 50%.

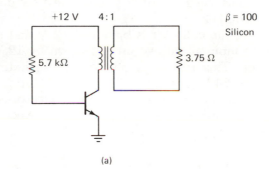

(a)

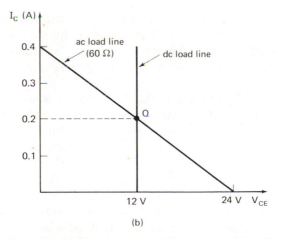

(b)

Figure 7-9 (a) Transformer-coupled class A amplifier. (b) with an ideal transformer, the dc load line is vertical but the ac load line is the reflected impedance. The voltage and current signal swings are now doubled and the efficiency is 50% maximum.

This example illustrates that under *ideal* conditions, the efficiency of the class A amplifier can be increased to 50% by using transformer coupling.

It is interesting to note that V_{CE} can reach 24 V or *twice* the supply voltage when the transformer coupling technique is used. This is possible because as I_C increases from 0.2 to 0.4 A, a voltage is developed across the primary of the transformer equal to 0.2 A $\times$ 60 Ω = 12 V. This voltage is positive at the V_{CC} side of the transformer. As the load line indicates, at this time I_C = 0.4 A, and V_{CE} = 0 V; the transistor is saturated.

On the negative half of the input cycle, I_C *decreases* from 0.2 to 0 A. This again causes a 12-V drop across the transformer, but this time of the *opposite* polarity. The transistor is now in cutoff and the collector voltage equals the 12 V of V_{CC} plus the 12 V developed across the transformer. This is 24 V.

In a sense, the collector is biased at 12 V and the ac input causes this voltage to momentarily go up and down by 12 V due to the transformer action. This results in momentary operating points of 0 V (saturation) and 24 V (cutoff).

The implication of this is that the collector-emitter *breakdown* voltage of the transistor must be selected to exceed *twice* V_{CC} when this coupling technique is used.

7.6   CLASS B AMPLIFIERS

The main drawback to the class A amplifier is that it draws power even in the absence of an input signal. This is a consequence of the bias point being in the center of the *active* region. Even with transformer coupling, which eliminates the power wasted in R_C, power is still wasted heating the transistor ($I_{CQ} \times V_{CEQ}$).

As Fig. 7-10a illustrates, a class B amplifier is one in which the Q point is biased at *cutoff*. As the input signal goes positive, the base current increases and the operating point moves into the active region.

Because the bias point is in cutoff, no power is consumed except when the input signal is applied. The average dc power level is much lower than that of the class A amplifier and the efficiency is improved.

There is one problem, however (isn't there always?). When the input signal goes *negative*, the operating point moves further into cutoff and no output occurs. Figure 7-10b illustrates typical input voltage and output current waveforms.

If the purpose of the amplifier is to produce an amplified *copy* of the input waveform, this scheme clearly does not work.

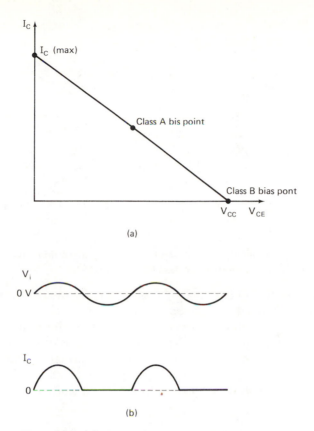

Figure 7-10 (a) Comparing class A and class B bias points, (b) when biased class B, the collector current flows only during the positive half cycle.

Push–Pull

For the class B amplifier to be practical its must employ a *push-pull* amplifier configuration. A simple push–pull circuit is shown in Fig. 7-11a. This circuit is also referred to as a complementary output stage because Q_1 and Q_2 are *opposite* but *matched* transistor types. It should be clear to you that Q_1 will conduct for positive excursions of the input signal while Q_2 conducts for the negative half of the input cycle. In a sense, Q_1 "pushes" while Q_2 "pulls," and vice versa.

When attempting to experimentally verify that the circuit in Fig. 7-11a actually does work, a new problem arises. As the input signal falls below 0.6 V, Q_1 stops conducting, but because the input signal must go below –0.6 V to cause Q_2 to begin conducting, it too is OFF. In fact, a

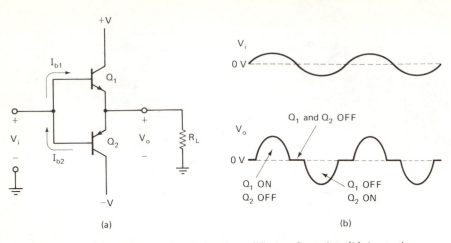

(a)   (b)

Figure 7-11 (a) A simple push-pull class B amplifier configuration. (b) due to the V_{BE} drops, crossover distortion occurs in the output signal.

deadband exists for $-0.6 \text{ V} < V_i < 0.6 \text{ V}$, and Q_1 and Q_2 are *both* OFF in this range. This results in *crossover distortion,* as shown in Fig. 7-11b.

Crossover distortion is solved by modifying the circuit to operate *class AB,* as illustrated in Fig. 7-12. Note that this circuit uses *complementary emitter followers.* No voltage gain is possible, but due to the current gain of the emitter follower, power gain is achieved. The following example will illustrate why this circuit is not quite class A or class B.

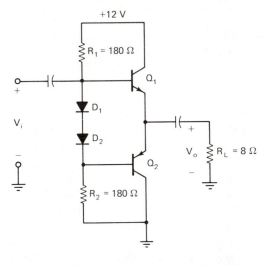

Figure 7-12 Class AB complementary emitter-follower output stage. The diodes help prevent crossover distortion.

Example 7-5 ————————————————————————————————

Determine the dc operating point for Q_1 and Q_2 in the complementary amplifier stage shown in Fig. 7-12. Estimate the efficiency under ideal conditions.

Solution (a) DC. If the base currents of Q_1 and Q_2 are small, they may be ignored, in which case R_1, R_2, D_1, and D_2 are all in series. $I_{R_1} = (12 \text{ V} - 1.2 \text{ V})/360 \, \Omega = 30$ mA. The emitter voltage of Q_1 is found as $+12 \text{ V} - (0.03 \text{ A})180 \, \Omega - 0.6 \text{ V} = 6 \text{ V}$. Then, $V_{B_1} = 6.6 \text{ V}$, and $V_{B_2} = 5.4 \text{ V}$.

It might appear that the collector currents of Q_1 and Q_2 could be exceedingly large because there are no collector resistors. However, diodes D_1 and D_2 are in parallel with the base-emitter junctions of these two transistors. Because the base-emitter voltage of a transistor determines its collector current, the collector currents will *also* be 30 mA if the diode's IV curves match the transistor's. This requires specially matched *npn* and *pnp* transistors called *complementary pairs*.

Because the collector current is not quite 0 A, this bias point is part class A and part class B. For this reason the circuit is called a *class AB* amplifier. Figure 7-13a illustrates the class AB bias point.

(b) AC. As the input signal goes positive, Q_1 conducts and Q_2 turns OFF. The output voltage now swings positive. When Q_1 saturates, the output voltage reaches its maximum value.

As the input signal goes negative, the opposite occurs. Q_1 turns OFF and Q_2 conducts. The output voltage now goes negative. Figure 7-13b illustrates these two cases.

For this particular circuit, the emitter voltage swings from the 6-V bias point *up* 6 V to 12 V and *down* 6 V to 0 V. This requires 6 V/8 Ω = 0.75 A of peak collector current.

(c) Efficiency. The output power is found as $0.75 \text{ A}(0.707) \times 6 \text{ V}(0.707) = 2.25 \text{ W}$. The dc power flows in pulses and only when Q_1 conducts (this ignores the small bias current needed to eliminate crossover distortion). The average value of I_{C_1} can be found from calculus to be $0.318 \times I_{peak} = 0.318(0.75 \text{ A}) = 0.239 \text{ A}$. The dc power is then $0.239 \text{ A} \times 12 \text{ V} = 2.87 \text{ W}$. Finally, the efficiency is $P_{o(ac)}/P_{i(dc)} = 2.25 \text{ W}/2.87 \text{ W} = 78.4\%$.

A realistic class B or class AB amplifier will have an efficiency less than the 78.4% calculated in this last example. This is because the output signal swings will be limited by $V_{CE(sat)}$ to less than 6 V and the dc bias point consumes a small amount of additional dc power. However, the efficiency of the circuit is much improved compared to the class A amplifier.

A disadvantage of class B or AB operation when compared to class A is that the typically large signal swings cause more *distortion* than the smaller signal swings of class A. This is because the transistor's characteristic curves become nonlinear as saturation and cutoff are approached for large signal operation.

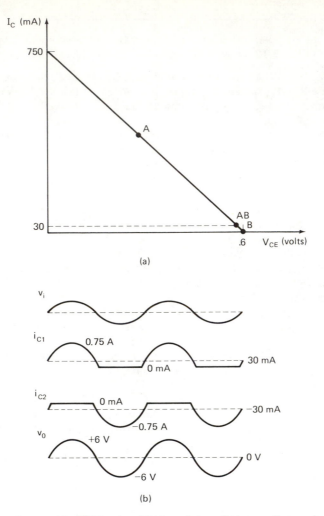

Figure 7-13 (a) The class AB bias point results in a small amount of collector current with no input signal. This is helpful in removing crossover distortion. (b) typical waveforms for the circuit in Fig. 7-12.

Transformer Coupled Push–Pull

The basic push–pull circuit shown in Fig. 7-11 is only possible if complementary transistors are available. Although there are many such transistor pairs available today, this was not always true.

It is possible to build a push–pull circuit using transistors of the same type if a means is developed to *split the phase* of the input signal.

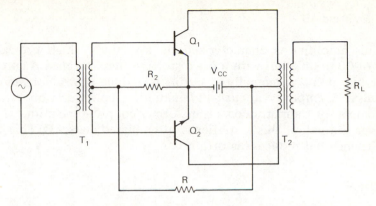

Figure 7-14 A transformer-coupled push-pull output stage. Complementary transistors are not required.

This is precisely what a transformer does. Figure 7-14 illustrates a *transformer-coupled push–pull amplifier* stage.

This circuit uses voltage divider bias to establish a small class AB bias current. This prevents crossover distortion, as discussed previously. The input signal is converted to two phases by T_1, and Q_1 and Q_2 conduct on alternate phases of the input signal as before. T_2 is necessary to reconstruct the output voltage produced by Q_1 and Q_2 and couple it to the load resistor.

This circuit in particular and transformer coupling in general have several *disadvantages*. Transformers capable of handling reasonable power levels are quite *bulky* and *heavy*. In addition, the cost may be high when compared to capacitive coupling techniques. Most practical amplifiers today use a form of the complementary stage shown in Fig. 7-12.

7.7 AMPLIFIER CLASSES SUMMARIZED

Amplifier classes are usually characterized by their operating points relative to the cutoff condition.

Class A

This amplifier is characterized by its low efficiency of typically 20 to 30% and good linearity with low distortion. It is basically a *small-signal* type of amplifier. The dc bias point is in the middle of the active region.

Class B and AB

This amplifier is characterized by a relatively high typical efficiency of 60 to 65% but with more distortion than the class A amplifier. The dc bias point is generally at cutoff to reduce power consumption. Typically, it employs a *push–pull* amplifier configuration requiring complementary transistors or a transformer coupling technique. Due to crossover distortion, this amplifier is often biased class AB. This lowers the efficiency but reduces distortion.

Class C

This amplifier is *not suitable* for audio applications. This is because the bias point is well into cutoff. No push-pull arrangement can reconstruct the input waveform. It is useful as an *rf* (radio frequency) amplifier where a parallel LC circuit (*tank* circuit) is reenergized by pulses of current as the class C amplifier briefly conducts. This is analagous to pushing a pendulum at the right instant to sustain its motion. Class C amplifiers consume little dc power because they conduct only briefly and have efficiencies from 65 to 85%.

Class D

This amplifier type might be called a *switching amplifier* because the output transistors switch between cutoff and saturation at a high frequency. The incoming signal modifies a pulse train forming a sequence of variable width pulses at a high frequency. These pulses turn the output transistors ON and OFF. A filter is used to extract the low frequency input signal riding on the pulse train. Because transistors consume little power when saturated or cutoff, the efficiency is high, often > 95%.

There have been other experimental amplifier types aimed at reducing dc power consumption when the input signal is at a low level. These include amplifiers with *dual* sets of output transistors and power supplies. At low signal levels, only one set is ON and consuming power. At higher levels, both sets are activated.

───────────────── KEY TERMS ─────────────────

Multistage Amplifier This refers to an amplifier that uses more than one stage of amplification in an effort to improve overall amplifier performance. This is a common design method because single-amplifier stages often have low gain in order to maintain ac and dc stability.

Capacitive Coupling   This is a method for passing an ac signal from one stage of an amplifier to another. A single-coupling capacitor is used between amplifier stages and the method is characterized by its simplicity, dc isolation, and high pass frequency response.

Direct-Coupling   As its name implies, this coupling technique consists of directly connecting the output of one amplifier stage to the input of another. Special care must be taken to maintain dc bias levels. It has the advantage of allowing amplification even at dc but the disadvantage of amplifying Q point drift.

Large-Signal Amplifier   This is an amplifier characterized by signal swings between cutoff and saturation (large signal). Such amplifiers are further characterized by their distortion levels and efficiencies. Typical large-signal amplifiers may operate class A, B, or AB.

Efficiency   This refers to the ratio of ac power out to dc power in. All amplifiers have efficiencies less than 100%, with the wasted power being lost due to heating of the active devices and resistors.

Transformer Coupling   This coupling technique is characterized by an impedance transformation property. Because the impedance looking into a transformer is a reflected impedance depending on the turns ratio, it is possible to select a transformer to match the output impedance of the amplifier to the load resistance and thereby achieve maximum power transfer. Other properties include dc isolation and limited frequency response.

Push–Pull   This is a type of large-signal amplifier characterized by class B or AB operation in which separate transistors conduct on alternate halves of the input signal. Several variations of this circuit are possible including complementary stages and transformer coupled stages.

Complementary Transistors   These are a matched pair of *npn* and *pnp* transistors that have nearly identical current-voltage characteristics. They are a necessity in most push–pull circuits to minimize output distortion.

─────────────────── QUESTIONS AND PROBLEMS ───────────────────

7-1   A two-stage amplifier has the following properties. Stage 1: $k = 25$, $R_i = 20$ kΩ, $R_o = 5$ kΩ. Stage 2: $k = 0.7$, $R_i = 10$ kΩ, $R_o = 100$ Ω. If 0.1 V is applied to the input, determine the output voltage and current with a 500-Ω load resistor.

7-2   Express the loaded voltage gains of the two stages in Problem 7-1 in dB and show that the sum equals the overall amplifier gain in dB.

7-3   Referring to Fig. 7-15, determine V_o.

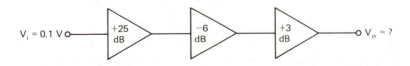

Figure 7-15

7-4   Determine the dc bias points for Q_1 and Q_2 in Fig. 7-16. Use the voltage divider bias method.

7-5   Calculate the overall voltage gain to the 1-kΩ load, input resistance and output resistance for the amplifier in Fig. 7-16. Assume $\beta = 175$ for both transistors.

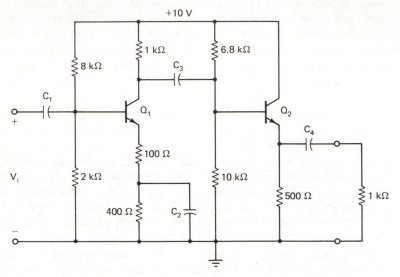

Figure 7-16

7-6   Sketch the overall amplifier equivalent circuit using the results from Problem 7-5 and determine the power gain in dB.

7-7   Estimate the maximum signal swing possible across the 1-kΩ load resistor for the circuit in Fig. 7-16. What power level does this represent for the 1-kΩ load?

7-8   Determine values for all capacitors in Fig. 7-16 if 100 Hz $\leqslant f \leqslant$ 10 kHz.

7-9   Refer to Fig. 7-16. Calculate the voltage gain of stage 1 directly driving the 1-kΩ load resistor (without the emitter follower). Compare your result to that obtained in Problem 7-5.

7-10 Design a two-stage amplifier that meets or exceeds the following: $R_i \geqslant 2$ kΩ, $A_V \geqslant 5$, $R_o \leqslant 100$ Ω, and output signal swing $\geqslant 1$ V peak-peak across a 10-Ω load resistance. Assume 1 kHz $\leqslant f \leqslant$ 20 kHz, $75 \leqslant \beta \leqslant 300$.

7-11 Estimate the dc bias at all nodes in Fig. 7-17. Assume $\beta = 200$ for both transistors.

7-12 Determine the overall voltage gain, input resistance, and output resistance for the amplifier in Fig. 7-17.

7-13 Calculate a value for the interstage coupling capacitor in Fig. 7-17. Assume 10Hz $\leqslant f \leqslant$ 1 kHz.

7-14 What are the advantages of direct coupling versus capacitive coupling?

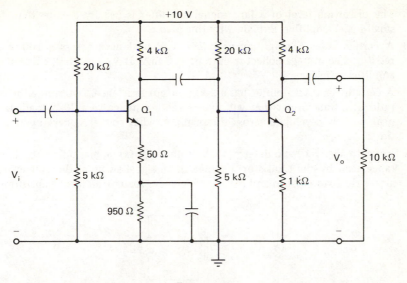

Figure 7-17

7-15 Describe the amplifier shown in Fig. 7-18 in terms of the individual amplifier stage types, coupling methods, and emitter resistor bypassing percentages.

7-16 Estimate the dc bias at all nodes for the circuit shown in Fig. 7-18. Assume $\beta = 100$.

7-17 Calculate the overall voltage gain for the amplifier shown in Fig. 7-18.

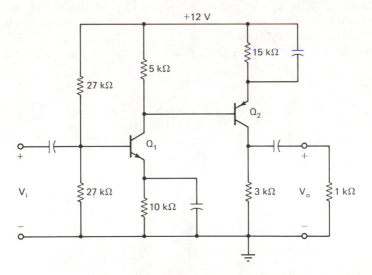

Figure 7-18

7-18 The distortion level of a large-signal amplifier is usually greater than for a small-signal amplifier. Explain why this is so.

7-19 A certain class A amplifier produces 4 V peak-peak across a 100-Ω load resistor. The average collector current is 5 mA. If $V_{CC} = 20$ V, calculate the amplifier efficiency.

7-20 A certain class A amplifier has a 2-kΩ output resistance. Determine the turns ratio of a transformer that will match this resistance to a 20-Ω load. If 4 V peak-peak is developed across the primary, what voltage is developed across the 20-Ω load?

7-21 Refer to Fig. 7-19 and determine the following: (a) dc bias point; (b) ac resistance seen by the transistor; (c) maximum signal swing at the collector and across the load resistor; and (d) the efficiency. Assume an ideal transformer.

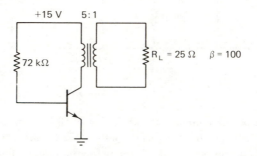

+15 V   5:1

72 kΩ

$R_L = 25\ \Omega$ $\beta = 100$

Figure 7-19

7-22 What is it that makes a class B amplifier more efficient than a class A amplifier? What special considerations must be made when designing a class B amplifier for audio purposes?

7-23 In Fig. 7-12, if $R_1 = R_2 = 380$ Ω and the Q_2 collector and R_2 resistor are returned to -12 V, recalculate the circuit operating point. Is the output coupling capacitor still needed?

7-24 Determine the dc operating point for all transistors in Fig. 7-20.

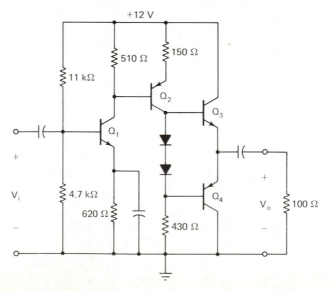

+12 V

510 Ω 150 Ω

11 kΩ

Q_2

Q_3

Q_1

V_i

4.7 kΩ

620 Ω

Q_4

V_o 100 Ω

430 Ω

Figure 7-20

Objectives

1. To gain experience in constructing, testing, and troubleshooting multistage amplifiers.
2. To compare the efficiency of several different amplifier designs.

Introduction   In this laboratory assignment you will construct a two-stage class A amplifier and measure its performance for various values of load resistance and coupling capacitance. Transformer coupling will also be examined using a class A amplifier. Finally, a class B push–pull circuit using complementary transistors will be tested. All amplifiers are compared for efficiency.

Components Required

Miscellaneous ¼-W resistors
Silicon *npn* transistors
Matched *npn–pnp* transistor pair
Audio transformer
Miscellaneous μF coupling capacitors

Part I: *Two-Stage Amplifier — Common Emitter and Emitter Follower*

STEP 1   Refer to Fig. 7-21 and calculate the dc bias expected at each node in the circuit.

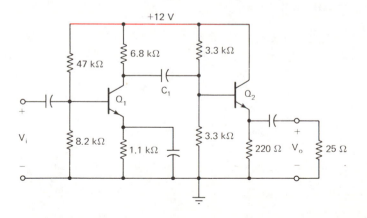

Figure 7-21   Class A amplifier used in Part I.

STEP 2 Set up this circuit and measure the quantities calculated in step 1. Explain any variations greater than 10%.

STEP 3 Estimate the overall voltage gain to the 25-Ω load resistance.

STEP 4 Measure the voltage gain for the following load resistances: 25 Ω, 100 Ω, 470 Ω, 680 Ω, and 1 kΩ.

Question 1 Why does the voltage gain increase for larger load resistances?

STEP 5 Using the 25-Ω load, measure the frequency response from 100 Hz to 100 kHz with C_1 = 10 μF and 1 μF. Graph each case using semilog paper.

STEP 6 At f = 1 kHz, measure the maximum signal swing across the 25-Ω load resistor and the total dc supply current. Calculate the overall amplifier efficiency by determining the ratio of ac load power to dc supply power.

Part II: *Transformer-Coupled Class A Amplifier*

STEP 1 Obtain an audio transformer and determine its turns ratio.

STEP 2 Place a 25-Ω load resistor on the secondary and use the method from previous labs to measure the primary reflected resistance.

STEP 3 Using $Z_p = (N_p/N_s)^2 \times Z_s$, does the turns ratio agree with the result in step 2?

STEP 4 Modify the circuit in Fig. 7-21 by removing the emitter-follower second stage and replacing it with the transformer-coupled stage shown in Fig. 7-22.

STEP 5 Measure the dc bias of this new output stage and calculate the dc resistance of the transformer.

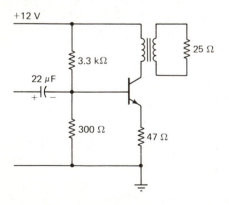

Figure 7-22 Transformer-coupled stage to be used in place of the emitter-follower stage in Fig. 7-21.

STEP 6 Measure the voltage gain to the collector and to the load resistor at $f = 1$ kHz.

STEP 7 Apply a 1-kHz input signal and determine the maximum output signal swing possible and calculate the ac output power. Measure the dc supply current and again calculate the amplifier efficiency as in step 6 of Part I.

Question 2 If the dc resistance of the transformer increases, will this increase or decrease the amplifier efficiency? Explain.

Part III: *Push–Pull Complementary Output Stage*

NOTE: The complementary output stage shown in Fig. 7-23 uses voltage divider bias compared to diode bias for the similar stage shown in Fig. 7-12. Ignoring the Q_3 and Q_4 base currents,

$$I_{R\,1} = \frac{V_{CC}}{2(R_1 + R_2)}$$

If each transistor is matched and conducts equally, then with no input signal, the output voltage should be $V_{CC}/2$. One *disadvantage* to this circuit is that as the temperature of Q_3 and Q_4 increase, their V_{BE}'s will decrease but the fixed bias will not. This may lead to a thermal

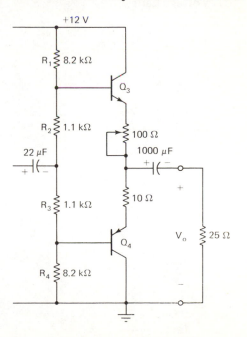

Figure 7-23 Complementary stage to be used in place of the emitter-follower stage in Fig. 7-21.

runaway situation. The diode bias in Fig. 7-12 will not have this problem if the diodes match the transistor V_{BE}'s and are at the same temperature.

STEP 1 Calculate the dc bias voltages and currents for the circuit in Fig. 7-23. Assume $V_{BE} = 0.65$ V. Are Q_3 and Q_4 biased class B or class AB?

STEP 2 Modify the circuit used in Part II by replacing the transformer stage with the complementary stage shown in Fig. 7-23.

STEP 3 Apply an input signal to the amplifier and adjust the 100-Ω pot to obtain the maximum undistorted output signal across the 25-Ω load resistor. Record this value and determine P_o.

STEP 4 Measure the voltage gain of the two-stage amplifier.

STEP 5 Measure the dc bias of the complementary stage and the overall amplifier dc current. Calculate the dc power and compute the efficiency.

Summary Compare the three amplifiers by preparing a chart for the case $R_L = 25$ Ω, and list the following:

(a) AC power out
(b) DC power in
(c) Efficiency
(d) Voltage gain

EIGHT

FIELD-EFFECT TRANSISTORS

As indicated in Chapter 3, the scientists at Bell Laboratories were actually working on a *field-effect* device when they discovered the bipolar transistor. Prior to this, all amplifiers were constructed using vacuum tubes, which are voltage-controlled devices. The field-effect transistor (FET) is actually the solid-state equivalent to the vacuum tube. Unlike the vacuum tube, however, there are two types of FETs, *n*-channel and *p*-channel. These two types are similar to the *npn* and *pnp* bipolar transistors.

Aside from the obvious advantages a solid-state device has over a vacuum tube, the FET also has several advantages when compared with the bipolar transistor. Among these are

1. Very high dc input resistance
2. No thermal runaway problems
3. Less noisy
4. No offset voltage (similar to saying $V_{CE(sat)} = 0$ V)

The main *disadvantages* to the FET are relatively low gain and bandwidth. In addition, high current FETs are not readily available (the VMOS FET discussed in Section 8.5 promises to change this).

In this chapter we study the physical principles of the field-effect transistor, various biasing circuits, typical amplifiers, and practical applications. Let us begin by examining the physics of the JFET transistor.

8.1 THE JUNCTION FIELD-EFFECT TRANSISTOR

A simplified diagram of an n-*channel junction field-effect transistor* (JFET) is shown in Fig. 8-1. This transistor consists of a bar of *n*-type

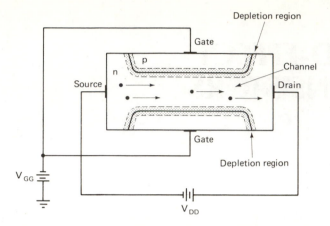

Figure 8-1 Pictorial of an *n*-channel junction field-effect transistor (JFET). Electrons enter at the source terminal, flow down the channel, and exit at the drain terminal. The gate voltage (V_{GG}) controls the channel resistance.

silicon with a *p*-type *gate* diffused into it. The two ends of the *n*-type bar are labelled *drain* and *source.*

With the biasing shown in Fig. 8-1, electrons leave the V_{DD} supply and enter the transistor at the source terminal. They then flow down the *n*-type channel, finally exiting at the drain terminal. In this instance, the channel region appears to be a simple resistor.

The gate supply voltage, V_{GG}, *reverse-biases* the gate-channel junction of the transistor. As explained in Chapter 1, a depletion region will exist in the vicinity of a reverse biased *pn* junction. This is due to the fact that all free carriers in this region are attracted to the reverse-bias potential, leaving only the immobile uncovered ions.

As would be expected, the depletion region in an FET extends into the channel region. This in turn causes the cross sectional area of the channel to *decrease* and in effect *increase* the channel resistance.

Figure 8-2 illustrates this effect by graphing the drain current and the V_{DD} supply voltage for various values of V_{GG}. Note that the characteristic curves are linear (straight lines), indicative of the channel resistance. As V_{DD} increases, I_D increases as expected by Ohm's Law ($I_D = V_{DD}/R_{channel}$). We should also note that the *value* of this resistance increases as V_{GG} becomes more negative, causing an increased gate-channel reverse bias and corresponding increased channel resistance. When $V_{GG} = -4$ V, much less current flows than when $V_{GG} = 0$ V for a given value of V_{DD}. This result implies a higher channel resistance for $V_{GG} = -4$ V and is consistent with the concept of a depletion region moving into the channel with increasing reverse bias.

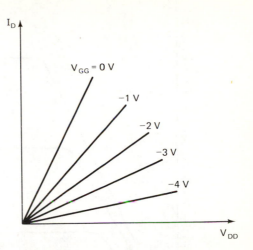

Figure 8-2 The effect of gate voltage on drain current. Larger values of gate voltage increase the channel resistance.

We are now in a position to see two of the main differences between a JFET and a bipolar transistor:

1. The JFET is *voltage* controlled. For a given value of V_{DD}, V_{GG} will control the channel current. The bipolar transistor is *current* controlled ($I_C = \beta I_B$).
2. Only *one* type of charge carrier is required for the JFET. The channel current is due entirely to electrons in this example. Since the gate is reverse biased it conducts no current. The FET is *unipolar*. As its name implies, the bipolar transistor requires *two* charge carriers (holes and electrons).

Characteristic Curves

The IV curves shown in Fig. 8-2 are not the whole story for the JFET. If the gate voltage is held constant while V_{DD} increases, the characteristic IV curves will initially follow a resistance curve (straight line) as previously explained. However, the curves will eventually fold over and become flat, as shown in Fig. 8-3. Further increases in V_{DD} will lead to a *breakdown* condition and a nearly vertical current–voltage characteristic.

The reason that the curves fold over can best be understood by studying the JFET equivalent circuit shown in Fig. 8-4a. The channel is represented as a resistance while the channel junction is represented by a series of diodes, all connected in parallel. Note that diode D_1 has 15 V reverse bias while diode D_n has 0 V reverse bias. Diodes between these two extremes have a reverse bias somewhere between 0 V and

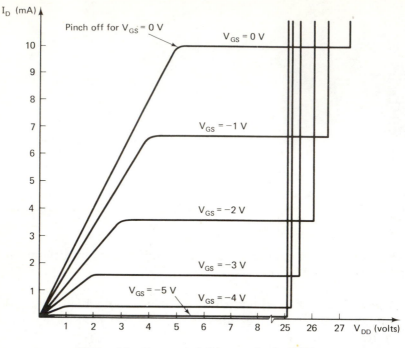

Figure 8-3 Characteristic IV curves for the JFET.

15 V due to the ohmic voltage drop along the channel resistance caused by the drain current. The significance of this is that the depletion region is *not uniform* in the channel. The maximum depletion region must exist at the *drain* end where the reverse bias is maximum.

For any given value of V_{GG}, as the channel current increases, the depletion region will move further into the channel. Eventually the channel will be constricted at the drain end such that further increases in current will cause the channel to close off altogether. At this point the channel current can no longer increase and further increases in V_{DD} *do not* result in increases in drain current. This is referred to as the *pinch-off* condition and at this point the characteristic curve becomes a flat line. Note that pinch-off does not mean 0 current but rather *constant* current.

Figure 8-4b illustrates that as V_{GG} increases, the depletion region moves further down and into the channel towards the source terminal. This results in an even higher channel resistance because less of the channel is available for conduction. Accordingly, as V_{GG} increases, the pinch-off drain current is correspondingly lower. This can be seen in Fig. 8-3. When $V_{GS} = 0$ V, pinch-off occurs when I_D is 10 mA. But when $V_{GS} = -3$ V, only about 1.5 mA of drain current is required for the pinch-off condition to occur.

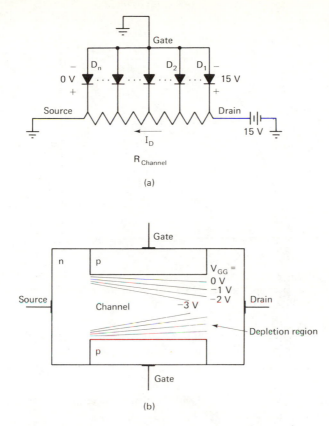

Figure 8-4 (a) The JFET channel is represented as a resistance and distributed parallel diodes. Due to the ohmic voltage drop along the channel, each diode has a different reverse bias. (b) as the value of V_{GG} increases, the effective channel area decreases.

As with most semiconductor devices, a *breakdown* condition also exists. For the JFET this occurs when the gate-drain junction is sufficiently reverse-biased to cause avalanche breakdown. When this occurs, a large current may flow limited only by external circuit resistances.

Figure 8-5 again illustrates the IV characteristics of the JFET. However, V_{GG} is now called V_{GS}, and V_{DD} called V_{DS}. Generally, the source terminal is chosen as the reference terminal and all voltages are expressed with respect to this terminal. This is similar to the convention followed for the bipolar transistor for which most voltages are expressed relative to the emitter terminal. Referring to Fig. 8-5, the following regions can be identified:

1. *Resistance region.* In this region the drain current increases linearly with V_{DS}.

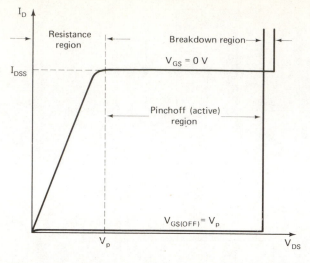

Figure 8-5 Characteristic curves for the JFET illustrating the operating regions and the definitions for I_{DSS}, V_p and, the $V_{GS(OFF)}$.

2. *Pinch off region.* As explained previously, the depletion region eventually pinches off the channel, causing the drain current to become constant. This is similar to the *active region* in the bipolar transistor.

3. *Breakdown region.* The JFET is not normally operated in this region because destructive currents may result.

Several important JFET parameters are also indicated in this figure. These are

1. I_{DSS}. Drain current for 0 gate voltage in the pinch-off region. This is the maximum drain current under normal conditions.

2. $V_{GS(OFF)}$. If V_{GS} is sufficiently large, the entire channel is pinched off and no current flows. This value is called $V_{GS(OFF)}$.

3. V_p. This is the value of drain-source voltage needed to cause the pinch-off condition when $V_{GS} = 0$ V. Actually, V_p is the value of V_{DS} needed to cause pinch-off without any help from V_{GS} ($V_{GS} = 0$ V). For this reason

$$V_p = V_{GS(OFF)} \tag{8-1}$$

This is because $V_{GS(OFF)}$ will pinch off the channel independent of the value of V_{DS}.

Example 8-1 ————————————————————————————————————

Refer to the data sheet for the 2N3821 in Fig. 8-6 and determine I_{DSS}, $V_{GS(OFF)}$, I_{GSS}, and V_p.

Solution The 2N3821 is an n-channel JFET and has an I_{DSS} of 0.5 mA minimum and 2.5 mA maximum when V_{DS} = 15 V, and V_{GS} = 0 V.
$V_{GS(OFF)}$ is –4.0 V maximum and is measured with I_D = 0.5 nA.
I_{GSS} is the gate *leakage current* and is specified as –0.1 nA with V_{GS} = –30 V.
Finally, V_p is *not* specified but by Eq. 8-1; V_p = $V_{GS(OFF)}$ = –4.0 V maximum.

Example 8-2 ————————————————————————————————————

Refer to Fig. 8-3 and estimate I_{DSS}, $V_{GS(OFF)}$ and V_p.

Solution When V_{GS} = 0 V, the drain current in the pinch-off region is 10 mA. This is I_{DSS}. $V_{GS(OFF)}$ is shown to be –5 V. Similarly, when the V_{GS} = 0 V curve begins to level off (pinch off); V_{DS} = 5 V. This verifies that V_p = $V_{GS(OFF)}$.

n-Channel versus p-Channel

If the n and p materials of the n channel device are reversed, a p-channel JFET results. In this case, holes are injected at the source terminal and travel down a p-channel to the drain. The gate voltage is now positive in order to keep the gate-channel junction reverse-biased. In fact, the only difference between p and n-channel JFETs is the polarity of voltage and currents. Figure 8-7 illustrates the schematic symbols for p- and n-channel devices, typical external biasing, and the physical device structure.

Amplification with JFETs

The amplification process with a JFET is similar to the bipolar transistor. In that device, a *small* change in base current leads to a *large* change in collector current and collector-emitter voltage. With the JFET, a small change in gate voltage leads to a larger change in drain-source voltage.

As was true with the bipolar transistor, a *load-line* method can be used to predict the JFET circuit operating point and amplifier gain.

Example 8-3 ————————————————————————————————————

Analyze the common-source JFET amplifier shown in Fig. 8-8a and determine: (a) the Q point, and (b) the voltage gain for a 1-V peak-peak sine wave input.

Solution The dc load line is determined by locating two points on the curves in Fig. 8-8b. When I_D = 0 A, V_{DS} = 10 V, and when I_D = 10 mA, V_{DS} = 0 V. This

SILICON N-CHANNEL
JUNCTION FIELD-EFFECT TRANSISTORS

. . . designed for audio amplifier, chopper and switching applications.

- Drain and Source Interchangeable
- Low Drain-Source Resistance —
  $r_{ds(on)} \leqslant 250$ Ohms (Max) — 2N3824
- Low Noise Figure — NF = 5.0 dB (Max) — 2N3821, 2N3822
- High AC Input Impedance — C_{iss} = 6.0 pF (Max)
- High DC Input Resistance — I_{GSS} = 0.1 nA (Max)
- Low Transfer Capacitance — C_{rss} = 3.0 pF (Max)
- JAN2N3821 and JAN2N3822 also Available

N-CHANNEL

JUNCTION
FIELD-EFFECT
TRANSISTORS

SYMMETRICAL
(Type A)

FEBRUARY 1971 — DS 5148 R2

*MAXIMUM RATINGS

Rating	Symbol	Value	Unit
Drain-Source Voltage	V_{DS}	50	Vdc
Drain-Gate Voltage	V_{DG}	50	Vdc
Gate-Source Voltage	V_{GS}	–50	Vdc
Drain Current	I_D	10	mAdc
Total Device Dissipation @ T_A = 25°C Derate above 25°C	P_D	300 2.0	mW mW/°C
Operating Junction Temperature	T_J	175	°C
Storage Temperature Range	T_{stg}	–65 to +200	°C

*Indicates JEDEC Registered Data.

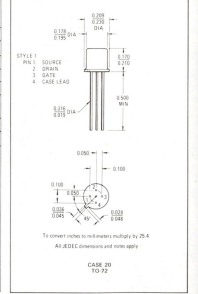

CASE 20
TO-72

Figure 8-6 Data sheets for the 2N3821, 22, 24 *n*-channel JFETs. (Courtesy Motorola Semiconductor Products Inc.)

*ELECTRICAL CHARACTERISTICS (T_A = 25°C unless otherwise noted)

Characteristic		Symbol	Min	Max	Unit
OFF CHARACTERISTICS					
Gate-Source Breakdown Voltage (I_G = –1.0 μAdc, V_{DS} = 0)		$V_{(BR)GSS}$	–50	–	Vdc
Gate Reverse Current		I_{GSS}			nAdc
(V_{GS} = –30 Vdc, V_{DS} = 0)			–	–0.1	
(V_{GS} = –30 Vdc, V_{DS} = 0, T_A = 150°C)			–	–100	
Gate-Source Cutoff Voltage		$V_{GS(off)}$			Vdc
(I_D = 0.5 nAdc, V_{DS} = 15 Vdc)	2N3821		–	–4.0	
	2N3822		–	–6.0	
Gate-Source Voltage		V_{GS}			Vdc
(I_D = 50 μAdc, V_{DS} = 15 Vdc)	2N3821		–0.5	–2.0	
(I_D = 200 μAdc, V_{DS} = 15 Vdc)	2N3822		–1.0	–4.0	
Drain Cutoff Current		$I_{D(off)}$			nAdc
(V_{DS} = 15 Vdc, V_{GS} = –8.0 Vdc)	2N3824		–	0.1	
(V_{DS} = 15 Vdc, V_{GS} = –8.0 Vdc, T_A = 150°C)	2N3824			100	
ON CHARACTERISTICS					
Zero-Gate-Voltage Drain Current(1)		I_{DSS}			mAdc
(V_{DS} = 15 Vdc, V_{GS} = 0)	2N3821		0.5	2.5	
	2N3822		2.0	10	
DYNAMIC CHARACTERISTICS					
Forward Transfer Admittance		$|y_{fs}|$			μmhos
(V_{DS} = 15 Vdc, V_{GS} = 0, f = 1.0 kHz)(1)	2N3821		1500	4500	
	2N3822		3000	6500	
(V_{DS} = 15 Vdc, V_{GS} = 0, f = 100 MHz)	2N3821		1500	–	
	2N3822		3000	–	
Output Admittance(1)		$|y_{os}|$			μmhos
(V_{DS} = 15 Vdc, V_{GS} = 0, f = 1.0 kHz)	2N3821		–	10	
	2N3822		–	20	
Drain-Source Resistance		$r_{ds(on)}$			Ohms
(V_{GS} = 0, I_D = 0, f = 1.0 kHz)	2N3824		–	250	
Input Capacitance		C_{iss}			pF
(V_{DS} = 15 Vdc, V_{GS} = 0, f = 1.0 MHz)			–	6.0	
Reverse Transfer Capacitance		C_{rss}			pF
(V_{DS} = 15 Vdc, V_{GS} = 0, f = 1.0 MHz)	2N3821		–	3.0	
	2N3822		–	3.0	
(V_{GS} = –8.0 Vdc, V_{DS} = 0, f = 1.0 MHz)	2N3824		–	3.0	
Average Noise Figure (V_{DS} = 15 Vdc, V_{GS} = 0, R_S = 1.0 megohm, f = 10 Hz, Noise Bandwidth = 5.0 Hz)	2N3821, 2N3822	NF	–	5.0	dB
Equivalent Input Noise Voltage (V_{DS} = 15 Vdc, V_{GS} = 0, f = 10 Hz, Noise Bandwidth = 5.0 Hz)	2N3821, 2N3822	e_n	–	200	nv/Hz$^{1/2}$

*Indicates JEDEC Registered Data.
(1)Pulse Test: Pulse Width ≤100 ms, Duty Cycle ≤10%.

MOTOROLA Semiconductor Products Inc.

BOX 20912 • PHOENIX, ARIZONA 85036 • A SUBSIDIARY OF MOTOROLA INC.

1316-5 PRINTED IN USA 3–71 IMPERIAL LITHO 821235   10M

DS 5148 R2

Figure 8-6   (cont.)

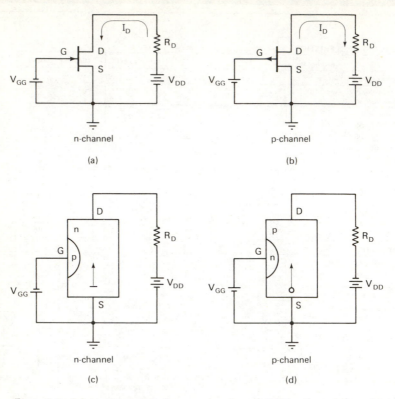

Figure 8-7 Schematic symbols for *n*- and *p*-channel JFETs. Current flow within the channel is illustrated in (c) and (d).

1-kΩ load line is shown in the figure. The Q point is found where this line crosses $V_{GS} = -1.5$ V (because the source terminal is at ground, $V_{GS} = -1.5$ V).

$$I_{DQ} = 4 \text{ mA} \qquad V_{DSQ} = 6 \text{ V}$$

The ac output signal is found by determining the *extreme* values of I_D and V_{DS} for a 1-V peak-peak input. Following the procedure used for bipolar transistors and referring to Fig. 8-8b, these are: $V_{GS} = -2$ V, $V_{DS} = 7.5$ V, $I_D = 2.5$ mA; and $V_{GS} = -1$ V, $V_{DS} = 4.5$ V, $I_D = 5.5$ mA.

The voltage gain is $(7.5 \text{ V} - 4.5 \text{ V})/1 \text{ V} = 3$.

You may be surprised at the low value of voltage gain. As discussed in the introduction, this is one drawback to the FET (and JFET). Unfortunately, this is a general result to be expected. This can be understood when studying the JFET IV curves. Volts on the gate control five to tens of volts at the drain. This represents gains on the order of tens.

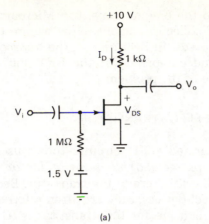

(a)

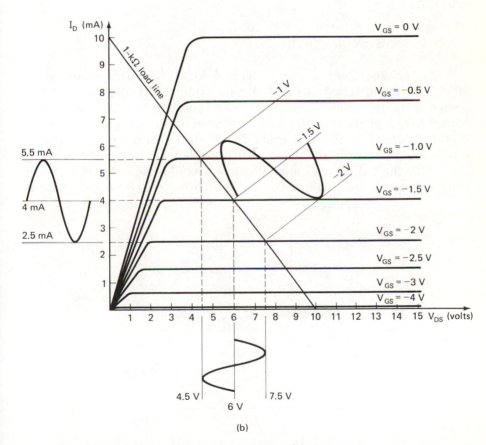

(b)

Figure 8-8 (a) Common-source JFET amplifier. (b) characteristic curves used for Example 8-3. The voltage gain is 3.

Compare this to the bipolar transistor. Microamps of base current control milliamps of collector current. Now we are talking about gains on the order of hundreds. In Section 8.4, the *transconductance* parameter is introduced and compared for the FET and bipolar transistor. This is the real source of the problem.

8.2   THE MOS FIELD-EFFECT TRANSISTOR

Most large-scale-integrated (LSI) circuits today use a form of MOS transistor. MOS means *metal-oxide semiconductor*, which is derived from the sandwich-like structure of the transistor. Because MOS transistors are also field-effect devices, they are often referred to as *MOSFETs*. Still another common name for this transistor is *IGFET* for *insulated gate* FET.

Enhancement Mode MOSFET

An enhancement mode *n*-channel MOSFET (also called an *E-MOSFET*) is illustrated in Fig. 8-9. The principle of operation for this device is quite different from the JFET, although both are voltage-controlled. A channel region exists between drain and source, as shown in the figure. For this *n*-channel device, the channel consists of the *p*-type region between the two *n* diffusions, labeled source and drain.

With the biasing indicated, electrons would like to travel from the source through the channel to the drain terminal. However, because

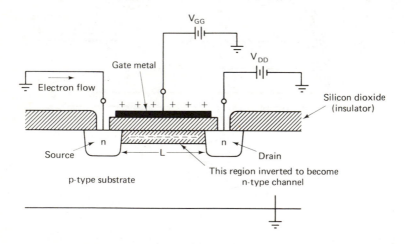

Figure 8-9   Enhancement mode *n*-channel MOS transistor (E-MOSFET). V_{GG} enhances the channel region between source and drain via "capacitor action." The channel length is shown as L.

the channel region is *p*-type, no current can flow and I_D = 0 A when V_{GG} = 0 V (unlike the JFET).

Now for the gate's role. The gate metal is insulated from the channel by a silicon-dioxide insulator (hence the name metal-oxide semiconductor) and actually acts as one plate of a *capacitor*. The channel is the other plate. As positive voltages are applied to the gate with respect to the channel or substrate, negative charges are attracted into the channel region. If the gate voltage is sufficiently large, the *p*-type region directly beneath the gate is *inverted* and becomes *n*-type. This is illustrated by the negative charges between source and drain in Fig. 8-9.

It is now possible for a drain current to flow through this *enhanced* channel region. Because the gate voltage controls this inversion layer and therefore the drain current, the MOSFET is a *voltage-controlled* device similar to the JFET.

There is one significant difference between the two, however. When the gate-source voltage of the JFET is 0 V, the transistor is conducting at its maximum (I_{DSS}). The E-MOSFET, on the other hand, requires a positive gate voltage to enhance the channel before conduction can occur. This device has 0 drain current when V_{GS} = 0 V.

Characteristic curves for the E-MOSFET are illustrated in Fig. 8-10. Note that I_D increases with increasing V_{GS} *opposite* to the JFET but exhibits a constant current active region *similar* to the JFET. As current flows down the channel, a *depletion region* is developed between the

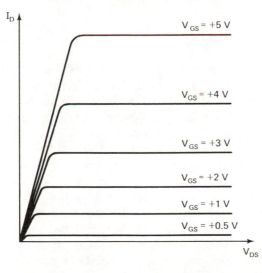

Figure 8-10 Characteristic curves for the E-MOSFET. When V_{GS} = 0 V, the transistor is not conducting (unlike the JFET).

reverse-biased p-type substrate and n-type channel. As was true with the JFET, this depletion region is maximum at the drain end but extends into the channel as I_D increases. For a given value of V_{GS}, the drain current will increase linearly (resistance region) until pinch-off occurs and will then remain nearly constant (pinch-off region).

It is interesting to note that the input to a MOSFET (the gate terminal) appears to be a capacitor. As such, it is an open circuit exhibiting extremely high input resistance. Values as high as *10,000 MΩ* are not uncommon at dc! In fact, dynamic MOS memories take advantage of this gate capacitance to store binary 1s and 0s with a minimum of circuit complexity (one-transistor memory cells).

One *disadvantage* to the MOSFET transistor is the extremely thin insulating layer between gate and channel (typically 4 to 5 one millionths of an inch!). This insulating layer is susceptible to "punch through," much as an ordinary capacitor can be destroyed by applying too large a voltage. When this occurs, the gate-channel becomes permanently shorted and the device is ruined. Unfortunately, it is very *easy* to develop large static potentials. For example, by simply shuffling your feet while walking across a carpet, potentials as high as 2000 to 4000 V are developed!

For this reason, MOS devices require special handling precautions, such as wrist straps to keep your body at ground potential. Similarly, soldering iron tips and test instruments should be well grounded. Typically, MOS devices are shipped with their leads *shorted* or placed in *conductive foam* to prevent the build-up of static charges.

Depletion Mode MOSFET

It is possible to manufacture the MOSFET with a built-in channel region. In this case the, transistor is already ON or conducting with 0 V of gate bias. This is illustrated in Fig. 8-11a for an n-channel device. As shown in this figure, the source and drain diffusions are simply extended so that a permanent channel now exists. A negative gate voltage is required to attract positive charges to the channel region and invert the n-type channel to p-type. This has the effect of *constricting* the channel between source and drain. If the gate voltage is sufficiently large, the channel can be pinched off completely.

Characteristic curves for an n-channel depletion mode MOSFET (D-MOSFET) are shown in Fig. 8-11b. Note that these curves are very similar to the n-channel JFET because negative values of gate voltage *deplete* the channel of carriers and turn OFF the transistor. The D-MOSFET is more versatile, however, because it can also be operated in the *enhancement mode* for *positive* values of V_{GS}. The enhancement and depletion regions of operation are indicated in the figure.

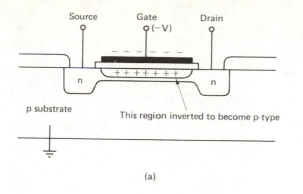

Source Gate Drain

(−V)

p substrate

This region inverted to become p-type

(a)

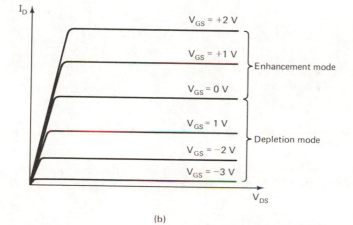

I_D

$V_{GS} = +2$ V

$V_{GS} = +1$ V

Enhancement mode

$V_{GS} = 0$ V

$V_{GS} = 1$ V

Depletion mode

$V_{GS} = -2$ V

$V_{GS} = -3$ V

V_{DS}

(b)

Figure 8-11 (a) Depletion mode MOSFET (D-MOSFET) has a built-in channel. (b) this transistor can be operated in the depletion or enhancement mode.

n-Channel and p-Channel

All of our comments so far have been directed at *n*-channel devices. By reversing the power supply polarities and directions of current flow, the same comments will apply for *p*-channel devices.

Schematically, *p*- and *n*-channel MOSFETs are illustrated in Fig. 8-12. Note that different schematic symbols are used for enhancement mode devices versus depletion mode devices. The middle terminal, labeled SUB, represents the MOSFET substrate that is usually connected to the source terminal internally. In all cases the arrow points in the direction of conventional current flow if the substrate is forward biased. Typical biasing schemes are also shown. It should be noted that opera-

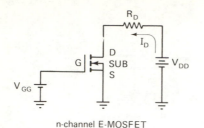
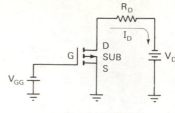

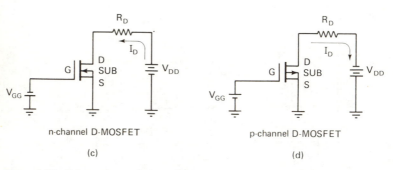

n-channel E-MOSFET

(a)

p-channel E-MOSFET

(b)

n-channel D-MOSFET

(c)

p-channel D-MOSFET

(d)

Figure 8-12 Schematic symbols and biasing circuits for the four possible MOS transistors.

tion in the enhancement mode requires only *one* polarity of supply voltage compared to *two* for depletion mode operation.

8.3   FET-BIASING CIRCUITS

When we discussed the bipolar transistor our approach was to sketch load lines on the characteristic curves to show that amplification was possible. We then concerned ourselves with establishing a stable circuit operating point.

Example 8-3 illustrated that amplification is possible with the FET. It also demonstrated how the circuit operating point could be found using a load line technique. We now examine biasing of the FET and the dc stability in more detail.

Transfer Curves

Figure 8-13 illustrates the JFET common source characteristic curves on the right half of the figure and what is called the *transfer* curve on the left half. The transfer curve plots I_D versus V_{GS} and is obtained by holding V_{DS} constant while V_{GS} is allowed to vary. We are

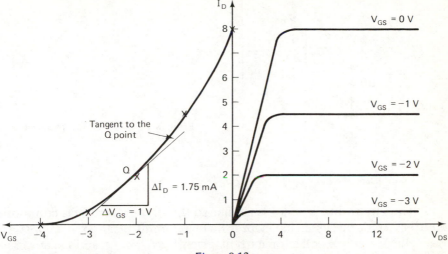

Figure 8-13

concerned about transfer curves obtained for values of V_{DS} in the *pinch-off* or *active-region* part of the characteristic curves. Because the characteristics are nearly flat in this region, the transfer curve obtained should be valid for most values of V_{DS} in the active region.

Why are the transfer curves important? The answer to this lies in realizing that the input voltage to the FET is V_{GS}, and this voltage in turn controls I_D, the output current. The transfer curves can thus be thought of as a graph of *output* (current) versus *input* (voltage).

Once the transfer curve is found, it is very easy to determine the dc operating point for an FET.

Example 8-4 ─────────────────────────────────

Verify the transfer curve shown in Fig. 8-13.

Solution Referring to the figure, note that the curve plots I_D versus V_{GS}. Transferring data from the common source characteristics in the active region: when $V_{GS} = 0$ V, $I_D = 8$ mA; when $V_{GS} = -1$ V, $I_D = 4.5$ mA, and so on.

Note that when $I_D = 0$ A, $V_{GS} = V_{GS(OFF)} = V_p \cong -4$ V. The resulting transfer curve is *parabola* shaped, with a maximum value of I_{DSS} where $V_{GS} = 0$ V, and a minimum value of 0 A where $V_{GS} = V_{GS(OFF)}$.

Example 8-5 ─────────────────────────────────

Determine the dc operating point of the circuit in Fig. 8-14. Use the transfer curve in Fig. 8-13.

Solution By inspection of Fig. 8-14, $V_{GS} = -2$ V. This must be true because $I_G = 0$ A due to the reverse-biased gate-source junction. Referring to the transfer curve, when $V_{GS} = -2$ V, $I_D = 2$ mA. Then at the Q point: $V_{GS} = -2$ V, $I_D = 2$ mA, $V_{DS} = 10$ V $- 2(1) = 8$ V.

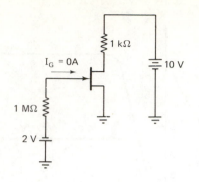

Figure 8-14 Fixed-bias common-source amplifier.

Fixed Bias

The circuit in Fig. 8-14 is similar to the simple fixed-bias common-emitter circuit discussed in Chapter 3. That circuit resulted in a constant base current and a collector current given by $I_C = \beta I_B$ and because the value of β can vary greatly from transistor to transistor, I_C is very *unpredictable* in this circuit. The fixed-bias circuit was shown to be unstable and, although simple to construct, a poor choice for a reliable amplifier.

Unfortunately, the same is true for the fixed-bias *common-source* amplifier in Fig. 8-14. The gate-source voltage is held constant at –2 V. It can be shown that the drain current is related to V_{GS} by

$$I_D = I_{DSS}(1 - V_{GS}/V_p)^2 \qquad (8\text{-}2)$$

This is similar to $I_C = \beta I_B$ for the bipolar transistor. Equation 8-2 indicates that I_D is related to I_{DSS} (I_D when $V_{GS} = 0$ V), and the ratio of V_{GS} to V_p, the pinch-off voltage.

The problem with this equation is that just as β typically varies between wide limits, so can I_{DSS} and V_p. Referring to Fig. 8-6, we see that I_{DSS} for the 2N3822 varies from a minimum value of 2 mA to a maximum of 10 mA. This means that for a given value of V_{GS}, we can expect a wide variation of values predicted by Eq. 8-2, depending on the actual values of I_{DSS} and V_p for the transistor used.

Example 8-6 ————————————————————————————————

Use Eq. 8-2 to construct minimum and maximum transfer curves for a JFET transistor with the following characteristics:

$$I_{DSS} = 4 \text{ mA (min) to } 10 \text{ mA (max)}$$
$$V_{GS(OFF)} = -3 \text{ V (min) to } -7 \text{ V (max)}$$

Solution The two curves are calculated using the minimum and maximum values for I_{DSS} and $V_{GS(OFF)}$ in Eq. 8-2, along with various values of V_{GS}. For example, when V_{GS} = 0 V, I_D (max) = 10 mA(1 - 0 V/7 V)² = 10 mA, and I_D (min) = 4 mA (1 - 0 V/3 V)² = 4 mA. The resulting curves are illustrated in Fig. 8-15.

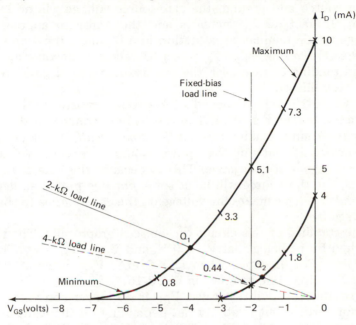

Figure 8-15 Using Eq. 8-2, maximum and minimum curves can be calculated using the extreme values of I_{DSS} and $V_{GS(OFF)}$. Fixed-bias and self-biased load lines are illustrated.

Example 8-7

Determine the extreme dc operating points possible for the circuit shown in Fig. 8-14, assuming the transfer curves in Fig. 8-15 apply.

Solution Referring to Fig. 8-14, V_{GS} is held constant at -2 V. The load line is therefore *vertical* at V_{GS} = -2 V. This is shown in Fig. 8-15. The two operating points are

$$I_D = 5.1 \text{ mA, and } V_{DS} = 4.9 \text{ V} \quad [10 \text{ V} - (5.1 \text{ mA} \times 1 \text{ k}\Omega)]$$

$$I_D = 0.44 \text{ mA, and } V_{DS} = 9.56 \text{ V} \, [10 \text{ V} - (0.44 \text{ mA} \times 1 \text{ k}\Omega)]$$

Example 8-7 should make it clear that fixed-bias FET circuits are as undesirable as they were with bipolar transistors.

Self-Bias

The dc stability problems of the bipolar transistor are solved by adding an emitter resistor to establish a *negative feedback* effect. In this case, the base current is no longer constant but a function of β. Ideally, the base current halves when β doubles.

A similar approach is taken with the FET. If a resistor is added between source and ground, the gate-source voltage will no longer be constant. In fact, as I_D increases and the transistor approaches the resistance region (similar to saturation in a bipolar), the source voltage will increase, causing V_{GS} ($V_G - V_S$) to decrease (become more negative) and tending to turn OFF the transistor. Again, *negative feedback* has been established.

In Fig. 8-16, we have added a 2-kΩ source resistor to the common source amplifier in Fig. 8-14. The gate is now connected through R_G directly to ground, making this a *self-biased circuit*. A unique property of this circuit is that only *one* power source is needed, yet the gate-source voltage is still negative. This is because the gate is at *ground* potential but the source will be at some *positive* potential, depending on I_D and R_S. This makes the voltage on the gate *relative* to the source (V_{GS}) *negative.*

The stability of this circuit can be seen graphically. The load line is sketched by realizing that $V_G = 0$ V, and $V_{GS} = V_G - V_S$. Then V_{GS} = $-I_D \times R_S$. For example, when $I_D = 2$ mA, $V_{GS} = -2$ mA $\times$ 2 kΩ = -4 V. A 2-kΩ load line is shown in Fig. 8-15.

Now the key point; because V_{GS} is no longer constant, the load line is not vertical but relatively *flat* depending on the value of R_S. As can be seen by inspecting Fig. 8-15, the minimum and maximum values of I_D (labelled Q_1 and Q_2) are much *closer* together than they were for the fixed-bias circuit. We can now be fairly confident that the transistor will be biased in the *active region* no matter what the specific I_{DSS} and V_p turn out to be.

Note that as R_S increases, the bias stability improves even further. This can be seen by sketching higher value load lines on Fig. 8-15. The larger the value of R_S, the flatter the load line and the smaller the variation in circuit operating point. However, the value of I_D decreases as R_S increases, restricting the output signal swing. The optimum value for R_S depends on the transistor involved and the output signal variation required.

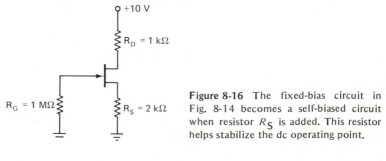

Figure 8-16 The fixed-bias circuit in Fig. 8-14 becomes a self-biased circuit when resistor R_S is added. This resistor helps stabilize the dc operating point.

MOSFET Biasing

When biasing a MOSFET, the Q point must be chosen to operate the circuit in the *enhancement mode* only or the *enhancement-depletion mode* corresponding to the two MOS transistor types.

Refer back to Fig. 8-9 (enhancement mode) and Fig. 8-11a (depletion mode) to refresh your memory on these two types of MOSFETs. The D-MOSFET can be operated in the depletion *or* enhancement modes, while the E-MOSFET device must be operated with positive values of V_{GS} (negative values for a p-channel device). This generally results in two different types of bias circuits.

A good bias point for the D-MOSFET would be $V_{GS} = 0$ V. This is in the center of the active region and results in operation in both enhancement and depletion modes as the ac input signal varies about the Q point. A circuit to establish the Q point at $V_{GS} = 0$ V is shown in Fig. 8-17.

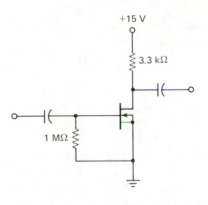

Figure 8-17 Bias circuit for an n-channel D-MOSFET. The bias point is at $V_{GS} = 0$ V and operation can be in the enhancement or depletion modes.

Example 8-8 ————————————————————————————

Assume $I_{DSS} = 2$ mA for the D-MOSFET shown in Fig. 8-17. Determine the circuit operating point.

Solution By definition, I_{DSS} is the drain current when $V_{GS} = 0$ V. Because $V_G = V_S = 0$ V in Fig. 8-17, $I_D = I_{DSS} = 2$ mA; $V_{DS} = 15$ V $- 2$ mA$(3.3$ k$\Omega) = 8.4$ V.

Of course the MOSFET can suffer from the same bias stability problems as the JFET. Figure 8-18 illustrates the transfer curves for the transistor in the previous example. Note that the dc load line (labeled $R_S = 0$ Ω) is *coincident* with the *center axes* and is vertical due to the fixed-bias condition ($V_{GS} = 0$ V). Although we calculated the bias point using the *typical* value of I_{DSS}, you can see that this current actually varies from 0.5 mA to 3.5 mA.

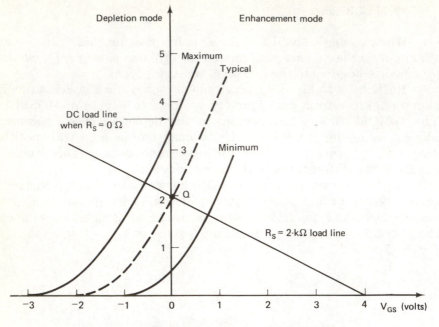

Figure 8-18   Transfer curves for a D-MOSFET.

An improved version of the D-MOSFET bias circuit is shown in Fig. 8-19. This circuit uses negative feedback via the 2-kΩ R_S resistor similar to the JFET self-biased circuit. When I_D = 0 A, V_S = 0 V, and $V_G = R_2/(R_1 + R_2) \times 15$ V = 4 V. Then V_{GS} = +4 V. This is point A in Fig. 8-18. Now, as I_D increases, V_{GS} decreases and follows the load line shown in Fig. 8-18. For example, when I_D = 2 mA, V_S = +4 V, and V_{GS} = 4 V – 4 V = 0 V. Note that the expected circuit operating point variation will be less than for the vertical load line case of the fixed-bias circuit due to the negative feedback.

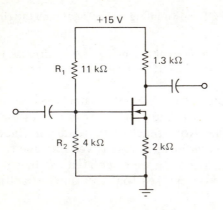

Figure 8-19 An improved D-MOSFET bias circuit. The operating point is stabilized due to the addition of resistor R_S.

Example 8-9 ─────────────────────────────────

Refer to the transfer curves and load line in Fig. 8-18 and determine the dc operating point for the circuit shown in Fig. 8-19.

Solution By inspection, the load line intersects the typical transfer curve at $I_D = 2$ mA, and $V_{GS} = 0$ V. This can be verified in Fig. 8-19. $V_G = 4/15 \times 15 = 4$ V. $V_S = I_D \times R_S = 2$ mA $\times 2$ k$\Omega = 4$ V. $V_{GS} = V_G - V_S = 0$ V. Finally, $V_{DS} = 15$ V $- 2$ mA $(1.3$ k$\Omega) - V_S = 15$ V $- 2.6$ V $- 4$ V $= 8.4$ V.

Hence we arrive at the *same* operating point as the simple circuit in Fig. 8-17 but with better stability to variations in I_{DSS}.

The circuit in Fig. 8-19 can also be used to bias E-MOSFETs. This is illustrated in Fig. 8-20. The transfer curve now exists only for *positive* values of V_{GS}. The Q point is shown on the typical curve as $I_D = 1$ mA, and $V_{GS} = +2$ V. Again, negative feedback helps to stabilize the circuit operating point.

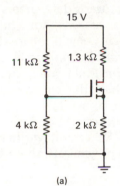

(a)

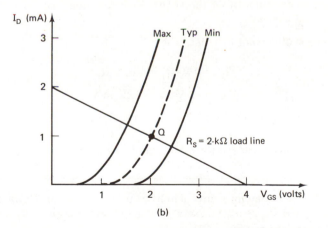

(b)

Figure 8-20 (a) *n*-channel E-MOSFET bias circuit. (b) transfer curves and load lines. In this circuit, V_{GS} must be positive.

The preceding sections have illustrated various types of FET bias circuits and techniques for analyzing these circuits. It is no coincidence that in all cases a *graphical method* was used. Unlike the bipolar transistor, there is no diode drop from which we can work to determine gate or source voltage. Equation 8-2 does relate I_D to V_{GS}, but in most cases V_{GS} cannot be found by inspection, as V_B or V_E can for a bipolar transistor. In fact, the gate-source voltage could be positive, negative, or 0 depending on the transistor type (see the load line in Fig. 8-18, for example!).

It is possible to set up *simultaneous equations* and another derived from the circuit under analysis. However, this can be difficult, and because I_{DSS} and V_p have wide tolerances, not too accurate.

Another method is to use Eq. 8-2 and the extreme values of I_{DSS} and V_p from the manufacturer's data sheets to construct maximum and minimum transfer curves, as was done in Example 8-6. Load lines sketched on these sets of curves will locate the *approximate Q* point.

Of course, when *troubleshooting* a circuit, the exact operating point is not of prime concern. Usually we want to know if the stage under test is good or bad so that we can quickly isolate and repair the actual problem. In this case, measurement of terminal voltages should tell us if the FET is biased properly and not defective.

When designing an FET circuit we may rely on the data sheets to construct a transfer curve or generate them from a curve tracer display.

Example 8-10 ──────────────────────────────────────

Obtain the transfer curve from the D-MOSFET common source characteristics in Fig. 8-21a.

Solution   Proceeding point by point, when

$$V_{GS} = 0 \text{ V}, I_D = 5 \text{ mA}.$$

$$V_{GS} = -2 \text{ V}, I_D = 1.25 \text{ mA}.$$

$$V_{GS} = +2 \text{ V}, I_D = 11.25 \text{ mA}.$$

and so on. The transfer curve is shown in Fig. 8-21b.

Example 8-11 ──────────────────────────────────────

Design a self-biased common-source amplifier using a D-MOSFET transistor biased at $V_{GS} = 0$ V, and $V_{DS} = +3$ V. Use a +5-V gate bias and assume $V_{DD} = 10$ V. The transfer curve is shown in Fig. 8-21b.

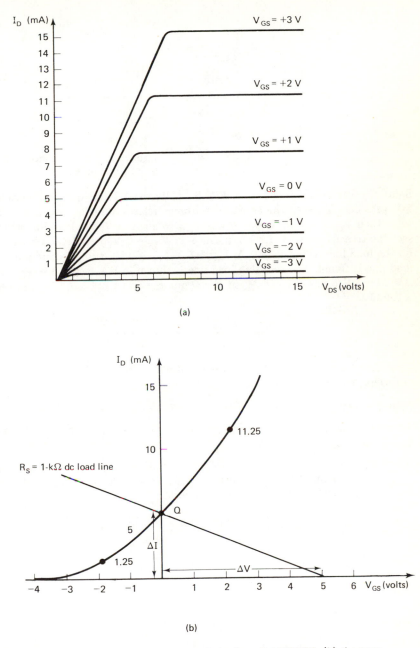

Figure 8-21 (a) Common source characteristics for a D-MOSFET. (b) the corresponding transfer curve.

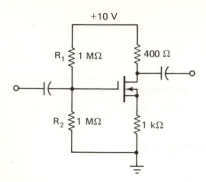

Figure 8-22 An n-channel self-biased D-MOSFET amplifier.

Solution The circuit is shown in Fig. 8-22. Resistors R_1 and R_2 will establish the 5-V gate bias. To keep the ac input resistance relatively high, choose $R_1 = R_2 = 1$ MΩ. Then $V_G = \frac{1}{2} \times 10$ V = 5 V. For V_{GS} to equal 0 V, the load line must intersect the transfer curve at $I_D = 5$ mA, where $V_{GS} = 0$ V. This determines one point on the load line. When $I_D = 0$ A, $V_{GS} = 5$ V ($V_S = 0$ V), and a second point on the load line is found. The resulting load line is shown in Fig. 8-21b. We see that $R_S = \Delta V/\Delta I = 5$ V/5 mA = 1 kΩ. V_S at the Q point = 5 mA $\times$ 1 kΩ = 5 V. Finally, R_D is found as

$$R_D = \frac{V_{DD} - V_{DS} - V_S}{I_D} = \frac{10 \text{ V} - 3 \text{ V} - 5 \text{ V}}{5 \text{ mA}} = 400 \text{ }\Omega$$

Another technique involves estimating the FET operating point by referring to the data sheet.

Example 8-12 ───────────────────

Design a p-channel self-biased JFET amplifier using a –12-V supply and given I_{DSS} = –6 mA, and $V_{GS(OFF)} = +5$ V.

Solution Choose a gate-source voltage of 2.5 V midway between $V_{GS(OFF)}$ and $V_{GS} = 0$ V. At this point, *assume* $I_D \cong \frac{1}{2}I_{DSS} = -3$ mA. Now, R_S is calculated as $R_S = V_S/I_D = -2.5$ V/–3 mA = 833 Ω. Choosing $V_{DS} = -6$ V,

$$R_D = \frac{V_{DD} - V_{DS} - V_S}{-3 \text{ mA}} = \frac{-3.5 \text{ V}}{-3 \text{ mA}} = 1.2 \text{ k}\Omega$$

The resulting circuit is shown in Fig. 8-23.

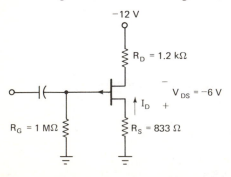

Figure 8-23 A p-channel JFET self-biased amplifier.

8.4 FET AMPLIFIERS

The FET will function as an amplifier because small changes in V_{GS} lead to larger changes in I_D and V_{DS}. This was illustrated *graphically* in Example 8-3.

When attempting to analyze the FET amplifier for ac we might take an approach similar to that used with the bipolar transistor. Determine an ac small-signal model, substitute and obtain the *ac equivalent circuit*, and, finally, develop the appropriate equations.

The FET Small-Signal Model

Beginning at the beginning, the input to an FET appears to be a *large resistance*. This may range from a few megohms for the JFET to hundreds of megohms for the MOSFET. If we restrict our small-signal model to low frequency, it would not be a bad assumption to say that looking into the gate-source of the FET there appears to be an *open circuit*.

The output characteristic curves for the FET are similar to those of the bipolar transistor and can be represented by a current source between drain and source terminals. Unlike the bipolar case, this current source depends on the input *voltage* instead of current. A change in V_{GS} causes a corresponding change in I_D, such that

$$g_m = \frac{\Delta I_D}{\Delta V_{GS}} \tag{8-3}$$

where g_m is the FETs equivalent of β but is referred to as the *transconductance*.

Finally, if the output curves of the FET are not perfectly flat, an *output* resistance is indicated. Call this r_o. The resulting small-signal model is shown in Fig. 8-24.

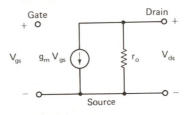

Figure 8-24 Small-signal model for the FET.

Example 8-13

Refer to the curve shown in Fig. 8-13 and determine g_m at the indicated Q point.

Solution The transconductance is determined graphically from the transfer curve by forming the triangle shown in the figure and applying Eq. 8-3.

$$g_m = \frac{\Delta I_D}{\Delta V_{GS}} = \frac{1.75 \text{ mA}}{1 \text{ V}} = 0.00175 \, \mho = 1750 \, \mu S$$

Note that the units of g_m are reciprocal ohms, formerly called mhos and now designated S for *Siemens*. Usually, g_m is expressed in microSiemens or milliamps-per-volt. In Example 8-13, 1.75 mA/V means a change of 1 V in V_{GS} will cause a 1.75 mA change in I_D.

Some transistor data sheets use y_{fs} in place of g_m and refer to this as the *forward transfer admittance*. These same data sheets may refer to the FET output resistance as y_{os}, the small-signal output admittance. The reciprocal of y_{os} is r_o in our FET small-signal model.

Example 8-14 ────────────────────

Refer to the data sheet for the 2N3821 *n*-channel JFET in Fig. 8-6 and determine the low-frequency small-signal model parameters.

Solution The transconductance is referred to as y_{fs} and is between 1500 and 4500 μS at 1 kHz (low frequency).

Similarily, the output resistance is listed as y_{os}, the output admittance. This is 10 μS maximum, meaning $r_o = \dfrac{1}{10 \, \mu S} = 100 \text{ k}\Omega$ minimum.

The Common-Source Amplifier

We are now is a position to apply the FET small-signal model in a circuit. Referring to Fig. 8-25a, a common-source self-biased JFET amplifier is illustrated. Note that the source resistor is bypassed. Following the rules developed in Chapter 5, the amplifier is redrawn in Fig. 8-25b with the FET replaced by its equivalent circuit.

Now the important ac characteristics of the amplifier can be determined.

1. *Input resistance.* By inspection,

$$R_i = R_G \tag{8-4}$$

2. *Output resistance.* Similarly, the output resistance is that seen by the load resistor looking into the drain terminal. In this case,

$$R_o = r_o \parallel R_D \tag{8-5}$$

3. *Voltage gain.* The output voltage is $-g_m \times V_{gs} \times (r_o \parallel R_D \parallel R_L)$, while the input voltage is V_{gs}. Taking the ratio of V_o to V_i results in

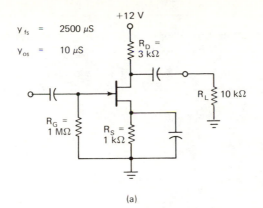

(a)

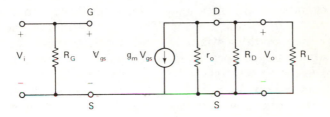

(b)

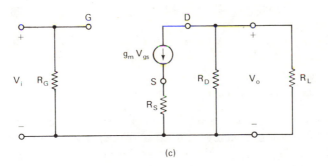

(c)

Figure 8-25 (a) The common source JFET amplifier, and (b) its equivalent ac circuit. (c) the ac equivalent circuit when R_S is unbypassed.

$$A_V = -g_m(r_o \parallel R_D \parallel R_L) = -g_m R_L' \qquad (8\text{-}6)$$

The minus sign indicates $180°$ phase reversal and

$$R_L' = r_o \parallel R_D \parallel R_L.$$

Example 8-15 ───

If I_D = 2 mA, determine the dc operating point of the circuit in Fig. 8-25a and calculate the ac quantities in Eqs. 8–4 through 8–6. Y_{os} = 10 μS and g_m = 2.5 mA/V.

Solution With I_D = 2 mA, V_S = 2 mA × 1 kΩ = 2 V, and V_{GS} = V_G - V_S = 0 V - 2 V = -2 V. V_D = V_{DD} - $I_D R_D$ = 12 V - 2 mA(3 kΩ) = 6 V. Finally, V_{DS} = V_D - V_S = 6 V - 2 V = 4 V.

Plugging into Eqs. 8-4 through 8-6 yields

$$R_i = 1 \text{ M}\Omega$$

$$R_o = 3 \text{ k}\Omega \parallel \frac{1}{y_{os}} = 3 \text{ k}\Omega \parallel 100 \text{ k}\Omega = 2.9 \text{ k}\Omega$$

$$A_V = -(2.5 \times 10^{-3}) \times (3 \text{ k}\Omega \parallel 100 \text{ k}\Omega \parallel 10 \text{ k}\Omega) = -5.6$$

This example confirms our previous conclusion that the FET *cannot* provide large voltage gains. The input resistance, however, is *much larger* than anything attainable with a bipolar transistor. And this brings up a good point. Because R_i is limited by R_G, why not choose R_G to be exceedingly large, say 5 MΩ or 10 MΩ?

The main reason for *not* doing this is that a small but definite *leakage current* does flow into the gate terminal (particularly for the JFET because its gate is a reverse-biased diode). If this current is 0.5 μA, this represents a 5-V drop across 10 MΩ. This would obviously upset the dc bias considerably. As a result, R_G is usually limited to a few megohms.

Finally, R_o is comparable to the bipolar case and limited to a few kilohms by R_D.

Using the Bipolar Formulas

We could now proceed to develop equations for the remaining FET amplifier circuits using the small-signal model just developed. However, by being a bit crafty, we may be able to save ourselves some work. You should have noted the similarity in ac equivalent circuits for the small-signal FET and the bipolar transistor. If we could change the βi_b current source in the bipolar small-signal model to a g_m term, we might be able to use the results from the previous chapters.

Example 8-16 ───

Determine g_m for a bipolar transistor in terms of β and r_{be}.

Solution The transconductance is by definition the ratio of change in *output current* to change in *input voltage*. For the bipolar transistor,

$$g_m = \frac{\Delta I_C}{\Delta V_{BE}} = \frac{\beta \, \Delta I_B}{r_{be} \, \Delta I_B} = \frac{\beta}{r_{be}} \tag{8-7}$$

The gain of the common-emitter amplifier was shown to be $-\beta R_L'/r_{be}$. If we substitute $g_m = \beta/r_{be}$, we obtain

$$A_V = -g_m R_L' \tag{8-8}$$

This is *identical* to Eq. 8-6 obtained for the FET! Now carrying this concept one step further, if we substitute g_m for each occurrence of β/r_{be} in the corresponding bipolar equations, we should obtain the FET equivalents.

Example 8-17 ───

Determine the voltage gain, output resistance, and input resistance to the amplifier in Fig. 8-25a if the source resistor is no longer bypassed.

Solution It was shown in Chapter 6 that a common-emitter amplifier with unbypassed emitter resistor has a voltage gain given by

$$A_V = \frac{-\beta R_L'}{r_{be} + (\beta + 1)R_E} \tag{6-7}$$

where $R_L' = R_C \parallel r_o \parallel R_L$. Substituting $R_S = R_E$ and dividing top and bottom by β, Eq. 6-7 becomes

$$A_V = \frac{-R_L'}{\dfrac{r_{be}}{\beta} + \dfrac{(\beta + 1)R_S}{\beta}} \cong \frac{-R_L'}{\dfrac{1}{g_m} + R_S} = \frac{-g_m R_L'}{1 + g_m R_S} \tag{8-9}$$

Plugging in the numbers,

$$A_V = \frac{-(2.5 \times 10^{-3})(3 \text{ k}\Omega \parallel 100 \text{ k}\Omega \parallel 10 \text{ k}\Omega)}{1 + (2.5 \times 10^{-3})(1 \text{ k}\Omega)} = -1.6$$

We can get into trouble if we blindly apply *all* the corresponding common-emitter equations, however. For example, the input resistance for the bipolar case is $R_B' \parallel [r_{be} + (\beta + 1)R_E]$. This equation does not contain a β/r_{be} term.

The solution here is to refer to the ac equivalent circuit. This is shown in Fig. 8-25c. From this we see that R_i is again R_G and R_o is equal to R_D (neglecting r_o). Therefore,

$$R_i = R_G = 1 \text{ M}\Omega$$

and

$$R_o = R_D = 3 \text{ k}\Omega$$

Note that Eq. 8-9 is the general voltage gain equation for an FET common-source amplifier. If R_S is bypassed, Eq. 8-9 becomes $-g_m R_L'$ ($R_S = 0 \ \Omega$) as obtained in Eq. 8-6.

Example 8-17 also completes the picture regarding the effect of the source resistor on the FET circuit. Similar to R_E in the common-emitter amplifier, R_S causes a *negative feedback* effect, stabilizing the dc bias but restricting the ac signal swing and reducing the ac voltage gain.

It is also interesting to note that although the basic gain equation for the common source or common-emitter amplifier is $-g_m R_L'$, the FET has much lower values of gain than the corresponding bipolar amplifier. This is due to g_m. For the JFET, g_m is typically 1 to 4 mA/V. But for the bipolar transistor,

$$g_m = \frac{\beta}{r_{be}}$$

and typical values might be $\beta = 100$, and $r_{be} = 3$ kΩ, resulting in $g_m = 100/3$ k$\Omega = 33$ mA/V. This represents an increase in voltage gain on the order of *ten* with the bipolar transistor. Physically, this means the FET is not as *sensitive* to changes in the input voltage as the bipolar transistor.

Example 8-18 ───

Determine a *general formula* for the common drain or source-follower circuit shown in Fig. 8-26. Calculate the gain if $g_m = 4000$ μS, and $y_{os} = 2$ μS.

+10 V

800 kΩ 2 kΩ

Figure 8-26 The common-drain or source-follower connection.

Solution   For the emitter follower we had

$$A_V = \frac{(\beta + 1)R_L'}{r_{be} + (\beta + 1)R_L'} = \frac{R_L'}{\dfrac{r_{be}}{\beta + 1} + R_L'} \cong \frac{R_L'}{\dfrac{1}{g_m} + R_L'}$$

Multiplying through by g_m, we obtain

$$A_V = \frac{g_m R_L'}{1 + g_m R_L'} \tag{8-10}$$

where $R_L' = R_S \parallel R_L$ for the source follower connection. Plugging in numbers,

$$A_V = \frac{(4 \times 10^{-3})2 \text{ k}\Omega}{1 + (4 \times 10^{-3})2 \text{ k}\Omega} = 0.89$$

The source follower is, of course, analogous to the emitter follower and exhibits similar properties. The input resistance is large (R_G), the voltage gain is just less than 1, and the output resistance can be shown to be small, $(1/g_m) \parallel R_S$.

The important common-source and source-follower equations are summarized in Table 8-1.

TABLE 8-1

	Common Source	Source Follower
A_V	$\dfrac{-g_m R_L'}{(1 + g_m R_S)}$	$\dfrac{g_m R_L'}{(1 + g_m R_L')}$
R_i	R_G	R_G
R_o	$r_o \parallel R_D$	$\dfrac{1}{g_m} \parallel R_S$

8.5 THE VMOSFET

Until recently, the FET has been limited to *small-signal* applications due to its low transconductance and relatively high drain-source resistance limiting channel current. Of course, the *MOSFET* has found wide application in complex integrated circuit technology such as microprocessors and semiconductor memories.

A new type of MOSFET transistor recently developed is called the *vertical MOSFET* or *VMOSFET*. This MOS transistor has a typical transconductance of 250 mA/V (compared to 5 to 30 mA/V for a JFET or MOSFET and 30 to 50 mA/V for a bipolar transistor) and maximum drain currents in excess of 10 A for selected devices.

A cross-sectional view of a VMOSFET is shown in Fig. 8-27. One of the striking differences about this transistor is that the drain connection is on the *bottom* of the device, as opposed to the *top* surface in a normal MOSFET. This results in a vertical path for drain current. Operation is similar to the MOSFET in that application of a positive voltage to the gate terminal attracts electrons under the gate metal and in particular to the *p*-region, separating source and drain. With sufficient gate voltage, the channel becomes inverted and a current may flow between source and drain.

It is not obvious why the VMOSFET is an improvement over the

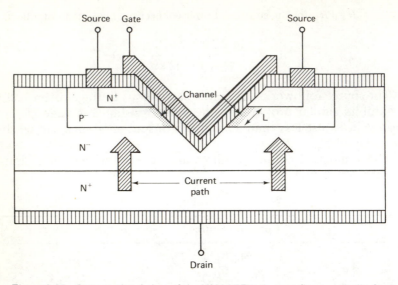

Figure 8-27 Cross-sectional view of the VMOSFET. Current flows vertically from the drain through the enhanced channel region to the source contact.

conventional MOSFET. However, the *length* of the channel, shown as *L* in Fig. 8-27, is now controlled by doping densities and diffusion times similar to the base width in a bipolar transistor. A shorter channel means less channel resistance and a greater chance that all carriers entering the channel at the source end will exit at the drain terminal without disappearing due to recombination. This in turn means that *higher transconductances* are possible. Note that the length of the channel in a normal MOSFET is determined by the dimensions of a photographic mask, which does not allow short channel lengths. Refer to Fig. 8-9 to compare the conventional channel length with the VMOSFET's.

In most other ways, the VMOSFET is similar to a normal MOS device. The device shown in Fig. 8-27 is an *n*-channel enhancement mode transistor and has characteristic curves very similar to those in Fig. 8-10. *P*-channel devices are also possible. The input resistance is again large, as expected, due to the insulated gate design. The maximum operating voltage is *higher* than that of a comparable MOSFET due to the lightly doped N⁻ region closest to the channel. Typical values are 80 V compared to 20 to 30 V for standard MOSFETs. Switching times from ON to OFF are considerably *faster* due to reduced capacitance between gate-source and drain-source caused by the smaller gate-source and gate-drain areas characteristic of the "V" groove construction.

In summary, the VMOSFET is a *high-current, high-voltage* enhancement mode transistor capable of fast switching times and large-signal amplifier applications.

VMOSFET Circuit Applications

A typical *large-signal* application of a VMOSFET is shown in Fig. 8-28. This is a 4 W class A audio amplifier using transformer coupling to a speaker load. As is characteristic of a class A amplifier, the VN66AF transistor is biased ON even without the presence of an input signal. In this case, the gate voltage is approximately 4 V, resulting in a bias current of about 0.5 A.

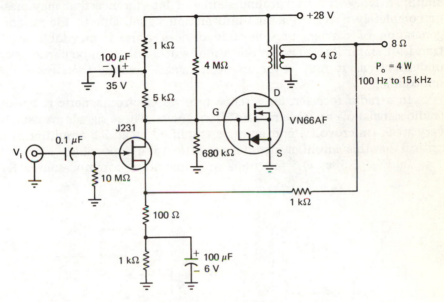

Figure 8-28 4-W class A VMOSFET audio amplifier. (Courtesy Siliconix, Inc.)

In the following chapter, we see how TRIACs and SCRs can be used to turn high current ac devices such as lamps and motors ON or OFF. Figure 8-29 illustrates a similar application using a VMOSFET.

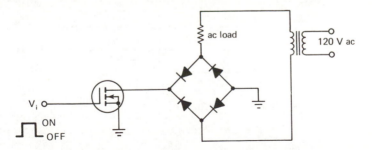

Figure 8-29 DC control of an ac load using a VMOSFET. (Courtesy Stone and Berlin from *Design of VMOS Circuits With Experiments* © Howard W. Sams & Co., Inc.)

255

A small low-powered *dc* input voltage can switch the VMOSFET ON, allowing current to flow through the bridge and activate the ac load.

8.6 FET CIRCUIT APPLICATIONS

In communications work, the FET is often preferred over the bipolar transistor. One of the reasons for this is that the FET is a *low-noise* amplifier. Noise is a background source of interference that may mask or completely cover the actual information signal. Due to the random generation of carriers in solid-state devices, noise is inevitable in all transistor amplifiers. In communications work, noise can be particularly undesirable as it may cover up weak signals, making their reception impossible.

In a radio receiver, an antenna receives electromagnetic radiation (radio signals) over a wide range of frequencies. These signals are usually very weak (microvolts) and must be amplified by an RF amplifier. It is critical that this amplifier introduce as little noise as possible.

In Fig. 8-30a, Q_1 functions as a low-noise common-source RF

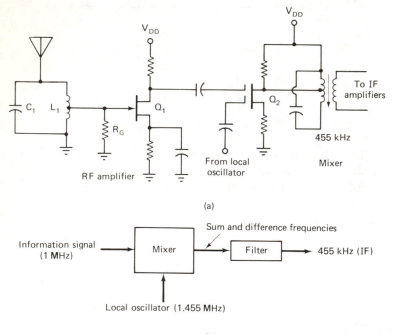

(a)

(b)

Figure 8-30 (a) The JFET (Q_1) is used as a low-noise amplifier and the dual gate FET (Q_2) as a mixer. (b) the mixer produces sum and difference frequencies, which are filtered to produce the IF signal at 455 kHz.

amplifier. L_1 and C_1 allow the receiver to be tuned to one particular frequency only. Q_1 then amplifies this signal and passes it to Q_2, a *dual-gate RF mixer.*

Nearly all radio and television receivers use frequency mixers to convert incoming RF signals to an intermediate frequency (IF). This simplifies the design of the radio and television because all circuits following the mixer need work at the IF frequency only, independent of the actual incoming signal frequency.

The principal of the electronic mixer is that two sine waves mixed together will result in two output signals, one with a frequency equal to the *sum* of the incoming frequencies, and the other the *difference*.

Figure 8-30b illustrates an incoming 1-MHz signal. The mixer receives this signal plus a 1.455-MHz sine wave from the *local oscillator*, producing two output signals with frequencies

$$f_{sum} = 1.455 + 1 = 2.455 \text{ MHz}$$

$$f_{diff} = 1.455 - 1 = 0.455 \text{ MHz}$$

If the stages following the mixer are tuned to 455 kHz, only this signal will pass. And if we design the oscillator to always be 455 kHz above the incoming signal (by making it variable and mechanically linking it to the C_1 tuning control), the output of the mixer will always be 455 kHz.

The dual-gate FET (Q_2) is particularly well suited to this application. This transistor has two separate gate diffusions, which conveniently allows mixing of the gate signals but with no dc interaction between the inputs. In addition, both inputs are very high impedance, minimizing any loading effects. Because the FET is a *square-law device* (I_D is proportional to V_{GS}^2), it introduces less distortion than the bipolar transistor, which is an exponential device. This distortion could possibly result in frequencies other than the sum and difference being produced and cause subsequent interference with the information signal. This is particularly true as the amplitude of the incoming signal increases. The FET will tend to pass large signals with less distortion than a bipolar transistor.

Another common application is to use an FET input stage as a *buffer* for a bipolar gain stage. This is illustrated in Figure 8-31. The voltage gain of the FET stage is low but its input resistance is extremely high, preventing the driver from being loaded down.

As a final example, consider the *logic inverter* shown in Fig. 8-32. This circuit uses both *p*-channel and *n*-channel E-MOSFET transistors and is commonly referred to as a *CMOS (complementary MOS) inverter.* A positive input voltage, greater than the switching threshold of the *n-*

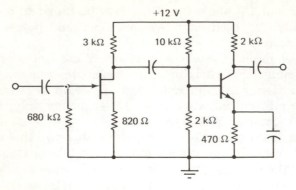

Figure 8-31 High-input-resistance two-stage amplifier.

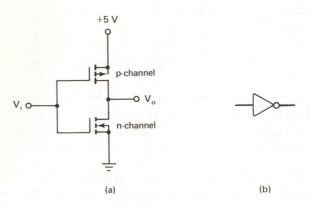

Figure 8-32 (a) Complementary MOS (CMOS) logic inverter.
(b) logic symbol.

channel device, turns this transistor ON but turns the *p*-channel transistor OFF. This causes the output voltage to be near 0 V (a logic 0). When $V_i = 0$ V, the opposite occurs and the *n*-channel transistor goes OFF while the *p*-channel goes ON. The output voltage now equals 5 V (a logic 1). The output voltage is always the logical opposite of the input.

A unique characteristic of the CMOS inverter is that there is never a path from V_{DD} to ground because both transistors are never ON at the same time. This means *power consumption* is extremely low, usually measured in microwatts.

─────────────────── KEY TERMS ───────────────────

FET Field-effect transistor. A voltage-controlled transistor characterized by high input resistance and an output current made up of majority carriers only (unipolar transistor). Several types of FETs have been developed, including the JFET, MOSFET, and VMOSFET.

JFET Junction field-effect transistor. This type of FET is characterized by a reverse-biased gate-channel junction. The device is normally ON and the channel must be depleted by the reverse-biased gate voltage to turn it OFF.

Pinch-off In an FET, when increases in V_{DS} no longer result in increases in I_D, the transistor is said to be operating in the pinch-off region. This corresponds to the active region for a bipolar transistor.

MOSFET Metal-oxide semiconductor field-effect transistor. This type of FET is characterized by an insulated gate, resulting in extremely high input resistance. MOSFETs are commonly broken down into two categories — D-MOSFETs and E-MOSFETs.

D-MOSFET Depletion-mode MOSFET. This transistor is normally ON and the channel must be depleted by the gate voltage to turn the device OFF. D-MOSFETs may also be operated in the enhancement mode.

E-MOSFET Enhancement-mode MOSFET. This transistor is normally OFF and the channel must be enhanced by the gate voltage to turn it ON.

Transfer Curves For an FET, these indicate how I_D will vary as V_{GS} changes. Transfer curves are very useful for determining FET dc circuit operating points.

g_m Transconductance. This parameter indicates the change in drain current for a 1-V change in gate-source voltage. Typically, g_m varies from 2 to 10 mA/V for the JFET, to 20 to 30 mA/V for the MOSFET, and as high as 250 mA/V for the VMOSFET.

VMOSFET Vertical MOSFET. This is a type of MOSFET characterized by a vertical path for current flow and a relatively short channel length compared to conventional MOSFETs. VMOSFETs have much higher current and voltage capabilities than either JFETs or MOSFETs.

————————————— QUESTIONS AND PROBLEMS —————————————

8-1 List two main differences between an FET and a bipolar transistor.

8-2 The FET has a high input resistance while the bipolar transistor is considered to have a low input resistance. Explain why this is so.

8-3 Redraw Fig. 8-1 assuming a p-channel device.

8-4 Explain the meaning of the following terms commonly found on an FET data sheet: I_{GSS}, I_{DSS}, $V_{GS(OFF)}$, y_{fs}, y_{os}.

8-5 What is the difference between the *resistance* region and the *pinch-off* region on the JFET characteristic curves? What are the corresponding regions for the bipolar transistor?

8-6 What is $V_{GS(OFF)}$ for the transistor whose characteristic curves are shown in Fig. 8-8?

8-7 Why is the FET sometimes called a "square-law" device?

8-8 When a MOSFET is damaged by a static discharge, what actually takes place?

8-9 Show characteristic curves for a p-channel D-MOSFET and E-MOSFET. Show schematic symbols for each.

8-10 Construct a transfer curve from the JFET characteristic curves in Fig. 8-3.

8-11 Show the schematic diagram of a fixed-bias n-channel JFET circuit using $V_{GG} = -3$ V, and $V_{DD} = +18$ V. Choose R_D such that $V_{DS} = 9$ V. Use the transfer curve from Problem 8-10.

8-12 Generate a set of maximum and minimum transfer curves using Eq. 8-2 and the following:

$$I_{DSS} = 6 \text{ mA (min) to } 12 \text{ mA (max)}$$

$$V_{GS(OFF)} = -5 \text{ V (min) to } -9 \text{ V (max)}$$

8-13 Use the transfer curves from Problem 8-12 and determine the maximum and minimum operating points for a self-biased n-channel JFET amplifier with $V_{DD} = +15$ V, $R_D = 1.5$ kΩ, and $R_S = 1$ kΩ.

8-14 What is the advantage of self-bias versus fixed-bias?

8-15 Refer to Figs. 8-17 and 8-18. If R_D is changed to 1.5 kΩ and the transistor operates on the maximum transfer curve, determine the new circuit operating point.

8-16 Refer to Fig. 8-22. If R_2 is changed to 600 kΩ, determine the new circuit operating point using the transfer curve shown in Fig. 8-21b.

8-17 For Problem 8-16, what value of R_S is needed to operate at $V_{GS} = 0$ V?

8-18 Design an n-channel E-MOSFET bias circuit and show the schematic diagram. Use the method illustrated in Example 8-12. Assume $I_{D(ON)} = 10$ mA when $V_{GS} = +4$ V, and $V_{GS(OFF)} = 0$ V. Use $V_{DD} = +10$ V.

8-19 Determine the value of g_m at $V_{GS} = 0$ V on the typical transfer curve in Fig. 8-18.

8-20 A certain common-emitter amplifier has $\beta = 150$ and $I_B = 15$ μA. Determine g_m.

8-21 Determine the input resistance, output resistance, and voltage gain of the common-source amplifier shown in Fig. 8-19 with R_S bypassed. Assume the typical transfer curve in Fig. 8-18 applies.

8-22 In Fig. 8-31, if $I_D = 2.5$ mA, determine the dc operating point at all nodes in the circuit. Assume $\beta = 125$.

8-23 Determine the input resistance, output resistance, and overall voltage gain of the two-stage amplifier in Fig. 8-31. Assume $\beta = 125$ and $g_m = 6$ mA/V.

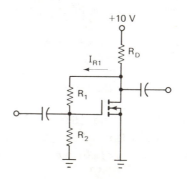

Figure 8-33

8-24 Bias the n-channel enhancement-mode common-source amplifier shown in Fig. 8-33 at the following Q point: V_{GS} = 3 V, V_{DS} = 5 V, I_D = 4 mA, and I_{R1} = 0.1 mA.

8-25 Illustrate a circuit using an n-channel VMOSFET as a relay driver. Assume the relay draws 100 mA at 5 V. Use V_{DD} = +12 V. Indicate the input voltage polarities needed to switch the relay ON and OFF.

8-26 What logic function is performed by the circuit in Fig. 8-34?

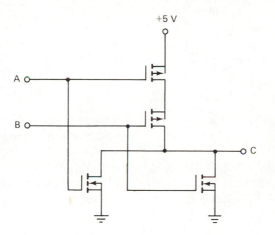

Figure 8-34

LABORATORY ASSIGNMENT 8:
FIELD-EFFECT TRANSISTOR

Objectives

1. Experience with the FET.
2. Practice in obtaining FET transfer curves.
3. Testing of various FET amplifier circuits.

Introduction   In this laboratory assignment you will experimentally determine the transfer curves of an n-channel JFET. A self-biased circuit will be designed and constructed. Load lines will be used to predict performance. Finally, ac characteristics will be measured, including the performance of a two-stage bipolar-FET (biFET) amplifier.

Components Required

Miscellaneous ¼-W resistors
n-channel JFET transistor
npn transistor
1-kΩ and 10-kΩ pots
Assorted coupling capacitors

Part I: *Transfer Curves*

STEP 1 Set up the circuit indicated in Fig. 8-35. Initially, adjust R_1 such that $V_{GS} = 0$ V, and adjust R_2 such that $V_{DS} = 6$ V.

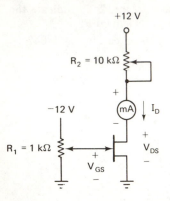

+12 V

$R_2 = 10$ kΩ

−12 V   (mA) ↓ I_D

$R_1 = 1$ kΩ

V_{GS}

V_{DS}

Figure 8-35 JFET test circuit for Part I of Laboratory Assignment 8.

STEP 2 Prepare a table of values for I_D and V_{GS} with V_{DS} constant at 6 V. Adjust R_1 such that V_{GS} varies in 0.5 V steps. Continue to increase the magnitude of V_{GS} until I_D has fallen to approximately 10 μA. Be sure to maintain V_{DS} constant at 6 V.

Question 1 At what value of V_{GS} do you think the FET is OFF? Is this $V_{GS(OFF)}$?

STEP 3 Determine a method to measure V_p (drain-source pinch-off voltage). Do your results compare to $V_{GS(OFF)}$ from Question 1? Should they? Explain.

STEP 4 Using graph paper, accurately sketch I_D versus V_{GS} using your data from step 2. Draw a smooth curve between the data points.

Optional If a curve tracer is available, display the common source characteristics using 0.5-V steps. Check the value of I_D for each value of V_{GS} with your data from step 2.

Part II: *Bias Circuits*

STEP 1 Select an appropriate Q point on your transfer curve obtained in Part I. Sketch a self-biased load line through this operating point.

STEP 2 Design a circuit consistent with your load line and with $V_{DS} = 6$ V. Don't forget the R_G resistor. Assemble the circuit and measure the resulting Q point. Double check any discrepancies greater than 10%.

STEP 3 Try several other transistors in your circuit and measure the resulting bias points.

Question 2 Comment on the stability of the self-biased JFET circuit you have constructed. How do you think it compares to similar bipolar circuits?

Part III: *AC Circuits*

STEP 1 Calculate the value required for an input coupling capacitor (assume $f = 1$ kHz), and then measure the voltage gain of the circuit from Part II.

STEP 2 Referring to the transfer curve, estimate g_m at the Q point and calculate A_V. Compare with the measured results in step 1.

STEP 3 Calculate the value required for a source-bypass capacitor (assume $f = 1$ kHz) and add to your circuit. Remeasure the voltage gain. Again check with calculations.

STEP 4 Refer to Fig. 8-31. Use your common-source amplifier from step 3 as the JFET stage and assemble this amplifier. Calculate a suitable value for this interstage coupling capacitor.

STEP 5 Measure A_V, R_i, and R_o for the overall amplifier.

Question 3 Draw the amplifier equivalent circuit for the biFET amplifier in step 4.

NINE

THYRISTORS AND OPTOELECTRONIC DEVICES

One is inclined to believe that bipolar and field-effect transistors are the whole story as far as solid-state electronics are concerned. But this is not true. In this chapter, a number of new solid-state devices are introduced along with *optoelectronic* versions of several already familiar components.

9.1 THE SILICON-CONTROLLED RECTIFIER

Basic Operation

The *silicon-controlled rectifier*, or SCR, was developed at the Bell Laboratories in 1956. This device is the best known member of a family of devices referred to as *thyristors* (thyratron-transistor) and represents the solid-state equivalent of the *thyratron* vacuum tube.

The SCR is a *high current switching device* having two states: ON and OFF. It is designed for control of ac and dc devices and selected parts have been designed to switch current values as high as 1600 A!

As illustrated in Fig. 9-1, the SCR consists of four alternating layers of *n*- and *p*-type semiconductor material. For this reason, it is often referred to as a *four-layer diode* or *pnpn diode*. The device has three terminals, labelled *anode*, *cathode*, and *gate*.

As we shall see, the SCR behaves nearly the same as a normal semiconductor diode except that it is triggered into conduction by a small current applied to the gate terminal. Without this trigger current, the SCR will not conduct.

Figure 9-2a illustrates an equivalent representation for the SCR, where it is shown to be two separate but connected *npn* and *pnp* transistors. This is shown schematically in Fig. 9-2b. Ignoring the gate

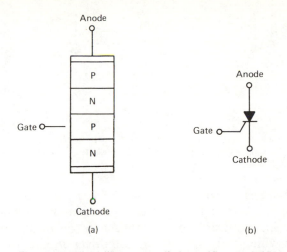

Figure 9-1 The silicon-controlled rectifier or SCR consists of four alternate *p* and *n* layers, shown in (a). The schematic symbol is shown in (b).

terminal, the *npn* base current is supplied by collector current from the *pnp*. Similarily, collector current of the *npn* supplies base current for the *pnp*.

If a voltage is applied positive to negative between anode and cathode with the gate terminal open, both transistors (and therefore the SCR) will be OFF. This is due to the "Catch 22" situation in which the *npn* cannot conduct unless the *pnp* conducts, and the *pnp* cannot conduct unless the *npn* conducts! With no way of getting base current to either transistor, they both must be OFF. This is illustrated in Fig. 9-3a.

However, if a brief *pulse* of current is applied to the gate terminal,

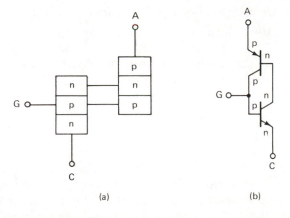

Figure 9-2 The SCR can be visualized as two separate transistors.

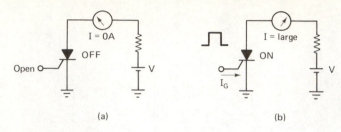

Figure 9-3   (a) If the gate terminal is left open, the SCR will not conduct. (b) applying a pulse of current to the gate terminal triggers the SCR ON, allowing current to flow.

the *npn* transistor can begin to conduct. This in turn allows base current to flow in the *pnp* and it too begins to conduct. This transistor in turn supplies more base current to the *npn*. A *regenerative* effect quickly takes place in which both transistors are driven into saturation and the SCR switches itself into the conducting state in 1 or 2 μs. In this condition, a very low resistance exists between anode and cathode and current is allowed to flow in the external circuit. This is illustrated in Fig. 9-3b.

It is interesting to note that as the SCR "fires," it regenerates its own gate current and *latches* itself into the conducting state. The external gate current is no longer needed. In fact, the gate terminal can even be *grounded* and the SCR will remain latched ON! This leads to the interesting question, "How do we turn it OFF?" Generally, there is only one way to do this and that is by "pulling the plug!" Actually, the anode to cathode current must be reduced below some minimum value (the holding current, I_H). This can be accomplished by

1. Turning OFF the power.
2. Shunting or shorting the anode–cathode with a low resistance.
3. Reversing the polarity of anode–cathode voltage.

Although at the moment these methods of turning the SCR OFF may seem inconvenient or complex to you, we will shortly see that in many circuits, this happens "automatically."

IV Characteristics

Figure 9-4 illustrates the IV characteristic curves for the SCR. A *family* of curves is shown for various values of gate current, I_G. As this current increases (I_{G2}, I_{G3}, in Fig. 9-4) the anode–cathode voltage needed to fire the SCR decreases accordingly. Note that it is possible

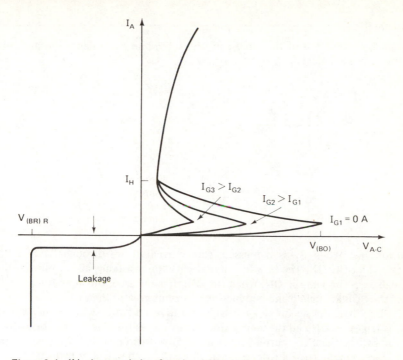

Figure 9-4 IV characteristics for the SCR. Indicated are the holding current, I_H, the forward breakover voltage, $V_{(BO)}$, and the reverse breakdown voltage, $V_{(BR)R}$.

to turn the SCR on *without* gate current. This occurs when the anode–cathode voltage is sufficiently high to cause an *avalanche breakdown* mechanism to occur, causing the SCR to latch itself on without external gate current. This voltage is called the *forward-breakover voltage* and is indicated as $V_{(BO)}$ in Fig. 9-4. Often, the SCR may simply be specified as "200 V 1 A", meaning $V_{(BO)}$ is 200 V and the maximum on state current (I_T) is 1 A.

One important fact not to be missed from the IV curves is that the SCR will only conduct for *positive* values of anode–cathode voltage. This should also be apparent from the two-transistor equivalent circuit in Fig. 9-2. Also note that the SCR exhibits a normal silicon diode breakdown characteristic for reverse bias. The value of $V_{(BR)}$ (the reverse breakdown voltage) is generally the same as $V_{(BO)}$.

The *holding current* is also indicated in Fig. 9-4. The value of I_H varies with I_T, the maximum on-state current. Typical values might be 5 to 10 mA for a 5-A SCR and 50 to 100 mA for a 25-A SCR. Generally, the holding current is less if the gate current is maintained instead of pulsed.

Example 9-1 ————————————————————————————————————

Explain the operation of the circuit in Fig. 9-5 when S_1 is depressed and released for (a) V_S = 12 V dc; (b) V_S = 12 V rms ac.

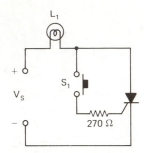

Figure 9-5 When S_1 is depressed, the lamp is latched ON when V_S is a dc source, but only lights at half brilliance when V_S is an ac source.

Solution (a) When S_1 is depressed, gate current flows through L_1 and S_1 and triggers the SCR ON. The now low anode–cathode resistance allows current to flow through L_1. The lamp is ON. When the switch is released, the lamp *remains ON* because the SCR is latched and conducting a current greater than I_H.

(b) When the ac input cycle goes positive and the switch is depressed, the SCR is triggered ON and the lamp lights. However, as the sine wave returns to 0 and goes negative, the SCR current must go to 0 A and fall below I_H (recall that the SCR won't conduct for a reverse bias). The SCR automatically turns OFF for each negative half cycle. The lamp is on but only at *half brilliance*. When the switch is released, the lamp will go OFF and remain OFF because the ac automatically reduces the anode current below I_H.

SCR Specifications

Several important specifications for the SCR have already been introduced. The following summarizes this information and includes standard nomenclature.

$V_{(BO)}$ —*forward-breakover voltage*. This is the minimum anode–cathode voltage that will trigger the SCR on in the absence of gate drive.

$V_{(BR)R}$ —*reverse-breakdown voltage*. This is the minimum value of reverse anode–cathode voltage that will cause the SCR to breakdown and conduct in the reverse direction.

I_T(rms) —*maximum rms on-state current*. This is the maximum anode current allowable in the on-state as limited by heat dissipation.

I_{GT} —*gate-trigger current*. This is the maximum gate current required to trigger the SCR ON.

V_{GT} —*gate trigger voltage*. This is the maximum, V_{GT}(max), or minimum, V_{GT}(min), gate–cathode voltage required to trigger the SCR ON.

I_H —*holding current*. This represents the minimum value the anode current can become without the SCR switching OFF.

Example 9-2

Refer to the SCR data sheet in Fig. 9-6 and compare the 2N880 with the 2N2348 at 25°C.

GENERAL ELECTRIC THYRISTOR AND DIODE CONDENSED SPECIFICATIONS

PHASE CONTROL SCR's
.5 TO 5 AMPERES

GE TYPE		C3	C103	C203	C5	C6	C7	–	C106	C107	C108
JEDEC		2N877-81[d]	–	2N5060-64	2N2322-29	–	2N2344-48	2N1595-99, A	–	–	–
ELECTRICAL SPECIFICATIONS											
VOLTAGE RANGE		30-200	30-200	30-400	25-400	25-400	25-200	50-400	15-600	15-600	15-600
FORWARD CONDUCTION											
$I_{T(RMS)}$	Max. RMS on-state current (A)	0.5	0.8	0.8	1.6	1.6	1.6	1.6	4.0	4.0	5.0
$I_{T(AV)}$	Max. average on-state current @ 180° conduction (A) @ T_C	0.32 @ 85°C	0.50 @ 25°C	0.50 @ 25°C	1.0 @ 85°C	1.0 @ 85°C	1.0 @ 55°C	1.0 @ 110°C	2.5 @ 30°C	2.5 @ 20°C	3.75 @ 30°C
I_{TSM}	Max. peak one cycle, non-repetitive surge current (A)	7	8	8	15	10	15	15	20	15	30
I^2t	Max. I^2t for fusing for > 1.5 msec (A^2 sec)	–	–	–	0.5	0.5	–	0.5	0.5	0.5	1
V_{TM}	Max. on-state voltage @ 25°C, 180° conduction, rated $I_{T(AV)}$ (V)	1.6	1.5	1.5	2.2	1.4	2	2	2.2	2.5	1.35
$R_{\theta JC}$	Max. internal thermal resistance, dc junction-to-case (°C/W)	80	125	75	10	10	–	–	10	10	10
I_H	Max. holding current @ 25°C (mA)	5	5	5	2	5	1	–	3	6	3
t_q	Typical turn-off time (μsec) @ max. T_J	15	15	15	40	40	20	40	40	40	40
	Maximum turn-off time (μsec @ 110 C)	–	–	–	–	–	–	–	100	100	100
$t_d + t_r$	Typical turn-on time (μsec @ 110°C)	1	1.4	1.4	1.4	1.4	1.4	1.2	1	1	1
di/dt	Max. rate-of-rise of turned-on current (A/μsec)	–	–	–	50	–	–	–	50	50	50
T_J	Junction operating temperature range (°C)	–65 to 125	–65 to 125	–65 to 125	–65 to 125	–40 to 125	–65 to 100	–65 to 150	–40 to 110	–40 to 110	–40 to 110
BLOCKING											
dv/dt	Typical critical rate-of-rise of off-state voltage, exponential to rated V_{DRM} @ max. rated T_J (V/μsec)	40	20	20	20	20	20	20	8	8	8
FIRING											
I_{GT}	Max. required gate current to trigger (μA) @ –65 C	300	500	500	350	–	75	–	–	–	–
	@ –40 C	–	–	–	–	–	–	–	500	–	500
	@ 25 C	200	200	200	200	1000	20	10,000	200	500	200
V_{GT}	Max. required gate voltage to trigger (V) @ –65 C	–	1	1	1	1	1	–	–	–	–
	@ –40 C	–	–	–	–	1	–	–	1	–	1
	@ 25 C	0.8	0.8	0.8	0.8	0.8	0.8	3	0.8	0.8	0.8
V_{GT}	Min. required gate voltage to trigger (V) @ 110 C	–	–	–	–	–	–	–	0.2	0.2	0.2
	@ 125 C	0.05	0.1	0.1	0.1	0.1	–	–	–	–	–
VOLTAGE TYPES											
Repetitive Peak Forward and Reverse Voltages											
	15	–	–	–	–	–	–	–	C106Q1	C107Q1	C108Q1
	25	–	–	–	2N2322 C5U	C6U	2N2344	–	–	–	–
	30	2N877	C103Y	2N5060 C203Y	–	–	–	–	C106Y1	C107Y1	C108Y1
	50	–	–	–	2N2323* C5F	C6F	2N2345	2N1595, A	C106F1	C107F1	C108F1
	60	2N878	C103YY	2N5061 C203YY	–	–	–	–	–	–	–
	100	2N879	C103A	2N5062 C203A	2N2324* C5A	C6A	2N2346	2N1596, A	C106A1	C107A1	C108A1
	150	2N880	–	2N5063	2N2325 C5G	C6G	2N2347	–	–	–	–
	200	2N881	C103B	2N5064 C203B	2N2326* C5B	C6B	2N2348	2N1597, A	C106B1	C107B1	C108B1
	250	–	–	–	2N2327 C5H	–	–	–	–	–	–
	300	–	–	C203C	2N2328* C5C	C6C	–	2N1598, A	C106C1	C107C1	C108C1
	400	–	–	C203D	2N2329* C5D	C6D	–	2N1599, A	C106D1	C107D1	C108D1
	500	–	–	–	–	–	–	–	C106E1	C107E1	C108E1
	600	–	–	–	–	–	–	–	C106M1	C107M1	C108M1
PACKAGE OUTLINE NO.		112	195.1, 228	263	101	101	101	101	173	173	173

* JAN & JANTX types available.
L. 2N885-89 available 20 mA max. I_{GT}
2. 2N2322A 28A available 20 mA max. I_{GT}

Figure 9-6 Typical SCR specifications. (Courtesy of General Electric Company Semiconductor Products Department.)

Solution

	2N880	2N2348
$V_{(BO)} = V_{(BR)R}$	150 V	200 V
$I_T(\text{rms})$	0.5 A	1.6 A
I_{GT}	200 μA	20 μA
$V_{GT}(\text{max})$	0.8 V	0.8 V
I_H	5 mA	1 mA

DC Circuit Applications

In most dc circuits, the SCR is used in place of a switch, eliminating the contact sticking, burning, and wear inherent in mechanical switches. Figure 9-7 illustrates such an application in an *emergency lighting* system. This circuit is designed to maintain a dc voltage across the 6-V lamp even in the event of an ac power failure. As such, it could be used for any application where a battery backup is essential.

The transformer and diodes, CR_2 and CR_3, form a conventional full-wave center-tapped dc power supply. The lamp is connected directly across this output and is normally powered by the ac main. Capacitor C_1 charges to the peak value of 6.3 V rms or 8.9 V and holds the SCR cathode at this value. Because the battery keeps the anode at 6 V, the SCR must be *OFF*. Diode CR_1 and resistor R_1 provide a charging path for the battery when ac is present.

If the ac power is lost, the cathode voltage will drop, allowing the SCR to be triggered ON by the battery through R_3. The battery now maintains the lamp through the conducting SCR. When ac power is restored, the SCR again becomes reverse-biased, automatically turning itself *OFF*.

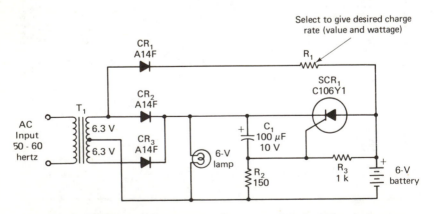

Figure 9-7 Emergency lighting system. (Courtesy of General Electric Company Semiconductor Products Department.)

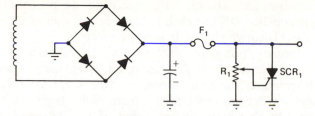

Figure 9-8 Using the SCR as an electronic crowbar. If the output voltage exceeds a preset value the SCR fires, blowing the fuse but protecting the load circuit.

Another dc circuit application of the SCR is illustrated in Fig. 9-8. This circuit is referred to as an *electronic crowbar*. For normal conditions, R_1 is adjusted such that the SCR is biased OFF. If the power supply voltage should momentarily surge and exceed a value determined by the setting of R_1, the SCR will be triggered ON, effectively placing a short circuit or "crowbar" directly across the power supply output. This, in turn, causes a large current to momentarily flow through the SCR, blowing fuse F_1 but protecting all equipment powered by the supply. Because it is not uncommon for one power supply to support several thousand dollars worth of electronics, a simple circuit of this nature may be a worthwhile investment.

As a final dc circuit application, consider the circuit shown in Fig. 9-9. This circuit functions as a *regulated battery charger*. Assume

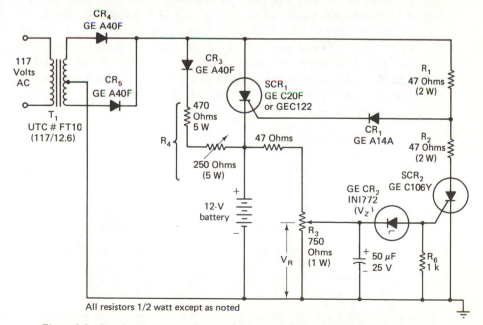

All resistors 1/2 watt except as noted

Figure 9-9 Regulated battery charger. (Courtesy of General Electric Company Semiconductor Products Department.)

initially a discharged battery. The pulsating dc available from the rectifier triggers SCR_1 ON for each half cycle and a substantial charging current can flow through this SCR and the battery.

As the battery potential rises due to the charging, the voltage tapped off by potentiometer R_3 also rises. Eventually, this voltage is large enough to trigger the zener diode and fire SCR2. At this point, the $R_1 R_2$ voltage divider prevents SCR_1 from being triggered ON and all charging stops. In this way, *overcharging* of the battery is prevented. Should the battery voltage drop below the preset value, SCR_2 will no longer trigger ON, allowing SCR_1 to resume charging the battery.

The 50-μF capacitor prevents false triggering of the zener diode, while resistor R_6 is required to prevent triggering of SCR_2 due to its own thermally generated leakage current. A *trickle* charge is maintained due to CR_3 and resistor R_4.

AC Circuit Applications

One of the more common SCR circuit applications is to control the amount of power absorbed by an ac load. This is done by varying that portion of the ac cycle, during which the load conducts. This is a very *efficient* way of controlling ac power and a much preferred method to using a series rheostat, which would have to dissipate large quantities of heat. A simple half-wave *phase control* circuit is illustrated in Fig. 9-10. Assume that the pot has been adjusted such that 120 V is required at the anode terminal to trigger the SCR into conduction. As illustrated in the figure, the SCR will conduct from 45° (170 V sin 45° = 120 V) to 180° or 135° per 360° cycle. If the load is an incandescent lamp, the lamp will appear to appear to burn at less than full brilliance and the circuit will function as a *light dimmer*. If the load is a motor, the motor will turn at less than full speed and the circuit now functions as a *motor-speed control*.

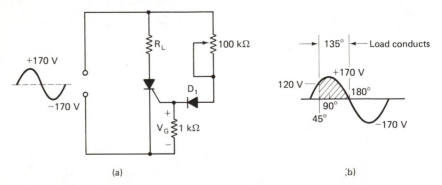

(a)   (b)

Figure 9-10   (a) SCR phase control circuit. (b) the load conducts for less than a full cycle, depending on the setting of the 100-kΩ pot.

Note that the SCR is turned OFF automatically when the ac input crosses 0 V because this forces the anode current to fall below I_H as the SCR becomes reverse-biased. During the reverse-biased time, the cathode is positive with respect to the anode. Because the gate terminal is returned to the anode, the gate–cathode is also reverse-biased during this time. Diode D_1 protects this junction against breakdown and possible damage by not allowing a reverse current to flow during this negative half cycle.

The conduction angle in this example is $135°$. As the pot is varied, what extremes are possible? When adjusted for minimum resistance and depending on the SCR, triggering could occur at nearly $0°$. This would allow a maximum of $180°$ of conduction. Minimum conduction occurs when the pot is adjusted to require the peak voltage of the sine wave before triggering. This allows $90°$ of conduction. These two cases are illustrated in Fig. 9-11. Note that conduction angles less than $90°$ are not possible (excepting $0°$) because all such triggering levels also occur *prior* to $90°$.

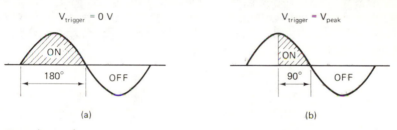

(a)   (b)

Figure 9-11 The circuit in Fig. 9-10 will conduct for (a) $180°$ maximum and (b) $90°$ minimum.

Example 9-3

Assume the $100\text{-k}\Omega$ pot in Fig. 9-10 has been adjusted to $84\ \text{k}\Omega$. Estimate the conduction angle for this circuit. Assume the SCR is a 2N881 and $R_L = 1\ \text{k}\Omega$.

Solution The 2N881 data sheet (Fig. 9-6) indicates a maximum of $200\ \mu\text{A}$ (I_{GT}) and $0.8\ \text{V}$ (V_{GT}) is required for triggering at $25°\text{C}$. Referring to Fig. 9-10, when $V_G = 0.8\ \text{V}$, the current through the pot will be

$$I_{pot} = I_{GT} + I_{1\,k} = 0.2\ \text{mA} + 0.8\ \text{V}/1\ \text{k}\Omega = 1\ \text{mA}$$

The total voltage required is then

$$V_i = 1\ \text{mA}(84\ \text{k}\Omega) + 0.6\ \text{V} + 0.8\ \text{V} = 85.4\ \text{V}$$

We now need to know what angle will cause V_i to be $85.4\ \text{V}$. That is,

$$170\ \text{V} \sin \Theta = 85.4\ \text{V} \quad \text{or} \quad \Theta = \text{INV} \sin (85.4/170) = 30°$$

The SCR will turn ON at approximately 30° and remain ON until the anode current falls below I_H, in this case 5 mA. Because I_H will occur when $V_i = I_H \times R_L = 5$ mA $\times 1$ kΩ = 5 V, the SCR will turn OFF when

$$170 \text{ V} \sin \Theta = 5 \text{ V} \quad \text{or} \quad \Theta = 178°$$

The total conduction angle is then $178° - 30° = 148°$.

Although the SCR can potentially conduct for 180° of an ac cycle, the circuit in Fig. 9-10 does not allow smooth phase control from 0 to 180°. We can improve this circuit with the addition of a capacitor and diode, as shown in Fig. 9-12. Although the circuit is beginning to appear more complex, its operation is similar to the previous phase control circuit. The addition of the capacitor means that the gate voltage will now *lag* the line voltage so that if, for example, the circuit was adjusted to trigger at 90° (V_{peak}), the gate voltage may not actually see this potential until 120° due to the time required to charge the capacitor. In this way, conduction angles less than 90° can be accomplished (60° in this example). In order that the capacitor does not retain its positive charge from one cycle to the next, diode D_2 conducts during the negative half cycle, forcing the capacitor to discharge before the new positive half cycle begins.

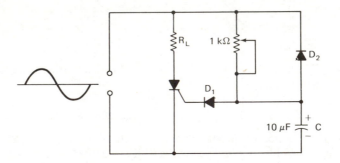

Figure 9-12 With the addition of a capacitor, the gate voltage lags the line voltage, allowing smooth phase control from 0 to 180°.

Now that we have phase control from 0 to 180°, you might think we should be satisfied. But of course we are not! Why not complete control over the *entire* 360° cycle? We might refer to this as *full-wave phase control* compared to the *half-wave phase control* we have been studying with the SCR. Full-wave phase control will require a new member of the thyristor family, called the *TRIAC*.

9.2 THE TRIAC

The word TRIAC is an acronym for *triode* (3-terminal) controlled *ac* switch. This device, developed by General Electric, has three terminals labeled *gate, main terminal one* (MT1), and *main terminal two* (MT2). Some books refer to these last two terminals as anode 1 and 2 respectively.

It is convenient to visualize the TRIAC as two SCRs connected in inverse parallel, as shown in Fig. 9-13a. Figure 9-13b illustrates the schematic symbol. As indicated in the previous section, the TRIAC will allow conduction for *both* possible voltage polarities (full-wave and half-wave conduction), and this should also be clear from the TRIAC equivalent circuit in the figure. The IV characteristics are very similar to those of the SCR but include operation in the third quadrant as well where the current and voltage are both negative.

TRIAC specifications are similar to SCR's. For example, a 400-V 6-A TRIAC will withstand 400 V between the two main terminals without triggering in the absence of gate drive. Its maximum on state rms current is 6 A. This same device might have a typical holding current of 50 mA, a maximum gate trigger voltage of 2.5 V, and a 50 mA gate trigger current. At present, TRIACs do not have as high a current capability as SCRs. Typical numbers are 50 to 100 A maximum for selected TRIACs, but over 1500 A for selected SCRs.

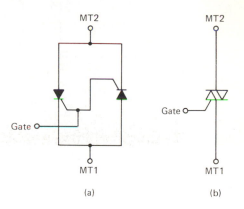

Figure 9-13 (a) The TRIAC equivalent circuit consists of two inversely-connected SCRs, (b) the schematic symbol.

TRIAC Triggering

Unlike the SCR, the TRIAC may be triggered into conduction in several ways. Figure 9-14 illustrates the four possible combinations. These are listed below.

	MT2	MT1	I_G	Fig. 9-14
(1)	+	−	+	a
(2)	−	+	+	b
(3)	+	−	−	c
(4)	−	+	−	d

Note that Fig. 9-14a and d correspond to the typical situation of an ac input, causing MT2 and MT1 to exchange polarities on alternate half cycles. In Fig. 9-14c and d, we see that it is possible for the gate current to be *negative* (flow *out* of the gate terminal). This is why full-wave conduction is possible.

Another interesting point we should note is illustrated in Fig. 9-14b. The *terminal* connections to the TRIAC are simply *opposite* to those in Fig. 9-14a and we might expect the TRIAC to trigger just as well as before. However, when connected in this manner, the TRIAC is much *less sensitive* to the trigger input. This indicates that the TRIAC is *not* symmetric and the choice of MT1 and MT2 is *not* arbitrary.

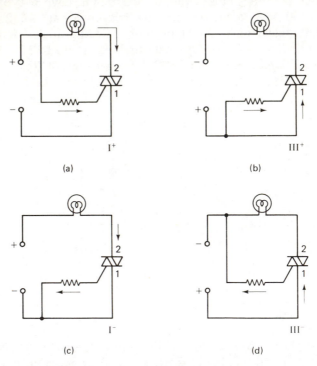

Figure 9-14 Various trigerring circuit combinations for the TRIAC. Each circuit is characterized by its quadrant of operation (I or III) and the direction of gate current (plus or minus).

When using a TRIAC, care should be taken to identify the terminals and then select a triggering circuit accordingly. Optimum triggering occurs for circuits in Fig. 9-14a, c, and d.

Circuit Applications

One of the main advantages that SCRs and TRIACs both enjoy is the replacement of *mechanical switching* devices and the associated contact bounce, arcing, and poor connections. Another advantage is that these solid-state switches always open at 0 current, eliminating the possibility of voltage transients due to stored *inductive* energy.

An example of such a circuit application is the *solid-state relay* shown in Fig. 9-15. When switch S_1 is closed, a small gate current, limited by resistor R_1, triggers the TRIAC into conduction, activating a possibly much larger load current. As long as S_1 remains closed, the load will conduct for both halves of the ac cycle. When the switch is opened, the TRIAC will not turn OFF until the ac input crosses 0 V, reducing the load current below the holding value.

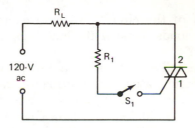

Figure 9-15 Solid-state relay. When S_1 is closed, the load conducts for the full 360° cycle.

One problem that is encountered when triggering TRIACs or SCRs is the *variability* of the trigger current and voltage from one device to another. This is similar to the β variations we experienced with bipolar transistors. In Example 9-3, we calculated a conduction angle based on data sheet values for I_{GT} and V_{GT}. In the real world these values might vary considerably, causing our calculations to be correspondingly in error.

Figure 9-16 illustrates one method of solving this problem. In this circuit, a new thyristor is introduced called the *DIAC*. This device is essentially a TRIAC without a gate terminal. In Fig. 9-16, as the ac input rises, capacitor C_1 builds up a charge and eventually reaches the breakdown voltage of the DIAC. When this occurs, the DIAC "fires," sucking the charge out of the capacitor and triggering the TRIAC. The DIAC essentially holds off triggering the TRIAC until a large enough charge has been accumulated in C_1. In this way, the TRIAC can be *guaranteed* to trigger despite device-to-device trigger variations.

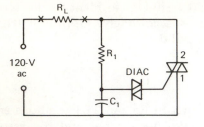

Figure 9-16 TRIAC phase-control circuit using a DIAC to ensure reliable
triggering.

In principal, the circuit in Fig. 9-16 will allow variable phase control from 0 to 360° if R_1 is replaced with a variable resistor. One of the problems with this circuit, however, is that the capacitor may not *discharge* fully after the DIAC conducts. This leaves a residual voltage of opposite polarity on the capacitor for the next half cycle. This may cause asymmetrical triggering of the TRIAC. For this reason, an *asymmetrical silicon bilateral switch* (ASBS) has been developed to replace the conventional DIAC.

Inductive Loads

All of the circuits we have examined up to this point have used *resistive* loads. What if the load is *inductive*, as it typically might be for a motor?

The inductor presents a new problem for the SCR or TRIAC due to its basic nature of wanting to maintain a constant current flow. Consider the circuit illustrated in Fig. 9-17. Two SCRs are being used to control an inductive dc load. This inductance may cause a current to flow in the SCR greater than I_H even though the voltage polarity has reversed. In this case, the SCR may *not* turn OFF as expected when the line voltage passes through 0 V. Circuits of this type should use a diode, D_1 in Fig. 9-17, to allow the inductor to discharge its current around the SCR.

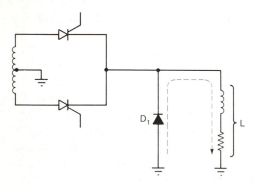

Figure 9-17 An inductive dc load prevents the SCR current from falling below
I_H when the output voltage passes
through 0 V. D_1 shunts this inductive
current to ground.

For inductive ac loads, a similar problem occurs. The load current and voltage are now out of *phase* with V leading I. This is shown for a TRIAC circuit in Fig. 9-18. As the current crosses 0 A the TRIAC must turn OFF and the nonzero line voltage at this instant appears across the TRIAC. Because all semiconductor devices have associated capacitances, this voltage also appears across the TRIAC *junction capacitance*, C_j. This in turn requires an additional TRIAC current to charge this capacitance. If this current exceeds I_H, the TRIAC may not turn OFF.

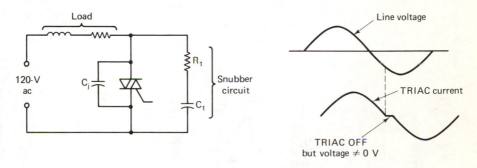

Figure 9-18   An inductive load causes the TRIAC voltage and current to be out of phase. The snubber circuit prevents the voltage across the TRIAC from building up too rapidly and possibly triggering the TRIAC into conduction.

In fact, this problem may exist even without an inductive load. When the SCR or TRIAC is OFF, any *sudden change* in anode–cathode voltage may cause a charging current $[i = C(dv/dt)]$ sufficient to trigger the device into the conducting state. For this reason, manufacturers rate how fast the voltage may change across the anode–cathode without triggering the device ON. This is called the *critical rate of rise of off-state voltage*, dv/dt. Typical numbers might range from a low of 20 V/μs to a high of 200 V/μs.

One method of getting around this problem for a particular inductive load is to select a device with a higher dv/dt rating. Another is to add the *snubber circuit*, consisting of R1 and C1 shown in Fig. 9-18. The RC time constant of this circuit limits the rate of rise of the off-state voltage. Values for R and C depend on the load, line voltage, and TRIAC used.

9.3 LIGHT-EMITTING DIODES

Optoelectronics generally refers to solid-state devices that *emit* light, *detect* light, or *transmit* electrical signals via a light path (opto-couplers). The remainder of this chapter will deal with specifications of these devices and typical circuit applications.

The Nature of Light

Light energy is actually a form of *electromagnetic radiation* similar to a radio wave. Each of the primary colors from blue to red has its own characteristic frequency and wavelength. Referring to Fig. 9-19, we see that *blue light* is near the high end of the visible light frequency spectrum because it has the shortest wavelength.

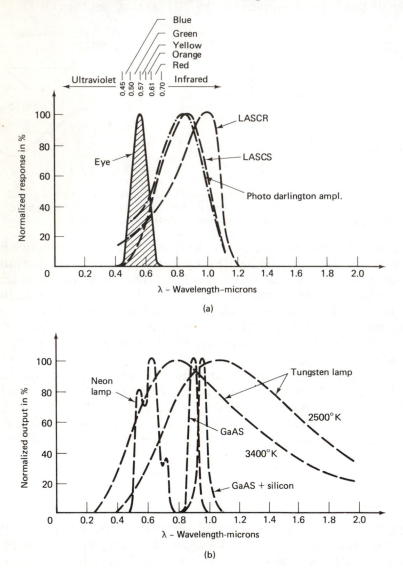

Figure 9-19 Spectral distributions of light-sensitive and light-emitting devices. (Courtesy of General Electric Company Semiconductor Products Department.)

Example 9-4 ————————————————————————————————

Calculate the extreme values of frequency for light in the visible spectrum ranging from 0.4 μm (microns) (blue) to 0.7 μm (red).

Solution   Recall that wavelength and frequency are related by the speed of light so that

$$f = c/\lambda \qquad\qquad (9\text{-}1)$$

where λ is the wavelength and c is the speed of light (3×10^{10} cm/s). Because 1 μm equals 10^{-4} cm, we can determine

$$f \text{ (blue)} = 3 \times 10^{10}/0.4 \times 10^{-4}) = 7.5 \times 10^{14} \text{ Hz}$$

and

$$f \text{ (red)} = 3 \times 10^{10}/0.7 \times 10^{-4}) = 4.3 \times 10^{14} \text{ Hz}$$

To appreciate these numbers, note that 10^9 Hz = 1 GHz, which is considered a microwave frequency. Blue light has a frequency of 750,000 GHz!

The LED

The LED or *light-emitting diode* is actually a *pn* diode that emits light in the visible spectrum when *forward*-biased. In reality, all semiconductor diodes are LEDs; however, the frequency of the emitted light is visible only for certain semiconductor materials.

When a semiconductor diode is forward-biased, a current flows and the resulting conduction electrons pick up an amount of energy such that they are now said to reside in the conduction band, as explained in Chapter 1. It is likely that some of the lower energy conduction electrons may fall into a nearby hole, leaving the conduction band and returning to the valance band. This process is called *recombination*. When recombination takes place, energy is given off, some in the form of *heat* and some in the form of light. Now here is the critical part. The *wavelength* of this light is *inversely* related to the bandgap energy (distance between conduction and valence bands) of the semiconductor material. This means that the larger the bandgap, the smaller the wavelength and the higher the frequency of the emitted light.

Typical red LEDs are made using gallium-phosphide (GaP), which has a band gap of nearly 2 eV (electron volts). Table 1-2 indicated that the bandgap of silicon is 1.1 eV. This means that the light emitted by a silicon diode will be even *lower* in frequency than that of red light. A check of Fig. 9-19 indicates that red is already the lowest visible frequency. For this reason silicon diodes cannot be used as LEDs.

IV Characteristics

The IV characteristics of the LED have been previously presented for comparison with silicon and germanium diodes in Fig. 1-14. Most manufacturers recommend a forward-operating current of approximately 20 mA for optimum LED brilliance. A typical forward voltage drop at this current level is 1.6 V but may be as high as 3 V.

In the *reverse* direction, the LED will not emit light and in fact begins to breakdown at a very low voltage. A typical specification is $100\,\mu A$ of leakage current with 5-V reverse bias.

Circuit Applications

One of the major circuit applications for the LED is as a *logic-level* indicator in digital circuits.

Example 9-5 ───────────────────────────────

Show the preferred way to drive an LED using a TTL (transistor-transistor logic) inverter (7404).

Solution The schematic diagram is shown in Fig. 9-20a. A logic 1 applied to the TTL inverter causes its output to go to the logic 0 or low state. In this condition, the TTL output looks like a saturated transistor and will *sink* the 20-mA LED current. The LED is ON, indicating that the input "sees" a logic 1. Resistor R is needed to limit the LED current. Assuming $V_{CE(sat)}$ of the TTL gate is approximately 0.2 V,

$$R = \frac{5 - 1.6 - 0.2}{0.02} = 160\ \Omega$$

Figure 9-20b shows an alternate way of driving the LED that is *not* preferred. Now the logic gate must *source* the 20 mA. The problem is that the TTL output

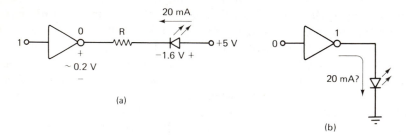

(a)

(b)

Figure 9-20 (a) Preferred method of driving an LED with a TTL gate. (b) this connection is not preferred because the gate may not be able to source the 20-mA LED current.

stage may not have the necessary drive capability when connected in this manner because it is specified to source only 40 μA at 2.4-V output in the high state.

Actually, the connection in Fig. 9-20a is also exceeding the rated gate specifications (16-mA maximum sink current). However, most TTL gates are conservatively rated and should handle this slight overload. If not, a higher current gate such as the 7406 could be used or two 7404's could be connected in parallel.

Often, LEDs are connected in a *seven-segment* format as shown in Fig. 9-21. Each segment is assigned a letter from a through g. By turning on combinations of these segments, the decimal numbers 0 to 9 can be displayed.

When using seven-segment displays, it is important to note whether they are *common-cathode* (Fig. 9-21b) or *common-anode* (Fig. 9-21c).

Some LED displays use a *dot matrix* concept, which allows alpha and numeric digits to be displayed. In all cases, a *digital-decoder* circuit is required to activate the proper LED segments or dots.

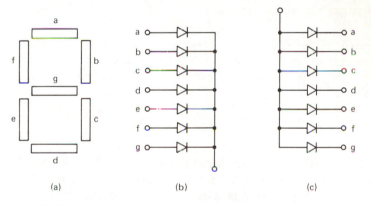

(a)   (b)   (c)

Figure 9-21 LEDs may be connected in a seven-segment format (a) in which all cathodes are common (b), or all anodes are common (c).

Example 9-6

Illustrate how a transistor can be used to turn an FND 503 seven-segment common-cathode LED display ON or OFF (a common requirement in *multiplexed* displays).

Solution The schematic diagram is shown in Fig. 9-22. An *n*-channel VMOSFET is used to sink the individual segment currents. At 20 mA/segment, this amounts to 140 mA when all seven segments are ON. Closing the switch turns OFF the transistor by grounding the gate and breaks the path from cathodes to ground through the VMOSFET. This is an ideal application for the VMOSFET, with its high-current-handling capability compared to conventional MOSFETs.

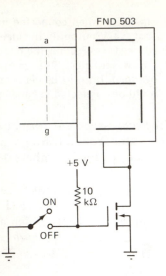

Figure 9-22 ON-OFF control of an FND 503 seven-segment display using a VMOSFET.

9.4 LIGHT SENSITIVE DEVICES

Light sensitive devices act as *photodetectors.* Semiconductor photo-detectors consist of *photocells*, whose resistance varies with the light intensity, and *photojunction* devices, which generate currents in response to the incident light.

Photocells

If a piece of pure silicon is connected to a voltage source a current will flow governed by *Ohm's Law.* This current will depend directly on the number of *free electrons* available to act as charge carriers in the silicon semiconductor. One way to increase this number is by the application of heat. Of course, this leads to the general thermal instability problems characteristic of all semiconductor devices.

The application of *light* is another way of increasing the free electron concentration. Light can be thought of as tiny particles of energy called *photons.* As these photons strike the valence electrons, some electrons may receive enough energy to leave the valence band and move to the conduction band. Because a hole is left behind in this process, the effect is identical to that of heat, namely, the creation of electron-hole pairs.

Externally, this effect is detected as a decrease in resistance between the terminals of the photocell. For this reason, photocells are often referred to as *photoresistive cells* or *photoresistors.*

Because a more intense light can be thought of as a greater number of light photons striking the surface in unit time, this results in a correspondingly greater *decrease* in cell resistance. Typically, dark-to-light resistance variations might be 100 kΩ in the dark and 10 kΩ in normal room light.

Although silicon could in principal be used as a photocell material, in practice it is not. This is because in silicon there are already a great number of thermally-generated electron-hole pairs available at room temperature, even in the dark. This has the effect of *swamping out* any increases due to incident light photons. For this reason, photocells generally use semiconductor materials with energy band gaps greater than silicon's 1.1 eV. Typical materials are cadmium sulfide (CdS) and cadmium selenide (CdSe).

Another interesting point is that the light photon's energy depends on its frequency and therefore the *color* of the incident light. CdS is most responsive to green-yellow light, while CdSe is sensitive to red light.

Example 9-7 ——————————————————————————

Illustrate a "night light" circuit that will activate an ac-operated light at darkness.

Solution The circuit is shown in Fig. 9-23. This circuit is nearly identical to Fig. 9-16 except for the addition of the photocell. During daylight hours the low resistance of the photocell prevents the capacitor from charging sufficiently to trigger the DIAC, and the TRIAC and light are both off. As darkness occurs the resistance of the cell rises, eventually allowing normal TRIAC triggering. Resistor R_1 allows for adjustments of the darkness triggering level.

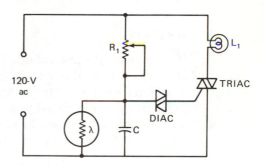

Figure 9-23 Darkness-activated night light. The low daylight resistance of the photo-cell prevents triggering of the DIAC until nightfall.

Photodiodes

Photojunction devices include the *photodiode*, *phototransistor*, and *photo SCR* also called the light-activated SCR (LASCR). All of these devices operate from a common principal.

Let's consider the photodiode first. When a *pn* diode is reverse-biased, current is limited to a leakage value due to the few minority carriers present (holes in *n*-material or electrons in *p*-material) that actually see this reverse bias as a forward bias. However, as we have just seen, light *photons* of the proper energy can create electron-hole pairs in a semiconductor. Because these electron-hole pairs are *exactly* the source of leakage current, an increased reverse current flow results in a photodiode subjected to light.

It is interesting to note that the photodiode is essentially the *opposite* of an LED. The LED, when *forward-biased*, emits light of a color dependent on the semiconductor material. The photodiode, when *reverse-biased*, exhibits an increased current flow dependent on the incident light color and intensity. In fact, when the LED is reverse-biased, it will act as a photodiode sensitive to light of the same color that it emits.

It is also interesting to remove all bias from the photodiode and illuminate its *pn* junction. Now, even though there is no reverse bias, electron-hole pairs will still be created. Any holes created in the depletion region will be swept into the *p* material by the built-in-voltage of the junction. Similarly, electrons created will be swept into the *n* material. The consequence of this is that the *n* side of the diode charges negative while the *p* side charges positive. If a resistor is connected across the diode, a current will flow. The photodiode is now acting as a *photovoltaic* cell or *solar* cell.

The typical photodiode has a poor efficiency when used as a solar cell, but when constructed with a large surface area and combined in series and parallel with other cells, relatively large power outputs are possible.

Phototransistors

One problem with the photodiode is that its photocurrent is usually quite small, generally in the microamp range. Although an *external* transistor could be used to amplify this current, we might note that when a bipolar transistor is biased in the active region, its base-collector junction is reverse-biased. We could use this junction as a photodiode, generating an effective base current for the transistor. The collector current would then be $I_C = (I_B + I_p)\beta$, where I_B is any external base current and I_p is the light or photon current. A phototransistor is illustrated in Fig. 9-24. When the light levels are quite low, it may be

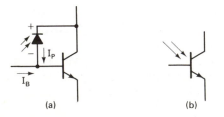

(a)   (b)

Figure 9-24 (a) Phototransistor with collector-base junction emphasized. (b) the schematic symbol.

desirable to use a *photodarlington*. The collector current is then approximately $(I_B + I_p)\beta^2$.

An interesting application of the phototransistor is the "flashlight communicator" shown in Fig. 9-25. When S_1 is in the receive position, incident light modulates the collector voltage of the phototransistor. This signal is in turn amplified by the common-emitter amplifier Q_2 and phase split by Q_3. Finally, Q_4 and Q_5 are complementary transistors operating in push–pull to drive the speaker.

In the transmit mode, the speaker acts like a microphone and its output follows the same path traced by the incoming signal. Now the complementary transistors modulate the *brilliance* of the flashlight bulb in accordance with the voice input. Intelligible conversations up to 100 ft have been reported!

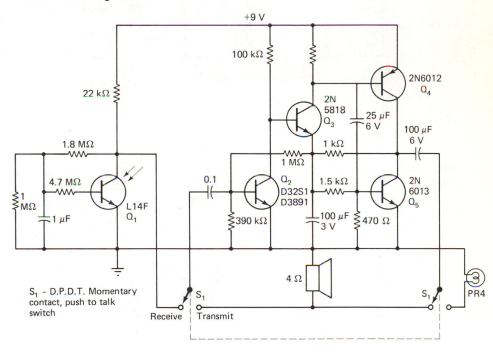

Figure 9-25 Flashlight communicator. (Courtesy of General Electric Company Semiconductor Products Department.)

Light-Activated SCRs

The light-activated SCR (LASCR) is similar to a normal SCR accept that gate current is supplied by incident light photons. When the anode to cathode of an SCR is biased at a voltage below $V_{(BO)}$, the middle *pn* junction (recall Fig. 9-1a) will be reverse-biased. This junction will be receptive to light and the resulting photon current can act as a gate current, triggering the SCR into conduction.

As might be imagined, there are numerous applications of such devices. One typical example allowing control of ac devices with full isolation is shown in Fig. 9-26. In this case, holding the beam of light on the two LASCRs will allow the load to conduct, while removal of the light will turn the load OFF. Applications might include automatic garage-door openers, remote control of various ac appliances, or remote television channel selection.

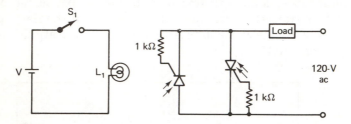

Figure 9-26   A fully isolated ac switch. When L_1 is ON, the load conducts.

9.5 OPTO-COUPLERS

The *opto-coupler* or *opto-isolator* is a relatively new member of the growing list of optoelectronic devices. Typically, the opto-coupler consists of an *infrared-emitting diode* (IRED) and *infrared-detecting phototransistor* (or photodarlington, photo-SCR, photo-TRIAC) both in the same six-pin dual-in-line integrated circuit package. The schematic representation is shown in Fig. 9-27.

When the IRED is biased to conduction (typically 10 to 20 mA), the infrared light emitted by the diode is detected by the phototransistor, turning it ON in proportion to the light intensity. For example, for the TIL 156, 4 mA of diode current will cause 40 mA of collector current, and 10 mA of diode current will cause 100 mA of collector current.

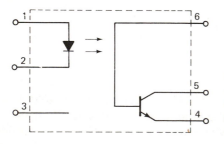

Figure 9-27 The opto-coupler consists of an infrared-emitting diode and photo-transistor. It is packaged in a six-pin dual-in-line package, with pin numbers as shown.

One of the primary advantages of the opto-coupler is the *isolation* between input and output circuits due to the light path between IRED and phototransistor. There is no necessity for a *common ground*.

Isolation is particularly important when low voltage dc circuits are used to control higher voltages and currents (particularly 120-V ac). Without the opto-coupler, a fault in the high-voltage circuit could possibly destroy the low-voltage control circuitry.

Two examples illustrating the opto-coupler's versatility are presented.

Example 9-8

Design a circuit to interface TTL (transistor-transistor logic) to an ASR-33 teletype (TTY) using a *20-mA current loop*.

Solution Although most computer terminals communicate using logic levels of +12 V and –12 V (*RS-232C*), there are a number of older teletype terminals using *current loops*. In this system, a logic 1 is represented by 20 mA of current flowing through a relay coil and a logic 0 as no current. Typical TTL output voltages are 0 V (logic 0) and 3 V (logic 1). A circuit that will convert from TTL voltage levels to 20-mA current loop is shown in Fig. 9-28.

The loop consists of the series connections of Q_3, Q_1, D_2, and the TTY printer and keyboard. Q_3 is biased in the active region as a current source to establish the 20 mA. Incoming data (labeled "serial out") turns D_1 ON and OFF, causing Q_1 to open and close the loop in unison with the digital data. These current pulses in turn cause the TTY printer to activate and type the proper characters.

In the receive mode, the TTY keyboard opens and closes the loop. This causes D_2 to turn ON and OFF. Q_2 converts the current pulses to TTL-compatible output voltages (labeled "serial in").

Note that the digital circuitry is completely isolated from the TTY and the current loop. All transmission between the two is via the light path within the opto-couplers.

The next example illustrates how the opto-coupler can be used to pull together the versatility of TTL digital control circuits with the high-current-switching capabilities of the TRIAC.

Example 9-9

Design a programmable timer circuit that will activate an ac load for a prescribed length of time in 1-s intervals. Use a photocell input.

Solution The circuit is illustrated in Fig. 9-29. The photoresistor (R_P) forms a voltage divider with variable resistor R_1. When covered, the resistance of R_P increases, placing a low voltage on the input of the first one-shot (74123A). This enables the one-shot to be triggered by pulses from the LM555 square-wave generator. Because the one-shot is retriggerable, its output goes high but never has a chance to time out due to the constant stream of input pulses from the 555.

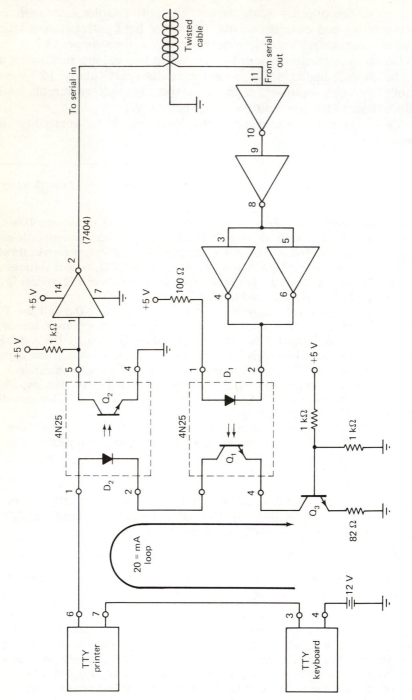

Figure 9-28 TTL-20-mA current loop interface.

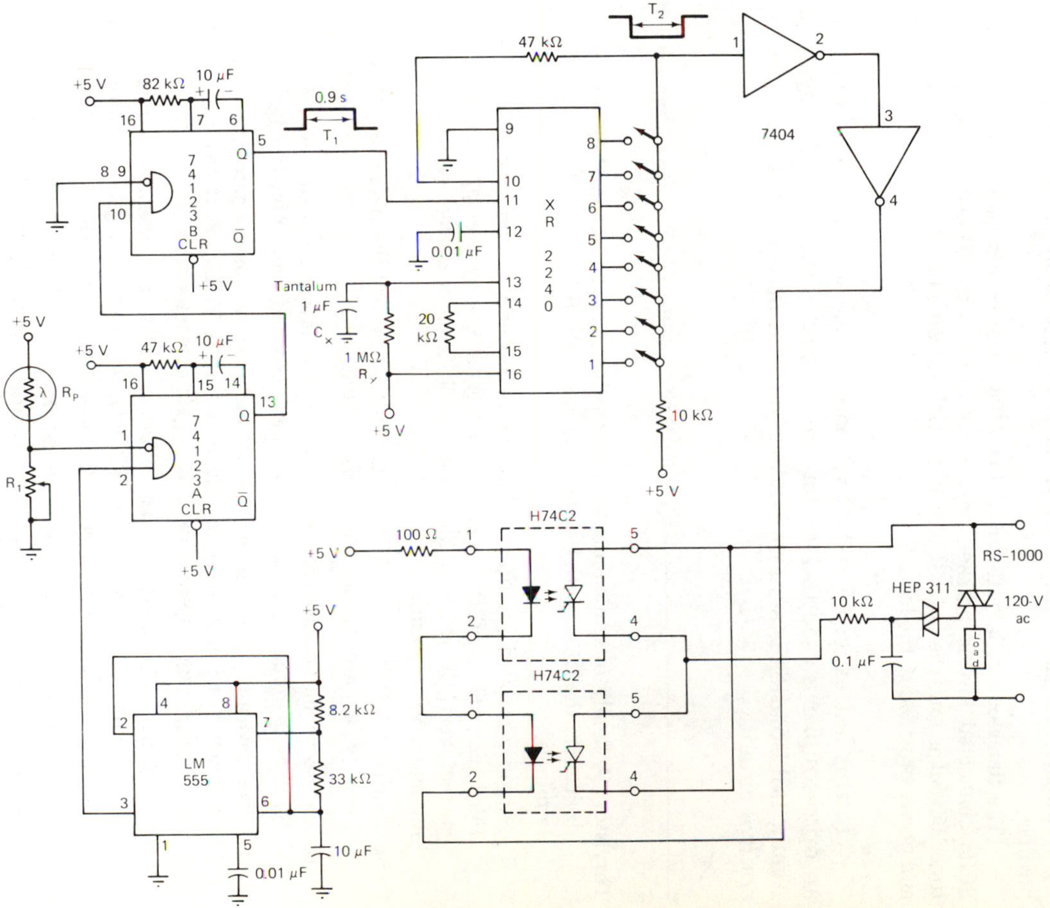

Figure 9-29 Opto-coupled programmable timer.

The rising edge of the first one-shot's output triggers the second one shot, which produces a brief 1 to 2-s pulse, activating the XR-2240 programmable timer. This circuit has a programmable time delay established by the eight SPST switches, connected as shown in the figure. Once activated, the output of the timer will go low for a specific time delay established by R_x and C_x and the binary value of the eight switches. For the example shown, accurate time delays from 1 to 255 s are possible.

The timer output is buffered by the 7404 inverters and then drives two opto-SCRs connected inverse parallel (equivalent to an opto-TRIAC). As long as the timer output is low, the opto-SCRs will be ON, allowing the TRIAC to be triggered and power provided to the load.

In this last example, a TRIAC handles the actual load current but an opto-coupler is used to facilitate dc control of the TRIAC triggering. Again, full isolation is achieved between the ac *power* circuit and the dc *control* circuit.

─────────────────────── KEY TERMS ───────────────────────

Thryistor   A family of semiconductor *pnpn* switching components. Representative devices are the DIAC with two terminals, and the SCR, and TRIAC with three terminals.

SCR   Silicon-controlled rectifier. Similar to a conventional *pn* diode but with a third terminal called the gate. Once triggered ON by the application of gate current, the SCR remains latched on until its anode current falls below some minimum holding value.

Phase Control   SCRs and TRIACs can be used to control the phase or portion of the ac cycle, during which the load conducts.

TRIAC   Also called bidirectional trigger diode, this device is similar to the SCR but allows conduction for both voltage polarities when appropriately triggered.

DIAC   This device is similar to the TRIAC but does not have a gate terminal. It is commonly used as a triggering element for the TRIAC as it ensures smooth TRIAC triggering despite variations in gate characteristics.

Solid-State Relay   This circuit replaces the conventional mechanical relay and moving contacts with an SCR or TRIAC and appropriate gate control circuit. Advantages include no moving parts, fast switching, and silent operation.

LED   Light-emitting diode. This semiconductor diode will emit light of a frequency proportional to its bandgap when forward-biased. Typical LEDs are made from gallium-phosphide and emit red light.

Photocell   This is a two terminal nonjunction semiconductor device. Its resistance decreases when exposed to light due to the generation of electron-hole pairs.

Photojunction Devices   These devices employ a reverse-biased *pn* junction. The leakage current of this junction depends on the intensity of the incident light. The resulting leakage current may act as a base current for a photo-transistor or gate current for a photo-SCR or photo-TRIAC.

Opto-Coupler   This is a six-pin integrated circuit employing an infrared photodiode and photojunction device. Main circuit applications involve isolation between the input and output circuits.

QUESTIONS AND PROBLEMS

9-1   Indicate two *different* ways that a TRIAC or SCR can be triggered into conduction.

9-2   When a TRIAC is used with ac, it is automatically turned OFF each half cycle. Explain why this is so.

9-3   A certain SCR is rated at 250 V, 5 A. Explain what this means. Could this SCR be used to control a 120-V rms ac device?

9-4   Explain the operation of the circuit in Fig. 9-5 if the SCR is replaced with a TRIAC and V_S = 12 V rms ac. Will L_1 burn as brightly as it does with an SCR?

9-5   For the circuit in Fig. 9-8, assume I_{GT} = 15 mA, V_{GT} = 1.5 V, R_1 = 1 kΩ, and the transformer is rated at 6.3 V rms. If the pot is adjusted to its center, at what voltage will the SCR trigger ON?

9-6   The circuit shown in Fig. 9-9 will not function if a filter capacitor is used across the rectified output. Explain.

9-7   Assume the zener diode in Fig. 9-9 is rated at 5.1 V and R_3 is adjusted so that 300 Ω remains across the capacitor. At what battery voltage will the circuit stop charging?

9-8   Indicate the *minimum–maximum* smooth conduction angles possible for circuits (a) through (c) in Fig. 9-30.

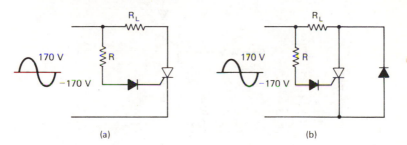

(a)   (b)

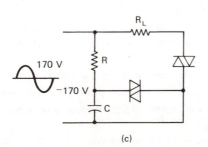

(c)

Figure 9-30

9-9 Refer to Fig. 9-30a. If $R = 68$ kΩ, $R_L = 1$ kΩ, $I_{GT} = 1$ mA, $V_{GT} = 1.25$ V, and $I_H = 30$ mA, determine the conduction angle.

9-10 Explain how *phase lag* can be introduced into an SCR triggering circuit to allow conduction angles less than 90°.

9-11 Refer to Fig. 9-15; which TRIAC triggering modes are used for this circuit?

9-12 Explain why the circuit in Fig. 9-16 will have more reliable triggering than the circuit in Fig. 9-15.

9-13 An SCR is used to control the current through a 120-V 500-Ω ac load. It is desired to have variable triggering from 10 to 90° on the ac waveform. Show the schematic diagram and determine a suitable value for the variable resistance. Assume $I_{GT} = 5$ mA, $V_{GT} = 1.5$ V, and $I_H = 10$ mA.

9-14 In Problem 9-13, what are the maximum and minimum conduction angles possible? Sketch the voltage waveforms across the SCR and load resistance for both cases.

9-15 Refer to Example 9-3; what value of resistance is needed for a 120° conduction angle?

9-16 Ultraviolet light frequencies are too _____ to be seen by the human eye, while infrared frequencies are too _____.

9-17 A transistor LED driver circuit is illustrated in Fig. 9-31. If $V_{CE(sat)} = 0.2$ V, determine values for R_1 and R_2 such that $I_F = 20$ mA. Assume $V_F = 1.6$ V.

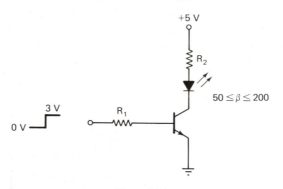

Figure 9-31

9-18 Determine which segments must be lit in a seven-segment display to indicate the numerals 0, 2, 6, and 9.

9-19 Hexadecimal numbers use the letters A through F to represent decimal numbers 10 to 15. Show how these letters can be represented in a seven segment format.

9-20 What is the main difference between a photodiode and an LED?

9-21 How does a photocell differ from a photojunction device?

9-22 What is the effect of increased light intensity on a photocell, photodiode, and phototransistor.

9-23 A phototransistor is to be used to drive a 360-Ω 10-mA relay coil. Assume the transistor will saturate when illuminated, with $V_{CE(sat)} = 0.2$ V. Show a schematic diagram using a 12-V dc source.

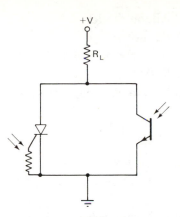

Figure 9-32

9-24 Explain the operation of the circuit in Fig. 9-32.

9-25 Show the schematic diagram of a solid-state relay similar to Fig. 9-15 but using an opto-coupler TRIAC driven by a TTL logic inverter. Assume the IRED requires 20 mA at 1.2 V. A logic 1 input should turn on the relay.

9-26 Refer to Fig. 9-28. Calculate the collector current of transistor Q_3. Why is this circuit considered a *current source*?

9-27 If the photocell in Fig. 9-29 varies from 1.5 kΩ to 50 kΩ (light to dark), to what value should the 10-kΩ potentiometer be adjusted if the 74123 will trigger with $V_i \leqslant 0.8$ V?

9-28 Calculate the IRED current for the opto-SCRs in Fig. 9-29. Assume $V_F = 1.2$ V.

_____   LABORATORY ASSIGNMENT 9:   _____
THYRISTORS AND OPTO-COUPLERS

Objectives

1. To compare and contrast the SCR and TRIAC
2. To measure and observe phase control waveforms
3. To gain experience with the opto-coupler

Introduction   In this laboratory assignment, you will assemble several SCR and TRIAC *control circuits* using ac and dc line voltages. Various types of phase control circuits will be built and conduction angles measured. An example of a dc motor speed control circuit will be assembled. Finally, a light-coupled amplifier and solid-state relay using opto-couplers will be tested.

Components Required

1 100-V 1-A SCR   1 10-μF capacitor
1 100-V 1-A TRIAC   1 0.002-μF capacitor

2 6.3-V lamps   4 silicon diodes

1 PBNO   1 dc motor

1 1-kΩ pot   1 opto coupler (4N25)

1 10-kΩ pot   1 7404

                             Miscellaneous capacitors and resistors

Part I: *SCR and TRIAC Triggering*

STEP 1   Set up the circuit shown in Fig. 9-5 using 6.3-V dc with the polarity shown. Depress and release the switch. Repeat with the power supply leads reversed.

STEP 2   Repeat step 1 with the SCR replaced by a TRIAC.

STEP 3   Repeat steps 1 and 2 using 6.3-V ac (it is not necessary to reverse the power supply leads in this case).

Question 1   Fill in the chart below, with bright, dim, or OFF for the state of the lamp.

Supply Voltage and Polarity	SCR	TRIAC
dc(+)		
dc(−)		
ac		

Question 2   For which cases was the lamp latched on? How did you turn it off? Explain.

Question 3   In step 3, why does the lamp burn *brighter* when the TRIAC is used?

Part II: *Phase Control*

A. SCR

STEP 1   Set up the circuit shown in Fig. 9-12 using 6.3-V ac and a 6.3-V lamp for R_L.

STEP 2   Apply power to the circuit and determine the *minimum* and *maximum* conduction angles. Record the waveforms across the lamp and SCR for the case of 90° conduction.

Question 4 What is the purpose of diode D_2? (*Hint:* Try removing it from the circuit and see what happens).

STEP 3 Add a diode in parallel with the SCR but with its anode connected to the SCR cathode. What affect does this have on the *maximum* and *minimum* conduction angles?

B. TRIAC

STEP 1 Assemble the TRIAC phase control circuit illustrated in Fig. 9-16. Use 12.6-V ac, a 10-kΩ pot for R_1, 0.002 μF for C_1, and two 6.3-V lamps in series for R_L. If a DIAC is unavailable it may be replaced with a 270 Ω resistor.

STEP 2 Adjust the pot and observe the minimum and maximum conduction angles. (You may find that the TRIAC does not trigger *evenly* on the positive and negative half cycles. This is why a DIAC is often used in this circuit). Be sure you can adjust the circuit to full ON and OFF (you may need to change the pot or capacitor values).

STEP 3 Replace the lamps with the *full-wave rectifier* circuit shown in Fig. 9-33. Observe the waveform across the 1-kΩ load resistor as the 10-kΩ pot is varied.

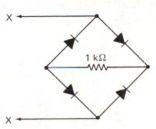

Figure 9-33 Full-wave rectifier circuit used to replace the two series lamps in step 3, Part II B.

STEP 4 If a small dc motor is available, use this in place of the 1-kΩ resistor. Again, observe the waveform across this load. Try bridging the motor with a large capacitor (500 to 1000 μF). Explain the result.

Part III: *Opto-Couplers*

A. Light-Coupled Amplifier

STEP 1 Using the circuit in Fig. 9-34 as a guide, calculate resistor values for the light-coupled amplifier shown. Bias the IRED at 10 mA, $I_C = 1$ mA (with IRED OFF), and assume $β = 100$.

STEP 2 Measure and record the dc bias voltages for your circuit *with* and *without* the IRED biased ON. Compare with calculations and explain any discrepancies.

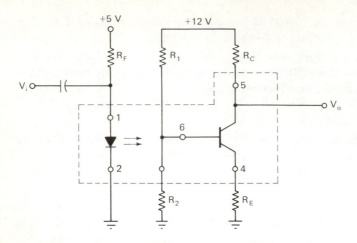

Figure 9-34   Light coupled amplifier using an opto-coupler.

STEP 3   Apply a small ac signal ($f = 1$ kHz) to the base of the photo-transistor through a suitable coupling capacitor. Measure A_V. Use an emitter bypass capacitor to achieve a voltage gain greater than 5.

STEP 4   Now apply the ac input to the IRED through a coupling capacitor and again measure A_V. Compare to your results in step 3.

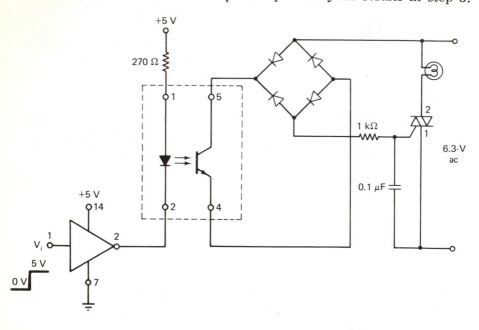

Figure 9-35   Solid-state relay circuit used for Part III B.

(*Note:* The IRED is a *forward-biased diode*, which means a large coupling capacitor will be required. Measure the gain using V_i at the IRED terminal).

Question 5   How do you account for the differences in gain measured in steps 3 and 4?

STEP 5   Measure the frequency response of the amplifier using *capacitive coupling* (step 3) and *light coupling* (step 4) from 100 Hz to 1 MHz. Graph on a common axes using semilog graph paper.

Question 6   What do you think will happen to the light-coupled amplifier if the IRED is not biased ON? Remove the 5 V and try it.

B. Solid State Relay

STEP 1   Set up the circuit shown in Fig. 9-35 using one gate in the 7404 hex inverter.

STEP 2   Note the condition of the load when $V_i = 0$ V; $V_i = 5$ V.

Question 7   Why is the bridge rectifier circuit necessary?

TEN

INTEGRATED CIRCUIT TECHNOLOGY

Most of us have come to take the seemingly day-to-day breakthroughs in electronics for granted. For this reason, I think it is appropriate to reflect on the amazing succesion of solid-state devices to come about since the birth of the transistor in 1947.

A brief history is presented in Fig. 10-2. Considering that the vacuum tube was the *only* active device available to designers for nearly 50 years, this list is truly impressive.

In this chapter we discuss the *integrated circuit* (IC), which came into being only 11 years after the very first transistor! In fact, the transistor was only in the limelight during the decade of the 1950's, having to give way to the integrated circuit in the early 1960's. Yet the integrated circuit is only the logical evolution of fabricating one transistor at a time, to the mass production of thousands of transistors all at once on a single piece of silicon.

This chapter explores the basic processes involved in fabricating integrated circuits. We discuss which components may or may not be readily integrated and put these together to form the differential pair and finally the fully-integrated amplifier or *operational amplifier* (op-amp), which is covered in the last two chapters.

10.1 THE PLANAR PROCESS

Figure 10-1 not withstanding, the integrated circuit is not manufactured by tiny little men with jack-hammers! In fact, the process is quite complex and delicate and not without its share of "black magic."

But before we begin, what exactly is an *integrated circuit* (IC)? We

Figure 10-1 January 1979 *Byte* cover.
(Reprinted courtesy Robert Tinney Graphics
and Byte Publications, Peterborough, N.H.)

might define it as a single chip or *die* of silicon (monolithic silicon) containing hundreds and perhaps thousands of individual transistors and resistors interconnected to perform some useful circuit function. In most cases we cannot actually see the silicon die and therefore tend to

1947 — First transistor (Bell Laboratories)

1953 — Unijunction transistor (General Electric)

1954 — Silicon junction transistor (Texas Instruments)
     — Commercial zener diodes (National Semiconductor)

1956 — Silicon-controlled rectifier (General Electric)

1957 — Tunnel diode (Sony)

1958 — Field-effect transistor (Teledyne Crystalonics)
     — First monolithic integrated circuit (Texas Instruments)

1960 — Plannar process (Fairchild)

1962 — MOSFET transistor (Fairchild)

1963 — Transistor-transistor-logic family (Texas Instruments)

1964 — 702 operational amplifier (Fairchild)

Figure 10-2 Some important dates in the history of electronic devices.

associate the IC with a particular package and associated leads. A typical such package is the 14-pin dual-in-line package (DIP) most of us are already familiar with, illustrated in Fig. 10-3. The actual silicon die is shown as a dashed line in this figure.

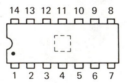

Figure 10-3 The 14-pin dual-in-line package (DIP). The dashed lines represent the relative silicon die size.

A Typical Processing Sequence

From the transistor's point of view, visualize an immense, flat, two-dimensional area of silicon. Into this area we must locate individual transistors consisting of alternate n and p type layers of silicon. The basic technique used is common to all integrated technologies and employs *photolithography* to define the individual bases, emitters, and collectors and a *diffusion process* to determine the doping type (p or n) and density.

Figure 10-4 illustrates the sequence of events in a typical processing step, for example, diffusing the emitters into the bases of all transistors. Each step is explained as follows and keyed to Fig. 10-4.

(a) The n- or p-type silicon is placed in a steam environment at 900 to 1200°C, causing a chemical reaction to occur and the subsequent growth of an SiO_2 (silicon dioxide) insulating layer over the

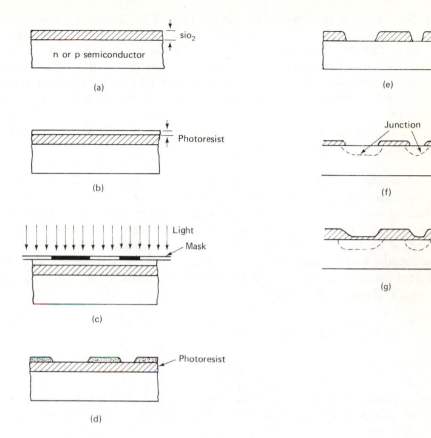

Figure 10-4 The typical sequence of events required to produce a *pn* junction in an integrated circuit. Refer to the text for an explanation of each step (a)-(g).

entire wafer of silicon. Alternately, a vapor-depositing method may be used.

(b) In preparation for the photography to follow, a thin layer of *photoresist* is spread over the wafer and allowed to harden.

(c) A glass plate called the *photo mask*, containing the specific transistor geometry to be etched, is placed over the wafer and the combination is briefly exposed to ultraviolet light.

(d) Those areas of photoresist exposed to light will *develop* and harden while those areas protected from the light by the mask will not. A subsequent dipping of the wafer into an etching solution will remove the *unexposed* photoresist.

(e) The wafer is now placed in a solution capable of etching the SiO_2 layer, and "windows" are opened, exposing the *n*- or *p*-type silicon layer below.

(f) The wafer is now ready for diffusion and is placed in a carefully controlled oven at 1000 to 1300°C with the selected doping impurity. The SiO_2 acts as a *mask* to the dopant, causing *pn* junctions to be formed only in the selected windows. Diffusion takes place over a period of time with deeper junctions resulting from longer diffusion times (refer to Section 1.3 for a review of semiconductor doping).

(g) Finally, a layer of SiO_2 is again grown over the wafer and the processing step is complete.

This *planar* fabrication process has many advantages. All transistors are fabricated at once and, because each mask may contain hundreds of individual circuits, the process is ideal for mass production. All dimensions are determined photographically, allowing extremely small individual components and the resulting miniaturization we have come to associate with integrated circuits. Finally, the SiO_2 insulating layer acts as a protection for the completed circuit, resulting in low leakage currents and high reliability.

10.2   A TYPICAL BIPOLAR PROCESS

The processing steps of the previous section are repeated for each new *p* or *n* diffusion and in a typical *bipolar* process may require as many as *seven* different masks. This requires very careful alignment from mask to mask and accurate documentation as the wafers proceed through the wafer fab area.

Figure 10-5 illustrates the top view and cross-sectional view of a bipolar transistor with collector pull-up resistor. Each of the individual junctions formed in this circuit follows the technique presented in the previous section. For example, the base mask is illustrated in Fig. 10-6.

As shown in Fig. 10-5, the IC is fabricated on a *substrate* of *p*-type silicon. The substrate's main purpose is to provide a solid mechanical base so the wafer can be handled without breakage. It is connected to the most negative circuit potential (usually ground) to maintain a reverse bias.

The *first* mask defines the n^+ *buried layer*, which is diffused into the *p*-type substrate. This layer will carry the collector current and the heavier doping provides a low collector resistance.

An *epitaxial n-type* layer is next grown over the entire surface by placing the wafer in a special *epitaxial reactor*. This *epi* layer will become the collector into which the bases and emitters are later diffused.

Because the epitaxial layer is common to all transistors, at this point all collectors are *shorted* together. It is therefore necessary to

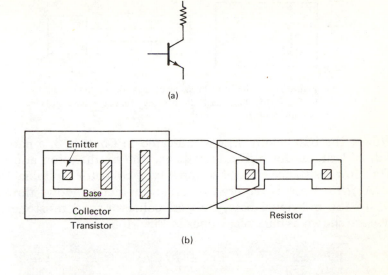

(a)

(b)

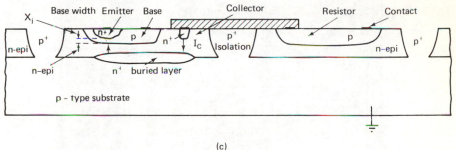

(c)

Figure 10-5

diffuse a *p-type isolation diffusion* through the epitaxial layer to isolate the individual transistors. This is the *second* mask.

The *third* mask defines all transistor bases and resistors. This is a *p*-type diffusion into the *n*-type collector. Note that a resistor is simply a base with contacts at both ends. Its value is determined by

$$R = \frac{\rho l}{A} \tag{10-1}$$

where ρ is the resistivity of the base diffusion, l is the resistor length, and A is the cross-sectional area. Large-valued resistors must be made long and narrow.

The *fourth* mask is the emitter diffusion (n^+) and collector contact diffusion. Note that the deeper the n^+ diffusion into the base, the

Figure 10-6 The third mask or base mask for the circuit
in Fig. 10-5. This mask defines the transistor base and
pull-up resistor.

narrower the base width, X_j. As indicated in Chapter 3, a narrow base
width is desirable for high β transistors. The n^+ diffusion also helps to
make a low resistance contact to the n-type collector (epitaxial layer).

A *fifth* mask is required to open contact windows through the
SiO_2 to the individual base, emitter, collector, and resistor contacts.
These are shown as dark black lines in Fig. 10-5.

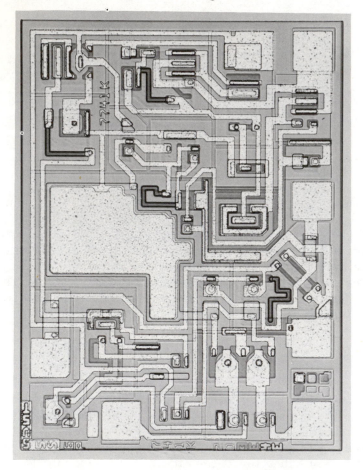

Figure 10-7   The μA 741 operational amplifier.
(Courtesy Fairchild Camera and Instrument Corporation)

The *sixth* mask is similar to the metal traces on a conventional printed circuit board. Aluminum is evaporated onto the wafer and selectively etched to define the proper interconnection of all components.

Sometimes the complete wafer is covered with a layer of glass (vapox) to protect the easily scratched aluminum. In this case, a *seventh* mask is required to open windows to the circuit *bonding pads* for the external package leads.

Figure 10-7 is a photograph of the popular 741-type *operational amplifier*. The bonding pads and aluminum metalization should be apparent. Closer scrutiny will reveal the individual transistors. Resistors are identified by the long snake-like patterns with contacts on each end.

10.3   OTHER INTEGRATED COMPONENTS

One unique characteristic of the integrated technology is that not all components can be integrated, and even those that can may not be compatible with existing processes. For example, can we make *pnp* transistors side by side with *npn* transistors?

pnp Transistors

For a number of reasons, *npn* transistors dominate electronic circuits today. They work off positive supply voltages, provide positive output voltages, and have improved frequency response compared to *pnp* transistors. For these reasons, the planar process has been refined for producing high performance *npn* transistors. However, some analog circuits require *pnp* transistors, *complementary output stages* for example. It is therefore desirable to be able to fabricate *pnp* transistors along with *npn* transistors.

Figure 10-8 illustrates two types of *pnp* transistors that can be fabricated using the conventional *npn* bipolar process. A *lateral pnp* transistor is shown in Fig. 10-8a. In this transistor, the emitter injects holes into the *n*-type base (epi region), which travel *laterally* (instead of *vertically*, as in the conventional *npn*) to the collector. The main problem with this device is the relatively large base width, limiting the value of β to a typical 5 to 50 range. This also results in poor frequency response and switching times.

A second method of fabricating a compatible *pnp* transistor is the *substrate pnp* shown in Fig. 10-8b. This device has a *vertical* current path and the base width is determined by the epitaxial thickness and *p* diffusion. Although higher values of β are possible compared to the lateral structure, the collector must be connected to the substrate (typically ground), limiting the usefulness of this transistor.

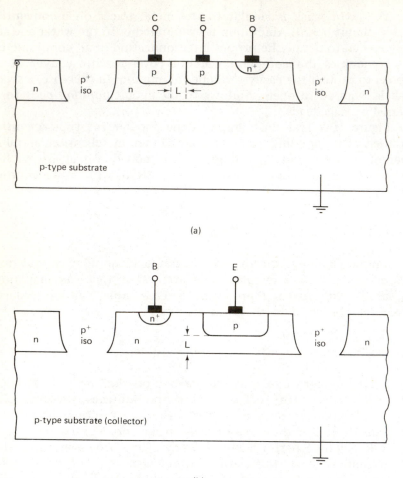

Figure 10-8 (a) In the lateral *pnp* transistor, current flows laterally through the base from emitter to collector. (b) the substrate *pnp* transistor has its collector at ground potential. *L* identifies the base width for both transistors.

It is possible to fabricate high performance *pnp* transistors but at the expense of added processing steps and complexity. For these reasons it is rarely done.

MOS Transistors

The MOS transistor structure was previously presented in Fig. 8-9. A study of this figure should reveal that the integrated circuit processing of this transistor is much simpler than the corresponding bipolar process. Only a *single diffusion* is necessary (for source and drain), and no buried

layer or epitaxial growth is required. In addition, one of the most significant advantages is that no *isolation diffusion* is required. This means
MOS transistors will require significantly less chip area and allow the
fabrication of more complex circuits (more transistors for a given die
size) than the bipolar technology. Indeed, most large scale integrated
(LSI) devices such as *microprocessors, computer memories,* and *electronic calculator chips* use MOS transistors.

   In general, the MOS and bipolar technologies are not compatible
from an IC processing standpoint. However, an increasing number of
devices are appearing using MOS and bipolar transistors on the same
chip (biMOS technology), indicating that this problem is being solved.

Capacitors

   As mentioned in Chapter 8, the MOS transistor is exactly a capacitor when looking into the gate terminal. An example of a metal-insulator-
semiconductor (MIS) capacitor is shown in Fig. 10-9. This device is
compatible with the conventional *npn* planar process and provides a
capacitance given by

$$C = \frac{\epsilon_r \epsilon_o A}{d} \tag{10-2}$$

where ϵ_r = relative dielectric constant of the insulator (2.7 to 4.2 for
           SiO_2)
       ϵ_o = permitivity of free space (8.85 $\times$ 10^{-14} F/cm)
       A = area in cm^2
       d = distance between the plates in cm

   The metal gate forms one plate while the n^+ diffusion forms the
other. The SiO_2 insulating layer acts as the dielectric.

Example 10-1 ──

Determine the value of a capacitor with an area 1.7 $\times$ 10^{-3} cm^2, distance between
plates of 1.5 $\times$ 10^{-5} cm, and dielectric constant of 3.0.

Solution   Plugging into Eq. 10-2,

$$C = \frac{(8.85 \times 10^{-14})(3)(1.7 \times 10^{-3})}{1.5 \times 10^{-5}} = 30 \text{ pF}$$

   To put this capacitor into perspective, a 30-pF capacitor is shown
in Fig. 10-7 as the large area in the center of the photograph. In
fact, in this case the 1.7 $\times$ 10^{-3} cm^2 capacitor area represents about
10% of the *total* chip area! And that is the point; although capacitors

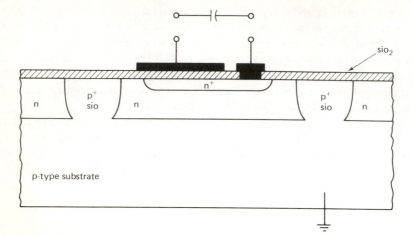

Figure 10-9 The metal insulator semiconductor (MIS) capacitor. One plate is the n^+ diffusion and the other plate is aluminum. SiO_2 is the insulating dielectric.

can be integrated, they require a great deal of chip area for relatively small capacitance values. For this reason, capacitors are infrequently designed into integrated circuits.

Summary

The development of the integrated circuit has required a rethinking of standard design practices. Previously, the *number* of components was kept to a minimum to reduce total system cost. In addition, the selection of component *sizes* (large-valued resistors for example), *transistor types*, and the use of bypass and coupling *capacitors* was of little significance in fabricating the product.

Now, however, all of these factors are of major significance. Large-valued components eat up chip area, meaning fewer chips per wafer and a greater likelihood of chip failure. Some components, like high performance *pnp* transistors and microfarad capacitors, simply cannot be had.

On the other hand, transistors fabricated side by side on the same piece of silicon exhibit excellent *matching* qualities. But even here there is a wide variation from wafer to wafer in absolute component values (resistor tolerances of ±30% are not uncommon) due to the variations in processing from one batch of wafers to another.

Despite the apparent difficulties, the enormous complexity possible on a tiny chip of silicon far outweighs all these inherent problems. Imagine replacing several circuit boards worth of electronics with a single plug-in package that may sell for only a few dollars!

A brief list of advantages and disadvantages that the integrated circuit designer must consider include

| *Advantages* | *Disadvantages* |

Advantages

1. Reduced size and weight
2. Low power consumption
3. Low cost
4. Greatly improved reliability and serviceability
5. Readily mass-produced
6. Excellent on-chip matching

Disadvantages

1. Low performance *pnp* transistors
2. Limited to pF-value capacitors
3. Resistor values generally limited to < 50 kΩ
4. Wide tolerance of component values
5. No integrated inductors.

Often the advantages and disadvantages are played against each other. For example, the *differential pair* is a commonly-used integrated-amplifier stage that requires *well-matched* transistors and resistors but *no* coupling capacitors. It is a more complex circuit, requiring as many as eight transistors, than would ordinarily be attempted with *discrete* parts, but because of the matching capabilities and miniaturization possible, it is a realistic circuit from an *integrated* standpoint. This circuit is discussed in Section 10.5.

10.4 THE MANUFACTURING SEQUENCE

We have talked in some detail about the wafer fabrication process but there is more involved than fabricating the wafers. Let's follow the typical *life cycle* of an integrated circuit from design through shipment to end user.

Design

After the marketing group determines that a market exists for a particular circuit, the *design engineers* then begin the design cycle. This may include a computer analysis of the circuit, breadboarding in the laboratory, design and actual placement of the components by a drafts-man, and sample runs through the wafer fab area for design evaluations and characterization for data sheets. If all goes well, the design is turned over to the production group.

Wafer Fab

As previously discussed, the circuit now follows a complex path through the *wafer fabrication area*, consisting of masking, diffusion, and visual inspections.

Wafer Test

Eventually, the completed wafers emerge to be tested in the *wafer sort* or *wafer test area*. Each wafer is placed on a computerized test station and functional and parametric tests are done automatically by the computer. Individual die on the wafer that fail any tests are *red-inked* to separate them from the good die. In addition, wafers may be tested and sorted for individual grades such as military or industrial quality parts.

An important parameter in the wafer sort area is *yield*. This number is usually expressed as a percent and represents the number of good die compared to the total die on the wafer.

Wafers with excessively high failures (low yield) may be pulled from the line and manually probed to determine the actual cause of failure. For example, in the bipolar process, the n^+ emitter diffusion could be too deep, causing the emitter to *short* through the base and into the collector, a *C–E short*. When this problem is detected, diffusion times can be adjusted accordingly.

Assembly

The tested wafers are now *scribed* (cut) and the good die are assembled into packages, leads bonded, and the package sealed. Care must be taken in this area not to mix units of various types as no part numbers are yet branded on the package covers.

Final Test and QA

Once packaged, the units are tested once again in the *final test area*. Ideally, yields at this point should be 100% but practically range from 5 to 10% below this value. Again, sorting may take place for particular grades of parts or unique customer requirements.

The quality assurance (QA) group may take selected samples of the tested product and retest these parts under carefully-controlled conditions of temperature to verify the quality of the product before shipment. Finally, the parts are branded and shipped to the individual customers.

Economics

Many of us have had the experience of buying an electronic calculator only to see the price decrease in the following months. The semiconductor industry is unique in this respect, because prices tend to *decrease* over time rather than increase. The peculiar economics here is related to the *yield* of the tested wafers.

Newly designed products may initially run at *low* yields in the wafer test area due to new processes being required, tight geometry requirements in the design, and general unfamiliarity with the product. As an example, assume a wafer has 100 potentially good die. If this wafer is tested with a 10% yield, only 10 good die are obtained. In order to recover the cost of manufacturing this wafer, the price of the completed circuit must be high. On the other hand, if this wafer is tested with a 50% yield, 50 good die will be obtained and the *same costs* can be recovered but at a substantial reduction in unit selling price.

Another factor influencing yield is the *die size*. Figure 10-10 illustrates two identically sized wafers, each with three defects. In Fig. 10-10b, these three defects represent a large portion of the total die on the wafer and lower the yield accordingly. In Fig. 10-10a, these same three defects have a much lower impact on the total good die realized from the wafer due to the much smaller die size. For this reason, it is desirable to make die size, and thus individual component sizes, as small as possible.

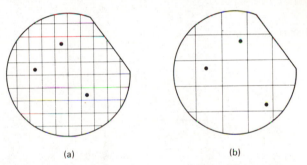

(a)   (b)

Figure 10-10 The effect of die size on wafer yield. In (a), the three defects have a minimal effect due to the small die size. But in (b), the three defects represent a substantial portion of the total die on the wafer.

It is expected that new products will run at low yields until most problems have been worked out, at which time prices will drop accordingly. Because the wafer test yield is never 100%, there are always some improvements that can be made and the decreasing price phenomena often continues over several years.

10.5 THE DIFFERENTIAL PAIR

The *differential pair* is an example of an amplifier designed with the integrated circuit in mind. It is *dc* or *direct-coupled*, requiring no capacitors, and in its "full blown" form as an integrated operational amplifier, it has a liberal sprinkling of transistors (more than 20). In fact, one common analog integrated circuit design practice is to *substitute* transistors for resistors wherever possible. This is because a transistor biased in its active region will have a large collector–emitter resistance due to its small-signal output resistance yet will take up less die area than the corresponding normal resistor.

DC Bias

The basic differential pair amplifier stage is shown in Fig. 10-11. For the moment, assume the two inputs, applied at the Q_1 and Q_2 bases, are at ground potential. Note that because of the $-V_{EE}$ supply, ground is *not* the lowest circuit potential. If the base currents are small, then there is a negligible voltage drop across R_{G_1} and R_{G_2}, the equivalent input generator resistances. Each base is then at 0 V. Both emitters must be at –0.6 V and

$$I_T = \frac{V_{EE} - 0.6 \text{ V}}{R_E} \tag{10-3}$$

or 950 μA for this example.

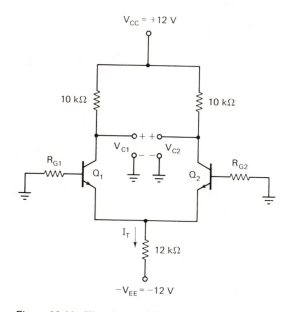

Figure 10-11 The basic differential pair amplifier stage.

Because both transistors are integrated on the same piece of silicon, they should match very closely, such that

$$I_{C1} = I_{C2} = \frac{I_T}{2} \tag{10-4}$$

or 475 μA a piece in this example. Finally, the collector voltage of each transistor is given by

$$V_C = V_{CC} - I_C R_C \tag{10-5}$$

This is 7.25 V. Note that Q_1 and Q_2 are both biased in the *active region*, with collector-base junctions reverse-biased and base-emitter junctions forward-biased.

AC Analysis

Although the amplifier in Fig. 10-11 has both inputs at ground potential, this is actually a *nonzero* input level because the supply voltage ranges between ±12 V (ground is 12 V *above* –12 V). Now if the output is defined as the voltage *between* the Q_1 and Q_2 collectors, we see that $V_o = 0$ V because $V_{C1} = V_{C2}$. In fact, if the transistors are matched and both inputs are at the *same* potential, the output will *always* be 0 V. This holds true whether the input is a dc voltage or a sine wave. This mode of circuit operation is referred to as the *common mode* and is a highly *desirable* property of the circuit. To understand this better, we need to examine the *differential mode* of operation.

In this case, two separate inputs, V_1 and V_2, are applied. Figure 10-12 illustrates how *superposition* can be used to determine the circuit's response. In Fig. 10-12a, only V_1 is active. Q_1 functions as a common-emitter amplifier, producing an amplified but 180°-out-of-phase signal at its collector and an in-phase signal at the common-emitter connections. Because the Q_2 base is at ground potential and an input is being applied to its emitter (the Q_1 and Q_2 emitters follow the Q_1 base), this transistor is functioning as a *common-base* amplifier. This causes a signal equal in amplitude to the Q_1 collector signal, but of opposite phase, to appear at its collector (recall that the common-base amplifier does not invert phase).

Figure 10-12b illustrates the opposite case where V_2 is active. Note that for both cases the two collectors are *out of phase* with respect to each other and the output voltage is therefore *nonzero*. If the voltage gains for Q_1 and Q_2 are equal, we can add the two effects to obtain the total circuit response.

$$V_o = V_{o1} + V_{o2} \qquad (10\text{-}6)$$

In this case, the output voltage due to V_2 is opposite in phase to that due to V_1 such that

$$V_o = AV_1 - AV_2 = A(V_1 - V_2) \qquad (10\text{-}7)$$

where A is the voltage gain.

A quick check reveals that when $V_1 = V_2$, the output is 0 V, as discussed previously, for the common mode. You should be able to see that the differential pair is actually a difference amplifier, amplifying the *difference* between the two inputs but giving no output to any signal common to both.

What types of signals may be common to both inputs? These may include various sources of noise, power supply variations, and, in general, anything but the signal we are trying to amplify. It is interesting to note that component variations and thermal drifts within the IC itself appear to be *common-mode* inputs and therefore tend not to be amplified. On the other hand, component *mismatching* will cause differential inputs to appear. But because matching of components within the IC is quite good, this effect should also be negligible. For these reasons, the differential pair is ideally suited for integration.

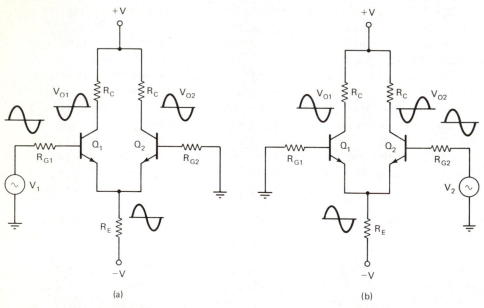

Figure 10-12 Superposition can be used to determine the effects of the two inputs V_1 and V_2. In (a), only V_1 is active, while only V_2 is active in (b). In both cases, the collector voltages are out of phase with respect to each other.

Refer to Fig. 10-12a. The differential pair is being driven by a single input V_1. Because Q_1 is acting as a common-emitter amplifier, its voltage gain is given by Eq. 6-7.

$$A_V = -\frac{\beta R_L'}{r_{be} + (\beta + 1)R_E}$$

In this case, R_L' is actually R_C, and R_E is the parallel combination of R_E and the resistance seen looking into the emitter of Q_2. As mentioned in Chapter 6, this resistance is very low, on the order of r_{be}/β. Because R_E is much larger than this, it can be ignored and Eq. 6-7 becomes

$$A_V = -\frac{\beta R_C}{r_{be} + (\beta + 1)\dfrac{r_{be}}{\beta}} = -\frac{\beta R_C}{2\,r_{be}} \tag{10-8}$$

This is the voltage gain from one input to a single collector output (single-ended output). When the output is taken *differentially* between the two collectors, the actual output is twice this value or

$$A_{DM} = -\frac{\beta R_C}{r_{be}} \tag{10-9}$$

This is called the *differential mode gain*, A in Eq. 10-7. Although Eq. 10-8 was derived with only V_1 active, the gain due to V_2 is the same because the two transistors are assumed matched and Eq. 10-7 can be used to predict the *total* output voltage.

Example 10-2

Calculate the expected output voltages for the differential pair in Fig. 10-11 for the following ac input/output configurations: (a) $V_1 = +3$ mV, $V_2 = +1$ mV, differential output; (b) same as (a) with single-ended output; (c) $V_1 = +3$ mV, $V_2 = 0$ V, differential output. Assume $\beta = 100$.

Solution   The three cases are shown in Fig. 10-13. Notice that the differential pair is replaced by its equivalent *symbol* indicating the two inputs and two outputs.
   (a) For this case, Eq. 10-7 becomes $V_o = A(3\text{ mV}) - A(1\text{ mV}) = A(2\text{ mV})$, where A is given by Eq. 10-9. From Eq. 10-4, we previously determined that $I_C = 475\ \mu\text{A}$, and therefore $r_{be} = 5.3$ kΩ. From this,

$$A_{DM} = \frac{-100(10\text{ k}\Omega)}{5.3\text{ k}\Omega} = -189$$

and $V_o = -189 \times 2$ mV $= -378$ mV.

(b) For this case, the output is taken at one collector only. The gain is then one half of the previous case and $V_o = -189$ mV.

(c) Because $V_2 = 0$ V, Eq. 10-7 becomes $V_o = AV_1 = -189 \times 3$ mV $= -567$ mV.

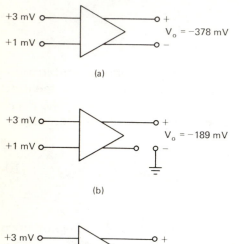

(a)

(b)

(c)

Figure 10-13 The output voltage of the differential pair for three separate output/input cases. (a) differential output and input. (b) differential input and differential output.

The Common Mode

Ideally, the total emitter current of the differential pair, I_T, should remain *constant* no matter what input voltage is applied. Figure 10-14 illustrates a common mode input. This input tries to turn both transistors on simultaneously, but a constant emitter current will not allow this to happen and the output voltage cannot change. The common mode gain is therefore 0.

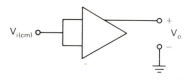

Figure 10-14 Common-made gain test circuit.

In reality, the emitter current is *not* constant and V_o can change slightly. The common mode gain is defined as

$$A_{\text{CM}} = \frac{\Delta V_o}{\Delta V_{i(\text{cm})}} \tag{10-10}$$

where $V_{i(cm)}$ and V_o are illustrated in Fig. 10-14 (note that this is a single-ended output). A differential pair is characterized by its *common-mode rejection ratio* (*CMRR*), expressed in dB.

$$CMRR = 20 \log(A_{DM}/A_{CM}) \qquad (10\text{-}11)$$

It is desirable that the common mode rejection ratio be large. In this way the amplifier will amplify *differences* between the two inputs but reject signals *common* to the two inputs (for example, 60 Hz hum).

Example 10-3 ———————————————————————

The following data is collected on a differential amplifier: (a) differential input = 5 mV, differential output = 0.6 V; (b) common mode input change from 0 V to 1 V, single-ended output changes from 8 to 8.02 V. Determine the *CMRR*.

Solution

$$A_{DM} = 0.6 \text{ V}/5 \text{ mV} = 120$$

$$A_{CM} = \Delta V_o/\Delta V_{i(cm)} = 0.02 \text{ V}/1 \text{ V} = 0.02$$

$$CMRR = 20 \log\left(\frac{120}{0.02}\right) = 75.6 \text{ dB}$$

What this literally means is that the common-mode response will be 75.6 dB below the differential mode response.

Figure 10-15 shows an improved differential pair, again taking advantage of the integrated technology. Q_4 functions as a *diode-connected transistor*. Because its base and collector are shorted, it appears to be a base-emitter diode but with lower ON resistance (more vertical slope) than a normal diode due to transistor action shunting most of the diode current through its collector. A reference current is established in this transistor, given by

$$I_R = \frac{20 \text{ V} - 0.6 \text{ V}}{20 \text{ k}\Omega} = 970 \text{ }\mu\text{A}$$

If Q_3 is identical in geometry to Q_4, and because it must see the same base-emitter voltage,

$$I_T = I_{C3} = I_R = 970 \text{ }\mu\text{A}$$

neglecting base currents. The remainder of the dc analysis proceeds as discussed earlier.

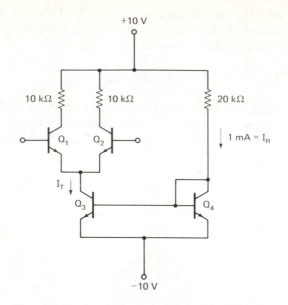

Figure 10-15 An important differential pair. Q_4 establishes a reference current, which is reflected into Q_3 if the two transistors are matched.

There are a number of advantages to this more complex differential pair.

1. The I_T current is a better approximation to a current source than the circuit with a single R_E resistor. This means improved *CMRR*.

2. Q_3 may be physically smaller than the large R_E resistor, saving on chip area.

3. Geometric scaling can be used to establish *multiples* of the reference current. If Q_3 was physically *twice* the size of Q_4, the reflected I_T current would *double* the value of I_R.

The Op-Amp

The *operational amplifier* or op-amp makes extensive use of the differential pair. A block diagram is shown in Fig. 10-16. The input stage employs bipolar or field-effect transistors in a *differential* configuration and should have high input resistance, low input bias currents, and a large common-mode rejection ratio.

Because the differential output of this stage is *not* referenced to ground, it is desirable to take the output *single-endedly* and establish a ground reference. This results in a dc level (the bias level of that

Figure 10-16 Block diagram of an operational amplifier. The op-amp has a differential input stage, level-shifting stages to ensure O V out with O V in, and a class B output stage for driving the load resistance.

collector) appearing in the output. For this reason, *level-shifting* stages are required to drop the output voltage to 0 V.

Finally, the *class B* output stage allows power to be supplied to low resistance loads efficiently and with reasonable signal swings.

A chip photograph of the μA741 op-amp is shown in Fig. 10-7 and its schematic diagram is shown in Fig. 10-17. As we shall see in the next two chapters, the modern op-amp comes close to being the ideal voltage amplifier, which can be characterized by

1. $R_i = \infty \; \Omega$
2. $R_o = 0 \; \Omega$
3. $A_V = \infty$

No better example of the advantage of the integrated technology need be presented than the schematic diagram in Fig. 10-17. This

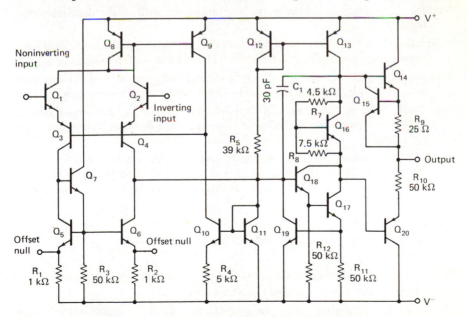

Figure 10-17 Schematic diagram of the μA741 op-amp.

obviously complex circuit would be impractical for all but the most critical of designs in a discrete form. Yet in integrated form, it sells for less than *50¢* and can be used for applications not even considered a few years ago.

KEY TERMS

IC   Integrated circuit. A single chip of silicon on which a number of transistors and resistors have been simultaneously fabricated to perform some useful circuit function.

DIP   Dual-in-line package. This refers to the type of package most integrated circuits are assembled in. Typical DIPs have 14, 16, 24, or 40 pins.

Diffusion   This refers to the process of doping a semiconductor to become *n*-type or *p*-type by placing the wafer in a furnace at a very high temperature with selected impurities.

Photolithography   A photographic process involving photoresist, photo masks, and ultraviolet light for defining the specific locations and geometries of integrated circuit components.

Wafer   A round, thin disc of silicon on which several hundred or thousand individual chips or die are simultaneously fabricated. Typical wafers are 5 in. in diameter.

Yield   After a testing operation, the percentage of good parts compared to the total tested.

Matching   The ability to fabricate similar components with nearly identical characteristics.

Differential Mode   This refers to a differential amplifier's response to the difference between its two inputs. It is desirable to have a high differential mode gain.

Common Mode   This refers to a differential amplifier's response to voltages common to both inputs. It is desirable to have a low common-mode gain.

Single-Ended   An output taken at a single collector is called a single-ended output. This is compared to the differential output that is taken between the two collectors.

Op-Amp   Operational amplifier. A highly-developed integrated circuit amplifier featuring a differential input stage and single-ended output stage. It has a high input resistance, low output resistance, and a large differential voltage gain.

QUESTIONS AND PROBLEMS

10-1   Why did the planar process lead to the development of the integrated circuit?

10-2   Explain why the β of a lateral *pnp* transistor is so low.

10-3   List two important properties of SiO_2 related to the manufacture of integrated circuits.

10-4 What is the purpose of the p^+ isolation diffusion in Fig. 10-5? Why do MOS transistors not need to be isolated?

10-5 Which do you think controls the basewidth of an *npn* transistor, photomask definition, or diffusion time? Explain.

10-6 Name the six masks required for the fabrication of the transistor in Fig. 10-5 and draw each one separately.

10-7 Calculate the length of a 50-kΩ 7-μm wide resistor. Assume $\rho = 0.1$ Ω-cm and a junction depth of 5 μm. Convert your answer to mils and compare with the size of the 741 die (56 mil^2). *Note:* 1 μm = 10^{-4} cm, and 1 cm = 394 mil.

10-8 Determine the area required for a 100-pF capacitor with a distance between plates of 0.15 μm and dielectric constant of 3.0.

10-9 The pinning of many integrated circuits is such that, if plugged in backwards, the ground and V_{CC} connections are reversed. Explain why this may damage the IC. Consider the substrate diode.

10-10 Explain why MOS transistors require substantially less chip area than bipolar transistors. What affect does this have on possible circuit complexity?

10-11 A wafer is tested and found to have 167 good die and 103 bad die. What is the wafer sort test yield?

10-12 Explain why larger die sizes will tend to have lower wafer test yields.

10-13 Recalculate the dc bias for the circuit in Fig. 10-11 if $R_{C1} = R_{C2} = 15$ kΩ, and $R_E = 18$ kΩ. $\beta = 100$.

10-14 Estimate the differential mode gain of the amplifier in Fig. 10-15. Assume $\beta = 100$.

10-15 Refer to Fig. 10-18. Determine the differential-mode gain, common-mode gain, and *CMRR* in dB.

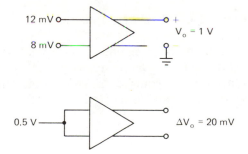

Figure 10-18

10-16 A differential amplifier with 100-dB CMRR has a differential mode gain of 100. If a 1-V common mode input is applied, what change in output voltage can be expected?

10-17 Refer to Fig. 10-19. Estimate the voltage at all nodes in this circuit.

10-18 Refer to Fig. 10-21. Estimate the voltages at all nodes in this circuit if both inputs are grounded.

10-19 The amplifier in Fig. 10-21 has a single-ended output. Which input will be in phase with respect to this output?

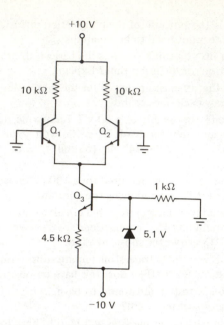

Figure 10-19

LABORATORY ASSIGNMENT 10:
THE DIFFERENTIAL PAIR

Objectives

1. To investigate the dc bias and ac performance of the differential pair.
2. To observe the matching of integrated transistors.

Introduction   In this laboratory assignment, you will assemble and test two forms of the differential pair. An integrated *npn* transistor array (CA3086) will be used for all transistors. This 14-pin DIP contains five *npn* transistors, two of which are connected as a differential pair.

Components Required

1 CA3086
Miscellaneous ¼-W resistors

Part I: *Standard Differential Pair*

A. DC

STEP 1   Refer to Fig. 10-20 and determine pin numbers for the differ-

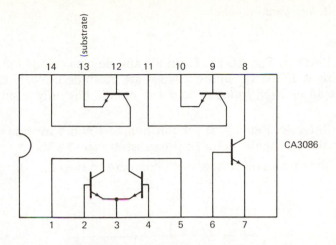

Figure 10-20   The CA 3086 14-pin DIP.

ential pair circuit illustrated in Fig. 10-11. Assemble this circuit on your breadboard. Be sure the substrate (pin 13) is connected to –12 V.

STEP 2   With both inputs grounded, measure all dc voltages in this circuit and compare with calculations.

Question 1   If $V_{C_1} \neq V_{C_2}$, does this mean Q_1 and Q_2 are not matched? What about the matching of the two 10-kΩ resistors?

B. AC

STEP 1   Connect a signal generator ($f = 1$ kHz) to the Q_1 base. Remember, the amplifier is *dc-coupled* and no coupling capacitor is required.

STEP 2   Measure the *single-ended* voltage gain to each collector. Observe both collectors simultaneously (two channels) and note the phases relative to V_i.

Question 2   The signal at the Q_1 collector is _____ phase with respect to the input, while the signal at the Q_2 collector is _____ phase with respect to this input.

STEP 3   "Float" the oscilloscope and measure the differential voltage gain between the collectors.

Question 3   Why must the oscilloscope float in step 3?

STEP 4   Indicate a method and measure the *common-mode* gain of the amplifier. Calculate the *CMRR* in dB.

A. DC

STEP 1 Refer to Fig. 10-21. Estimate the dc bias voltages at all nodes in this circuit if the Q_1 and Q_2 bases are grounded. *Note:* Q_5 is connected as an *emitter follower* and the circuit has only a single-ended output.

STEP 2 Refer to Fig. 10-20 for pin numbers and wire this circuit on your breadboard. Again, connect the substrate to –12 V.

STEP 3 Measure all dc voltages and compare to step 1.

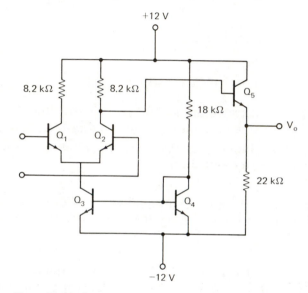

Figure 10-21 Improved differential amplifier used in Part II.

B. AC

STEP 1 Apply an ac input to the Q_1 base and measure the voltage gain of the amplifier.

STEP 2 Repeat step 1, applying the input to the Q_2 base.

Question 4 Often the inputs to an op-amp are labeled "inverting" and "noninverting," indicating the phase of the output for this particular input. In Fig. 10-21, the Q_1 base is the _____ input and the Q_2 base is the _____ input.

STEP 3 Measure the common mode gain of the amplifier. Compare to Part I.

Question 5 List the advantages of the circuit in Fig. 10-21 versus the circuit in Fig. 10-11. Are there any disadvantages?

ELEVEN

THE OPERATIONAL AMPLIFIER

A large portion of this text has dealt with *discrete* transistor amplifiers. We have discussed *npn* and *pnp* bipolar amplifiers, JFET amplifiers, MOSFET amplifiers, and even light-coupled amplifiers. Using any of these circuits first requires a careful selection of the circuit operating point for optimum dc stability and ac signal swing. Next, ac considerations are made and bypass and coupling capacitors chosen to obtain desired input and output resistances and amplifier gains.

All of these steps require time and often an experimental approach is necessary to obtain the optimum design. Contrast this with the *operational amplifier* (op-amp). Its dc bias is set internally and generally requires no adjusting. Just "plug it in and go." The voltage gain is set by the ratios of external resistors. For a 1% accurate voltage gain, use 1% resistors.

In this chapter we treat the op-amp as a new *three-terminal device*. The classic op-amp circuits are presented and techniques for analyzing and designing circuits are covered. Lest you think the op-amp is a perfect device, some of the more significant limitations, as they affect circuit response, are covered in the last section.

11.1 BASIC CONCEPTS

As discussed in Chapter 10, the op-amp is actually a highly-sophisticated multistage integrated amplifier (see Fig. 10-17). If we were to concern ourselves with the *internal* circuitry of the op-amp, we could become quite confused and actually miss one of the key points of its designers. That is, the op-amp was purposely designed as a complex circuit so that the user would have a minimum number of problems in applying it. What the op-amp designer has done is given us a component whose device-to-device variations are insignificant. The circuit's function is controlled by *external* components and not those internal to the op-amp. What must be mastered is an understanding of the device from an *external* terminals viewpoint.

The Power Connections

Figure 11-1 illustrates the symbol commonly used for the op-amp. The pin numbers shown are those for the μA741 8-pin DIP. Note that unlike most of the amplifiers we have discussed previously, there are *two* supply voltages required, labelled V+ and V-. These split or dual voltages allow the output voltage of the op-amp to swing negative as well as positive. They also mean that ground is *not* the lowest circuit potential. There is no set value for these voltages and their choice is based upon the output voltage swings desired and available voltage sources. Usually V+ = V-, but this is not essential and in some cases a single supply can be used. Typical values are ±12 V. In all cases, care must be taken not to exceed the maximum values established by the manufacturer. As indicated in Fig. 11-2, these are ±18 V for the μA741 and LM301A types and ±8 V for the CA3130.

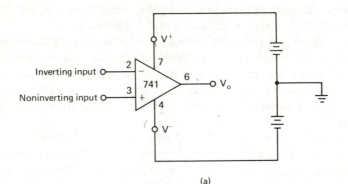

(a)

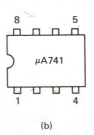

(b)

Figure 11-1 (a) The symbol commonly used for the op-amp. The pin numbers are shown relative to its 8-pin package in (b).

Symbol	Description	μA741 (bipolar)			LM301A (bipolar)			CA3130 (biFET)		
		Min	Typ	Max	Min	Typ	Max	Min	Typ	Max
A_{OL}	Open loop voltage gain		200,00			220,00			320,000	
I_{SC}	Output short-circuit current (mA)		25			26			22	
R_i	Input resistance (MΩ)	0.3	2.0		0.5	2.0			1.5×10^6	
V_{io}	Input offset voltage (mV)		1.0	5.0		2.0	7.5		8	15
I_B	Input bias current (nA) 25°C		80	500		70	250		0.005	0.05
I_{OS}	Input offset current (nA) 25° C		20	200		3	50		0.0005	0.03
	Slew rate (V/μs)		0.5			10			10	
CMRR	Common-mode rejection ratio (dB)	70	90		70	90		70	90	
V^+, V^-	DC supply voltage (volts)		+18			±18			±8	

Figure 11-2 Specifications of selected op-amps.

One point that is sometimes confusing about the op-amp is its ground. Where is it? In general, neither V^+ nor V^- is at ground potential. An inspection of Fig. 11-1 reveals that ground is there; it is the common reference point for the two supply voltages. It is *not* connected directly to the op-amp however. In general, all voltages are still measured with respect to ground, but we can expect negative as well as positive measurements.

The Output

As indicated in Fig. 10-17, the output voltage of the op-amp is taken single-endedly from a class B complementary output stage. This type of stage provides excellent drive capability, even to low resistance loads. Maximum signal swing is limited by saturation of the output transistors, Q_{14} and Q_{20} in Fig. 10-17; Assuming a $V_{CE(sat)}$ of approximately 1 V,

$$+V_o(\text{max}) = + V_{sat} \cong V^+ - 1 \text{ V} \tag{11-1}$$

and

$$-V_o(\text{max}) = -V_{sat} \cong V^- + 1\text{V} \tag{11-2}$$

The output of the op-amp is also protected against *short circuits*. Again referring to Fig. 10-17, excessive current through resistor R_9 will bias Q_{15} on stealing base drive from Q_{14} and limiting the output current.

Example 11-1

Assume a 741 type op-amp has ±12-V power supplies. Determine the maximum output voltages possible and the minimum value of load resistance without current limiting.

Solution   From Eqs. 11-1 and 11-2, $\pm V_{sat} = \pm 11$ V. Referring to Fig. 10-17, current limiting occurs when $V_{R9} = 0.6$ V, and Q_{15} begins to conduct. This requires $I = 0.6$ V/25 Ω = 24 mA. The smallest load resistor is therefore 11 V/24 mA = 458 Ω.

The Inputs

The op-amp is actually a *difference amplifier*. It has a very large differential gain, usually called the *open-loop gain* (A_{OL}). The output voltage is given by

$$V_o = A_{OL} \times E_d \tag{11-3}$$

where E_d is the *difference voltage* applied between the two inputs. In equation form,

$$E_d = V_i(+ \text{ input}) - V_i(- \text{ input}) \qquad (11\text{-}4)$$

These (+) and (–) inputs are referred to as the *noninverting* and *inverting* inputs respectively. As suggested by Eqs. 11-3 and 11-4, inputs applied to the positive input will make the output go positive, while inputs applied to the (–) input will make the output go negative. The actual polarity of the output voltage depends on the net *difference* between the two inputs.

Example 11-2 ———————————————————————————

Refer to the specifications for the μA741 in Fig. 11-2 and determine the maximum value of E_d that will *not* saturate the output. Assume ±12-V power supplies.

Solution With ±12-V power supplies, the output voltage is limited to approximately 11 V before saturation occurs. The maximum differential input E_d can now be found from Eq. 11-3.

$$E_d = V_o/A_{OL} = 11 \text{ V}/200{,}000 = 55 \text{ } \mu\text{V}$$

It is hard to appreciate how small 55 μV is. Most oscilloscopes have a minimum vertical sensitivity of 1 mV. We are talking about 0.055 mV! If you were to set this op-amp up in the laboratory and apply a test input, you would have a difficult time avoiding saturation let alone seeing the input signal!

You might wonder then, what good is this extremely high but rather unwieldy voltage gain? The answer is that the op-amp is almost *never used* in a linear application without *feedback*. As we shall see, feedback lowers the voltage gain considerably but makes the amplifier much easier to work with.

In the following sections, we study the op-amp operating in one of two modes: *linear* and *nonlinear* operation.

Nonlinear operation occurs when feedback is *not* used and the op-amp switches between $+V_{sat}$ and $-V_{sat}$. This is similar to the bipolar amplifier switching between saturation and cutoff and suggests square wave outputs and digital applications.

When feedback is used to limit the output voltage to values between $\pm V_{sat}$, the op-amp is operating *linearly*. In this case the output voltage is determined by Eq. 11-3, which leads to one of the key points in analyzing an op-amp circuit.

If an op-amp is operating linearly, its differential input voltage E_d can be assumed 0 V.

This is a valid assumption because linear operation restricts the output voltage to values less than V_{sat} and Example 11-2 demonstrated that the *maximum* input signal under these conditions is only 55 μV. In the next sections, we develop this idea further and you will see the real significance of this assumption.

An Ideal Amplifier

In many if not a majority of circuit applications, the op-amp behaves as a nearly ideal amplifier. As indicated in Fig. 11-2 for the μA741, the *input resistance* at either input terminal is typically 2 MΩ. This means we can assume negligible input current will be drawn from the input source.

Its *output resistance* is typically less than 100 Ω and even less than this when feedback is used. Therefore, low resistance loads can be driven without affecting the voltage gain (see Example 11-1 for a slight qualification).

The *open-loop voltage gain* is extremely large. So large, in fact, that only microvolts of input are needed to cause full output. This makes the op-amp particularly useful for voltage-sensing circuits (comparators) and enhances feedback effects when used linearly.

The op-amp will amplify dc as well as ac, making it useful in low frequency circuits as well as dc switching applications.

Although not perfect, the op-amp is quite an improvement over the discrete amplifiers we have discussed previously in this text. A summary of these advantages includes

1. DC bias established internally, "just plug it in and go"
2. DC coupled, the op-amp will amplify ac or dc voltages
3. Output dc voltage may go negative as well as positive
4. Large input resistance so there is no input loading
5. Low output resistance so the gain is not affected by R_L
6. Differential input, only the difference between the inputs is amplified
7. Common-mode response is typically 90 dB down
8. Open-loop gain is extremely large, allowing feedback to establish accurate closed loop gains.

11.2   LINEAR CIRCUITS

Linear op-amp circuits are those in which the output voltage is governed by Eq. 11-3; $V_o = A_{OL} \times E_d$. In order for this equation to be valid, the output voltage must be *less* than V_{sat}. For example, if $E_d = 1$ V, it is *not* logical to assume $V_o = 1$ V $\times$ 200,000 = 200,000 V! Rather, the op-amp must be *saturated* and $V_o = +V_{sat}$. This is *nonlinear* operation.

The Inverting Amplifier

The basic *inverting amplifier* is shown in Fig. 11-3a. Note that a portion of the output voltage is fed back to the inverting (-) input through resistor R_F. This is called *negative feedback*. Also note that the input signal is applied through resistor R_1 and not directly to the op-amp inputs. In actuality, there are now *two* inputs: V_i and E_d. The voltage gain from V_i to V_o is called the *closed-loop gain*, A_{CL}. As we shall see, this gain is established by the feedback resistor R_F and resistor R_1.

The gain from E_d to V_o is the *open-loop gain*, A_{OL}. This is approximately 200,000 for the μA741 but may vary from device to

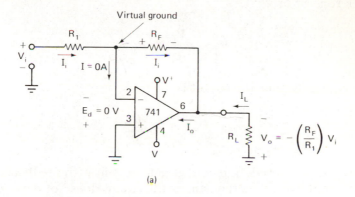

(a)

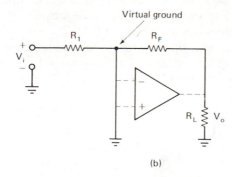

(b)

Figure 11-3 (a) The inverting amplifier. Because E_d is nearly O V the (-) input is at ground potential. Currents are shown for a positive input voltage. (b) the virtual ground is emphasized by showing a short circuit between the inverting and the non-inverting inputs. This is a key point for analyzing most linear op-amp circuits.

device. But this is precisely the advantage of negative feedback. As long as the output voltage is less than V_{sat}, E_d must be in the microvolt range $(E_d = V_o/A_{OL})$. This will be true even if A_{OL} changes by a factor of 10!

Referring to Fig. 11-3a, if $E_d = 0$ V, and the noninverting (+) input is grounded, the inverting input must also be ground potential. Figure 11-3b illustrates this point. Because this input is not actually *connected* to ground, it is called a *virtual* ground.

Input voltage V_i establishes a current in resistor R_1 given by V_i/R_1 (remember the virtual ground). Because the input resistance to either op-amp is very large, we can assume that all of this current also flows through resistor R_F. Again due to the virtual ground, the output voltage is the voltage developed across resistor R_F. As can be seen in Fig. 11-3a, a positive input voltage establishes a conventional current flow such that the voltage drop across R_F is *negative* with respect to the virtual ground. This voltage is equal to the input current (V_i/R_1) times the feedback resistor or

$$V_o = -\frac{V_i}{R_1} \times R_F \tag{11-5}$$

and the voltage gain of the circuit is

$$A_{CL} = \frac{V_o}{V_i} = -\frac{R_F}{R_1} \tag{11-6}$$

where A_{CL} is the closed-loop voltage gain or gain with feedback.

We might note that the gain of the circuit does not depend on the specific op-amp open-loop gain. In fact, as long as the assumption that $E_d = 0$ V is valid, Eq. 11-6 is valid. This simply requires that A_{OL} be large. A value of 100,000 will work as well as 200,000.

Also note that the closed-loop gain does not depend on R_L. As long as the current limit of the op-amp is not exceeded, the value of the load resistor is unimportant. This is quite a change from the discrete amplifiers previously studied.

Finally, the input resistance seen by V_i is just R_1 to ground (virtual ground).

$$R_i = R_1 \tag{11-7}$$

Example 11-3 ───────────────────────────────

Assume the inverting amplifier shown in Fig. 11-3a has $R_F = 10$ kΩ, $R_1 = 500$ Ω, and $R_L = 1$ kΩ. Calculate the output voltage, input resistance, and op-amp sink or source current if $V_i = +0.5$ V.

Solution Applying Eqs. 11-5 and 11-7,

$$V_o = -\frac{0.5 \text{ V}}{500 \text{ } \Omega} \times 10 \text{ k}\Omega = -10 \text{ V}$$

$$R_i = R_1 = 500 \text{ } \Omega$$

All currents have the polarities shown in Fig. 11-3a and the op-amp therefore *sinks* a total current of

$$I_o = \frac{V_i}{R_1} + \frac{V_o}{R_L} = \frac{0.5 \text{ V}}{500 \text{ } \Omega} + \frac{10 \text{ V}}{1 \text{ k}\Omega} = 11 \text{ mA}$$

Because this is within the typical current handling capability of the op-amp, there should be no problem driving this load.

You should be able to see that if the input polarity was reversed, the output voltage would be positive and the op-amp would have to source the 11 mA.

In summary, the inverting amplifier

1. Inverts the phase of the input signal
2. Has a voltage gain dependent only on resistor ratios (R_F/R_1)
3. Has an input resistance equal to R_1
4. Is not affected by R_L unless the current limit is exceeded.

The Summing Inverter

Figure 11-4 illustrates the *summing* or *adding inverter*. Note that the power connections (V^+ and V^-) have been left off the schematic

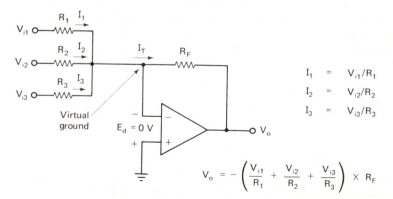

$$I_1 = V_{i1}/R_1$$
$$I_2 = V_{i2}/R_2$$
$$I_3 = V_{i3}/R_3$$

$$V_o = -\left(\frac{V_{i1}}{R_1} + \frac{V_{i2}}{R_2} + \frac{V_{i3}}{R_3}\right) \times R_F$$

Figure 11-4 The summing inverter. Due to the virtual ground, each input is independent of the others.

symbol, as is common practice. This circuit is nearly identical to the circuit in Fig. 11-3 except there are now three inputs instead of one. However, due to the virtual ground, all three inputs are *independent* of each other. Each contributes a current V_i/R, which is *summed* at the virtual ground and multiplied by R_F to determine the actual output voltage.

$$V_o = -\left(\frac{V_{i1}}{R_1} + \frac{V_{i2}}{R_2} + \frac{V_{i3}}{R_3}\right) \times R_F \qquad (11\text{-}8)$$

The virtual ground of the op-amp is particularly useful in this circuit. Without the summing point being at ground potential, one input could feed into the others, possibly swamping one or both out.

If R_1, R_2, and R_3 are made variable, each input can be adjusted to produce equal amplitude output signals. An audio mixer, for example, might have three inputs from three separate microphones. If one of the speakers is more soft spoken, the gain of that channel can be adjusted accordingly.

Example 11-4

The circuit in Fig. 11-4 has $R_1 = R_2 = R_3 = 1$ kΩ, and $R_F = 10$ kΩ. Determine the output voltage if $V_{i1} = +0.6$ V, $V_{i2} = 0.3$ V, and $V_{i3} = -0.4$ V.

Solution   When all the input resistors are equal, Eq. 11-8 becomes

$$V_o = -(V_{i1} + V_{i2} + V_{i3}) \times \frac{R_F}{R_1} \qquad (11\text{-}9)$$

and in this case, $V_o = -(0.6 \text{ V} + 0.3 \text{ V} - 0.4 \text{ V}) \times 10 = -5$ V

As Eq. 11-9 indicates, the summing inverter actually acts as an *adder* of the various inputs. This circuit, together with the inverting amplifier, which is actually a multiplier, provides the addition and multiplication functions. One of the earliest applications of op-amps was in *analog computers* to solve *differential equations*. Each arithmetic term in the equation is represented by a separate op-amp circuit performing addition, multiplication, integration, or some other mathematical function. All terms are summed using an adder circuit and its output represents the solution to the equation.

The Voltage Follower

The *voltage follower* is the op-amp equivalent of the *emitter follower*. The circuit is shown in Fig. 11-5a. Again, for linear operation, $E_d = 0$ V, and therefore $V_o = V_i$. As with the emitter follower, there is

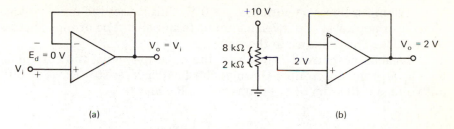

Figure 11-5 (a) The voltage follower connection. $V_o = V_i$. (b) the voltage follower provides a stable reference circuit unaffected by changes in load resistance.

no voltage gain ($A_{CL} = 1$) and the input resistance is large (typically 2 MΩ for the μA741).

An application of such a circuit is shown in Fig. 11-5b. In this case, the op-amp functions as a stable 2-V reference capable of sinking or sourcing 24 mA of current. Without the op-amp, the voltage divider would be subject to loading as soon as an external resistor was connected.

Example 11-5 ——————————

Calculate the minimum value of load resistance for the circuit in Fig. 11-5b. Assume a maximum allowable load current of 24 mA.

Solution Because the output is adjusted to 2 V, when 24 mA is drawn, $R_L(\text{min}) = 2\,V/24\,mA = 83\,\Omega$. Any resistor 83 Ω or larger will see a stable 2-V reference. In a sense, the circuit functions as a voltage regulator.

Other applications of the voltage follower include *buffering* a high output resistance source to a low resistance load. In this case, the large input resistance of the op-amp prevents loading of the input source while its low output resistance is suitable for driving low or varying resistance loads.

The Noninverting Amplifier

Figure 11-6 illustrates an amplifier which *does not invert* the input signal phase. Although this amplifier is noninverting, negative feedback is still employed. The input, however, is applied to the noninverting (+) input.

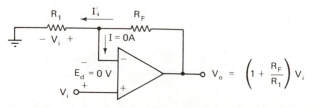

Figure 11-6 The noninverting amplifier. Because $E_d = 0$ V, V_i appears directly across R_1. There is no phase inversion.

Again, for linear operation, $E_d = 0$ V. This causes the input voltage V_i to also appear at the inverting input terminal and be dropped across resistor R_1. This is shown in the figure.

Because no current flows into the op-amp inputs, the current in R_1 (V_i/R_1) must also flow through the feedback resistor R_F. The output voltage is the sum of the drops across R_F and R_1.

$$V_o = V_{R1} + V_{RF}$$

Substituting $V_{R_1} = V_i$, and $V_{RF} = (V_i/R_1) \times R_F$,

$$V_o = V_i + \left(\frac{V_i}{R_1}\right) \times R_F = V_i\left(1 + \frac{R_F}{R_1}\right) \qquad (11\text{-}10)$$

Rearranging to obtain the closed loop gain,

$$A_{CL} = \frac{V_o}{V_i} = 1 + \frac{R_F}{R_1} \qquad (11\text{-}11)$$

As the conventional currents in Fig. 11-6 illustrate, a positive input voltage causes a positive output voltage and this circuit is referred to as a *noninverting amplifier*. Compared to the inverting amplifier, this circuit has nearly the identical voltage gain but does not invert the input signal phase and has a typically large input resistance limited by the op-amp itself instead of R_1.

Summary

By feeding a portion of the output voltage back to the input, the op-amp can be made to have manageable voltage gains established by resistor ratios. In this configuration, the input is no longer applied directly between the op-amp inputs (*open loop*) but rather through a resistor network that establishes the actual circuit gain (*closed-loop gain*). The general effect of this feedback is to *lower* the voltage gain but *stabilize* the circuit and provide more predictable performance.

11.3 NONLINEAR CIRCUITS

The op-amp is not always used as an amplifier. When operating *nonlinearly*, the following characteristics can be noted:

1. V_o is always at $\pm V_{sat}$.
2. $E_d \neq 0$ V, causing condition (1).

3. *Positive* feedback may be used.

4. The output signal is a *square wave*.

5. The op-amp is often connected in an *open loop* configuration.

The Voltage Comparator

The *voltage comparator* is a circuit that compares an unknown input voltage to a reference level and provides a corresponding output. This output voltage will indicate if the unknown input is above or below the reference level.

When used in the *open-loop* configuration, the op-amp fits this description quite well. In fact, any attempt to use the op-amp without feedback will probably result in a voltage comparator circuit rather than an amplifier!

Figure 11-7a illustrates the simplest possible op-amp voltage comparator. Because no feedback is used,

$$V_o = E_d \times A_{OL}$$

and if E_d exceeds 50 to 100 μV, the output will be at $\pm V_{sat}$. As an example, a time varying input signal is shown in Fig. 11-7b. Whenever this input exceeds +1 V, the reference level, the output switches to $-V_{sat}$. This is because $E_d = V[(+) \text{ input}] - V[(-) \text{ input}]$ and this is negative when V_u is greater than + 1 V.

When V_u is less than +1 V, E_d is positive and $V_o = +V_{sat}$. In this

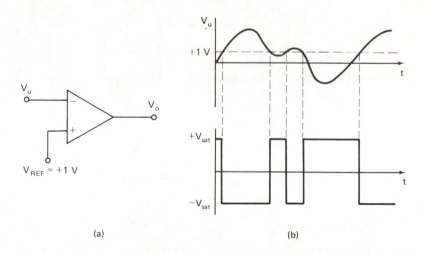

(a)   (b)

Figure 11-7 (a) A simple voltage comparator. If V_u differs from V_{ref} by more than 50 to 100 μV, the output voltage will be at$^+ V_{sat}$. (b) the output waveform when V_u varies with time.

way, the status of the V_u signal can be monitored. Whenever the output switches to $+V_{sat}$, the unknown signal has crossed $+1$ V going negative. The opposite is true when V_o switches from $+V_{sat}$ to $-V_{sat}$.

Example 11-6 ———————————————————————————————————

Design an op-amp circuit to convert a 1-V peak-peak sine wave to a 20-V peak-peak square wave without phase reversal.

Solution The circuit is shown in Fig. 11-8a. The reference voltage (ground) is applied to the inverting input while the sine wave is applied to the noninverting input. In this way, whenever the input signal goes positive, the output voltage switches to $+V_{sat}$. The opposite occurs for negative swings of the input voltage. Input and output waveforms are shown in Fig. 11-8b and c.

A common application for such a circuit is to use the voltage comparator to extract the 60-Hz information from the ac power line. First, a transformer is used to step the 120 V down to a safer level for the op-amp. This sine wave is then applied to an op-amp circuit similar to Fig. 11-8a. However, most digital circuits work with voltages of 0 V

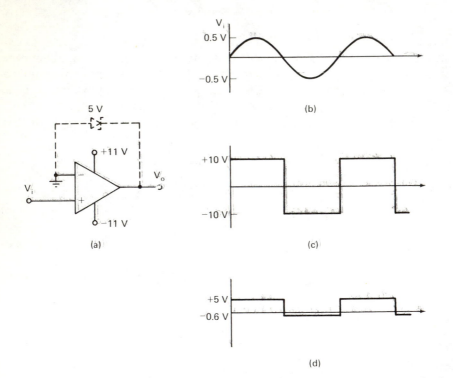

(b)

(a)

(c)

(d)

Figure 11-8 (a) The noninverting comparator. This circuit converts the input sinewave (b) to a 20-V peak-peak square wave (c) or a 5-V peak-peak square wave if a zener diode is used (d).

and +5 V. Even if the power supplies are reduced to ±5 V, the voltage comparator will still produce an undesirable negative output voltage. This problem is solved by adding a zener diode as shown by the dashed lines in Fig. 11-8a. With the zener diode in place, the output is limited to +5 V in the positive direction and –0.6 V in the negative direction. The corresponding output waveform is shown in Fig. 11-8d.

The circuit in Fig. 11-8a is also called a *zero-crossing detector* because its output switches each time the input sine wave crosses 0 V.

The Astable Multivibrator

An *astable multivibrator* is a circuit that free runs as a *square wave oscillator*. The op-amp equivalent is shown in Fig. 11-9a. Note that there is no input (free-running) and that both positive and negative feedback are applied.

When first energized, the capacitor is discharged and holds the inverting input at 0 V. Due to imbalances within the op-amp, the output voltage will be at some nonzero value. A portion of this voltage is fed back to the noninverting input through the $R_1 R_2$ voltage divider. Because only a small voltage across the inputs is sufficient to drive the op-amp into saturation, the output quickly switches to $\pm V_{sat}$.

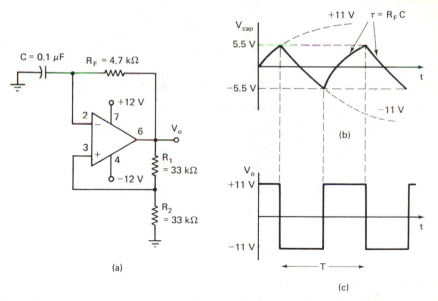

Figure 11-9 (a) Astable multivibrator. The circuit will free run as an oscillator, with typical wave forms shown in (b) and (c). The period of oscillation is T =

$$2R_F C \ln \left(1 + \frac{2R_2}{R_1}\right).$$

Assuming the output is at $+V_{sat}$ (+11 V), the capacitor begins to charge towards this value with a time constant given by $R_F \times C$. This is illustrated in Fig. 11-9b. As this charging is taking place, the op-amp itself is saturated at $+V_{sat}$ (+11 V) and the $R_1 R_2$ voltage divider is holding the noninverting input at

$$V(+) = +V_{sat} \times \frac{R_2}{R_1 + R_2} = 11 \text{ V} \times \frac{33 \text{ k}\Omega}{66 \text{ k}\Omega} = 5.5 \text{ V}.$$

Although the capacitor would like to charge to $+V_{sat}$, the op-amp will not let it. This is because the voltage on the inverting input will eventually *equal* the voltage on the noninverting input. Now when the capacitor voltage exceeds this by a tiny amount (typically 55 μV), the op-amp must switch to $-V_{sat}$. The capacitor now commences charging in the opposite direction, towards $-V_{sat}$.

Of course, when the output switched to $-V_{sat}$, so did the polarity of the feedback voltage on the noninverting input. This in turn means the op-amp won't let the capacitor charge to $-V_{sat}$ either. The resulting capacitor and op-amp output voltage waveforms are shown in Fig. 11-9b and c.

The period of this oscillator can be found (after a rather lengthy derivation) as

$$T = 2 R_F C \times \ln \left(1 + \frac{2R_2}{R_1} \right) \tag{11-12}$$

where ln represents the *natural log* (log base e, *not* base 10) function.

Example 11-7 ——————————————————————————————————

Estimate the frequency of oscillation for the circuit shown in Fig. 11-9a.

Solution   Plugging into Eq. 11-12,

$$T = 2 \times 4.7 \text{ k}\Omega \times 0.1 \text{ }\mu\text{F} \times \ln \left(1 + \frac{66}{33} \right)$$

$$= 0.94 \text{ ms} \times 1.1 = 1.03 \text{ ms}$$

The frequency of oscillation is $f = 1/T = 0.97$ kHz.

Example 11-8 ——————————————————————————————————

Recalculate the frequency of oscillation and peak-peak value of the capacitor voltage if R_2 is changed to a 10-kΩ resistor.

Solution   The peak value of the capacitor voltage equals the feedback voltage on the noninverting input. For this case,

$$V(+) = +11 \text{ V} \times 10 \text{ k}\Omega/(33 \text{ k}\Omega + 10 \text{ k}\Omega) = 2.56 \text{ V}$$

The capacitor voltage should swing between +2.56 V and –2.56 V, or 5.12 V peak-peak.

Because the time constant is the same, the frequency of oscillation should be higher as the voltage swings across the capacitor are less.

$$T = 0.94 \text{ ms} \times \ln\left(1 + \frac{20}{33}\right) = 0.45 \text{ ms, and } f = 2.2 \text{ kHz}$$

Finally, note that the op-amp itself establishes the peak to peak output voltage. Changing V^+ and V^- will change $\pm V_{sat}$ but will *not* affect the *frequency* of oscillation.

The Monostable Multivibrator

The *monostable multivibrator* is another example of a nonlinear application of the op-amp. This circuit is sometimes referred to as a *one-shot* because its output voltage switches states briefly but always returns to its original state.

The op-amp version of the one-shot is shown in Fig. 11-10 and typical waveforms in Fig. 11-11. In the stable state the output is at

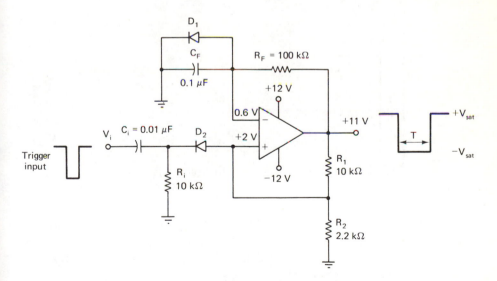

Figure 11-10 The op-amp monostable or one-shot circuit. A trigger pulse applied to the input causes an output of period $T = R_F C \ln\left(\dfrac{R_1 + R_2}{R_1}\right)$. Voltages on the diagram are indicated for the *stable* state.

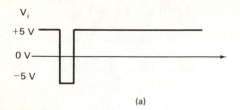

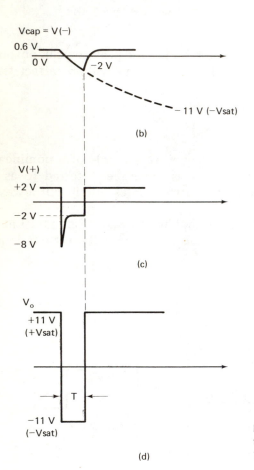

Figure 11-11 Voltage waveforms for the monostable multivibrator in Fig. 11-10.

$+V_{sat}$ (+11 V for the circuit in Fig. 11-10) and the inverting input is held at 0.6 V by diode $D1$. The $R_1 R_2$ voltage divider holds the non-inverting input at +2 V (2.2 kΩ/12.2 kΩ × +11 V). Thus E_d is positive and the output is biased at $+V_{sat}$.

The *negative* going edge of the trigger pulse must pull the *non-inverting* input below 0.6 V to initiate the output pulse. When this occurs the output will switch to $-V_{sat}$ and the $R_1 R_2$ voltage divider will cause E_d to become negative.

As soon as the output switches to $-V_{sat}$, capacitor C_F begins to charge towards this value. However, when it reaches -2 V $(R_2/(R_1 + R_2) \times -V_{sat})$ E_d switches and becomes positive again as the circuit reverts back to its stable state. The output pulse ends and $D1$ provides a fast discharge path for the capacitor in preparation for the next pulse.

The pulse width produced by the circuit is approximately

$$T = R_F C_F \times \ln(R_1 + R_2)/R_1 \qquad (11\text{-}13)$$

This equation is valid as long as the input time constant $R_i C_i$ is 4 to 5 times faster than the $R_F C_F$ time constant, as should normally be the case. In this case, and as illustrated in Fig. 11-11c, when the trigger occurs the noninverting input will quickly switch to -2 V before the inverting input has had a chance to fall below this level ensuring proper circuit operation.

Diode $D2$ prevents false triggering due to the *rising* edge of the trigger pulse.

Example 11-9 —————————————————————————————

Estimate the minimum trigger input amplitude and the pulse width for the monostable circuit shown in Fig. 11-10.

Solution   The trigger input must pull the (+) input to just below 0.6 V in order to cause the output to switch. In this case, the (-) input is biased at +2 V and V_i must therefore be at least 2 V peak-peak (1.4 V + 0.6 V across $D2$).

The pulse width is given by Eq. 11-13.

$$T = 100 \text{ k}\Omega \times 0.1 \ \mu\text{F} \times \ln 1.22 = 2 \text{ ms}$$

11.4   PRACTICAL CONSIDERATIONS

Although the op-amp designer has done a great deal to make application of the op-amp foolproof, there are instances when the op-amp does not behave as an *ideal* amplifier. It is the purpose of this section to point out these problem areas.

In most cases, careful circuit design can circumvent the problem. Occasionally, a higher quality op-amp may have to be selected. And even less occasionally, the design goal simply may not be achievable with an op-amp.

The problems to be discussed are related to use of the op-amp as a *dc* amplifier and as an *ac* amplifier. If it is desired to only amplify ac signals, *coupling capacitors* can be used and most of the dc-related problems, *bias currents* and *offset voltages*, will not be significant. In this case, ac-related problems such as *slew rate* and *bandwidth* must be considered.

Although it is a common assumption that no current flows into the op-amp input terminals, a small *bias current* must flow into each. Recall that the inputs to the op-amp are actually the base connections of two transistors connected in a differential arrangement similar to the circuit shown in Fig. 10-11. Even when these inputs are grounded, a current flows into each base because the emitters are at a negative potential due to $-V_{EE}$.

As an example, consider the circuit shown in Fig. 11-12. This is a capacitively-coupled noninverting amplifier. And it doesn't work! The output voltage will be at V_{sat}. The reason for this is that the coupling capacitor will *block* the dc base current for the noninverting input transistor. Accordingly, it must be biased *off* and the output stage saturated. A resistor to ground at this input would allow base current to flow and normal circuit operation.

The inverting amplifier would not have this problem because its bias current is supplied through R_1 and R_F. A capacitor blocking the path through R_1 would require the entire bias current to be supplied by the output stage itself through R_F.

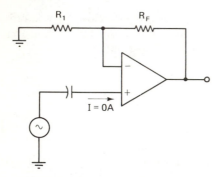

Figure 11-12 A capacitively-coupled noninverting amplifier. This circuit does not work because the capacitor blocks the path for the input bias current.

Now that we see that the op-amp must have bias currents, what is their effect? The basic difficulty with all dc-related op-amp problems is that the output voltage does not equal *zero* even though the input voltage is *zero*. This is actually an inherent problem in all *dc* amplifiers, as discussed in Section 7.3. What is worse, if the output voltage of one op-amp is fed to another, this second op-amp will amplify the dc *offset* voltage of the first, resulting in an even larger net dc output voltage.

Figure 11-13 illustrates an inverting amplifier with its input at ground potential. As previously mentioned, both op-amp inputs must be supplied with small bias currents. In this case, the noninverting input receives I_B^+ from ground while the inverting input receives I_B^- through

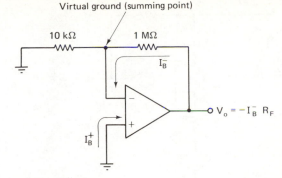

Figure 11-13 The effect of bias currents on the inverting amplifier. A voltage is developed due to the bias current flowing through R_F.

resistor R_F. Because the output is not saturated, $E_d = 0$ V, and a *virtual ground* appears at the summing point. For this reason, no current flows through the 10-kΩ resistor.

Example 11-10 ───────────────────────────────

Assume the op-amp in Fig. 11-13 is a μA741 and estimate the output voltage due to bias currents.

Solution The output voltage is $I_B^- \times R_F$, where $I_B^- = 500$ nA maximum for the μA741 (see Fig. 11-2). Then, $V_o = 0.5\ \mu$A $\times$ 1 MΩ = 0.5 V.

This example illustrates that the output voltage of an op-amp may not be 0 V even though the circuit input is at ground. The I_B^- bias current develops a voltage across R_F, which directly becomes the output voltage. Of course, one immediate solution to this problem should be obvious. Do not use large value (megohm) resistors in the feedback path.

Figure 11-14 illustrates a general op-amp circuit that could represent the inverting or noninverting amplifier as well as the voltage follower. In all cases, the input source is assumed adjusted to 0 V. After some algebra, the following equation can be derived for the output voltage due only to bias currents.

$$V_o = I_B^- R_F - I_B^+ R \left(1 + \frac{R_F}{R_1}\right) \tag{11-14}$$

where R represents the *source* resistance for the noninverting amplifier and R_1 includes the source resistance for the *inverting* amplifier. Note that the circuit in Fig. 11-13 has $R = 0\ \Omega$, and Eq. 11-14 then reduces to $V_o = I_B^- R_F$, as used in Example 11-10.

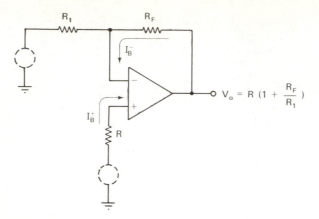

Figure 11-14 A general op-amp circuit illustrating the effect of the two bias currents.

Now a clever inspection of Eq. 11-14 reveals that if $R = (R_F \| R_1)$, V_o becomes

$$V_o = I_B^- R_F - I_B^+ R_F = R_F (I_B^- - I_B^+) = R_F I_{os} \qquad (11\text{-}15)$$

where I_{os} is the *offset current*. What we have done is to make V_o depend on the *difference* in bias currents. Because the transistors in an integrated differential pair are extremely well matched, this difference is quite small. The quantity $I_B^- - I_B^+$ is referred to as the *offset current* and is typically one fourth of the bias current value.

Example 11-11 ──

Modify the inverting amplifier shown in Fig. 11-13 to have a minimum output voltage due to input bias currents. Estimate the output voltage.

Solution The circuit is shown in Fig. 11-15. Resistor R, sometimes called the *current compensating resistor*, has been added to reduce the output voltage to $R_F I_{os}$. The value of R is $R = R_F \| R_1 = 1$ M$\Omega \| 10$ k$\Omega = 9.9$ kΩ. The output voltage is $V_o = R_F I_{os} = 1$ M$\Omega \times 0.2$ μA $= 0.2$ V where $I_{os} = 200$ nA maximum for the μA741.

What if 0.2 V is still too high for your application? Your best bet is to choose a higher quality op-amp. The LM301A op-amp has $I_{os} = 50$ nA. If this is still too high, a biFET op-amp such as the CA3130 can be used. This circuit employs an FET differential input stage to achieve very small input bias currents (typically 0.005 nA, as indicated in Fig. 11-2). With this circuit, I_{os} is 0.03 nA maximum, reducing V_o to 30 μV.

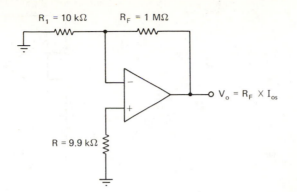

Figure 11-15 The inverting amplifier in Fig. 11-13 is improved by the addition of the 9.9-kΩ compensating resistor.

Input Offset Voltage

Even when the input bias currents are compensated for, the output voltage may still not be 0 V. The problem is due to *imbalances* within the op-amp. Transistors and resistors, though closely matched, are not *exactly* matched and, due to the large open loop gain possible with the op-amp, may cause a significant output voltage. The amount of voltage necessary at the input terminals to force the output back to 0 V is called the *input offset voltage*, V_{io}. Referring to Fig. 11-2, this ranges from 1 to 5 mV for the μA741.

A good example of this problem is illustrated in Fig. 11-16a. The effect of V_{io} is symbolized by a source connected to the noninverting input, but in reality both inputs are at ground potential and V_{io} is *internal* to the op-amp. Because the circuit is being used in an *open-loop* configuration, the output voltage is $V_o = A_{OL} \times V_{io}$, which will result in $\pm V_{sat}$, depending on the polarity of V_{io}. The point is, even though V_{io} is only millivolts, this is sufficient to drive the op-amp to saturation.

Figure 11-16b and c illustrate the same problem for the *closed-loop* cases. Note that when feedback resistors are used, the effect is less severe but still significant.

Example 11-12 ——————————————————————————————————

Estimate the output voltage of the amplifier in Fig. 11-13 due to input bias currents *and* input offset voltage. Assume a 741 type op-amp.

Solution In Example 11-10, the output was shown to be 0.5 V due to bias cur-

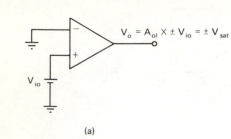

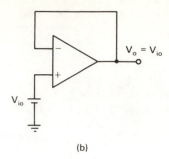

(a)   (b)

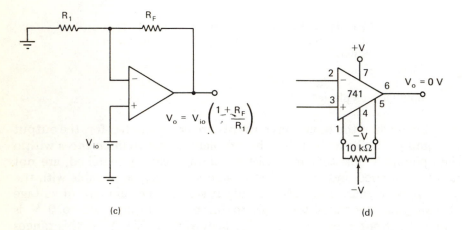

(c)   (d)

Figure 11-16 The effect of input offset voltage for (a) the open-loop case and (b) and (c) the closed-loop case. In (d), a 10-kΩ nulling potentiometer is used to reduce the outputs to O V.

rents. Referring to Figs. 11-16c and 11-13, the output voltage due to the input offset voltage is

$$V_o = \left(1 + \frac{1 \text{ M}\Omega}{10 \text{ k}\Omega}\right) \times 5 \text{ mV} = 101 \times 5 \text{ mV} = \pm 0.505 \text{ V}$$

Depending on the polarity of V_{io}, the output voltage may be 1.005 V, or –0.005 V.

What can be done about this problem? Most manufacturers have provided pins on the op-amp package to which a *nulling potentiometer* can be connected. An example for the 8-pin 741 is shown in Fig. 11-16d. This pot is adjusted until $V_o = 0$ V. The op-amp is then said to be *nulled* or *balanced*.

In summary, dc problems with the op-amp cause the output voltage to be nonzero even though the input is 0 V. Bias currents can be compensated for by the addition of a *compensating resistor* in series with the noninverting input. It is also good practice to avoid the use of large-value resistors in the feedback path. Input offset voltage can be *nulled* by the addition of an external potentiometer as indicated by the manufacturer.

Unfortunately, all of these solutions are valid at one temperature only and in general all affects will *drift* with time and temperature due to component aging and temperature sensitivity. In critical applications, components with low drift rates ($\mu V/^\circ C$, $nA/^\circ C$) should be selected.

The following sections detail ac problems of the op-amp.

Slew-Rate Limiting

Slew rate refers to how fast the amplifier output voltage can switch between its two extremes of V_{sat}. Referring to Fig. 11-2, the typical slew rate for the $\mu A741$ is 0.5 V/μs.

Example 11-13 ───────────────────────────────

If ±12 V power supplies are used with a 741 type op-amp, how long will it take for the output to switch from $+V_{sat}$ to $-V_{sat}$?

Solution   Assuming $\pm V_{sat} = \pm 11$ V, the output must switch a total of 22 V. At 0.5 V/μs, this will require 22 V/(0.5 V/μs) = 44 μs.

The slew rate of an op-amp is analgous to the *rise time* of a logic gate. However, logic gates can switch much faster (0 to 5 V in 5 to 10 ns). The cause of the slow switching time of the op-amp is an internal or external *compensating capacitor* designed into the circuit to stabilize frequency response and prevent oscillations (recall the 30-pF on-chip capacitor of the 741). When the output of the op-amp switches states, this capacitor must be charged. The problem is the op-amp has a limited amount of current to do this charging such that

$$Q = I \times t$$

and as the charge slowly builds up, so does the capacitor (and output) voltage;

$$Q = C \times V$$

Equating these two values of charge,

$$I \times t = C \times V$$

The rate at which the voltage builds up is called the *slew rate*.

$$\text{slew rate} = \frac{V}{t} = \frac{I}{C} \tag{11-16}$$

Example 11-14 ——————————————————————————————————————

Assume a certain op-amp can charge its internal compensating capacitor with a 100-μA current. Calculate the slew rate if the capacitor is 50 pF.

Solution Plugging into Eq. 11-16,

$$\text{slew rate} = \frac{100 \ \mu\text{A}}{50 \ \text{pF}} = 2 \times 10^6 \ \text{V/sec} = 2 \ \text{V}/\mu\text{s}$$

The affect of slew rate is to *distort* the output signal if the input signal demands a slew rate greater than the op-amp is capable of delivering.

As an example, a square wave input is shown in Fig. 11-17a. Due to slew rate limiting, the output waveshape appears as in Fig. 11-17b. Ultimately, a frequency may be reached where the output is totally unable to follow the input signal.

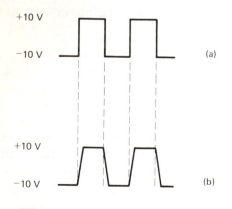

Figure 11-17 Slew-rate limiting distorts the input square wave (a) to become the waveform, shown in (b).

When the input is a *sine wave*, slew rate limiting is most pronounced as the sine wave crosses 0 because it is here that the sine wave changes at its *fastest* rate. For a sine wave of peak value E_p and frequency f, the op-amp must have a slew rate of at least

$$\text{slew rate (sine wave)} = 2\pi \ f E_p \tag{11-17}$$

Example 11-15 ——————————————————————————————————————

Determine the maximum operating frequency for a 741 type op-amp without slew-rate limiting for a sine wave input with (a) $E_p = 5$ V; (b) $E_p = 10$ V.

Solution   Rearranging Eq. 11-17

$$f = \frac{\text{slew rate}}{(2\pi \, E_p)}$$

Then for (a)

$$f(\text{max}) = \frac{0.5 \text{ V}/\mu\text{s}}{2\pi \times 5} = 15.9 \text{ kHz}$$

and for (b)

$$f(\text{max}) = \frac{0.5 \text{ V}/\mu\text{s}}{2\pi \times 10} = 7.96 \text{ kHz}$$

This example illustrates that slew-rate limiting is a *large signal* problem for the op-amp. The larger the output signal swing, the lower the frequency at which slew-rate limiting will occur. When $E_p = V_{sat}$, the slew rate limiting frequency in Eq. 11-17 is called the *full power* bandwidth.

For internally-compensated op-amps, there is little that can be done to improve the slew rate. Externally-compensated circuits can use *smaller* compensating capacitors but at the risk of poorer circuit stability. As was the case with bias currents, a higher quality op-amp can be selected.

Frequency Response

As discussed in Chapter 4, the voltage gain of all amplifiers will eventually fall off at higher frequencies. This is due to internal and external *capacitances* becoming active as frequency increases, shunting the signal to ground. In Section 4.5 it was shown that the rate of decrease in gain is *20 dB/decade* for one active capacitor.

Although it is desirable to have a large amplification bandwidth, this may also lead to potential oscillations at higher frequencies. An oscillator is by definition a circuit that generates its own input (and is thereby self-supporting). An amplifier can become an oscillator if its output signal has the *same phase* as its input and if the gain of the amplifier at this frequency is 1, thereby exactly regenerating itself. Because the op-amp already provides 180° of phase shift through its inverting input, the potential exists for some frequency to pass through the op-amp and acquire an additional 180° of phase shift. This meets the first requirement of an oscillator — 360° of total phase shift (input and output in phase). Depending on the gain at this frequency, instantaneous oscillations could result.

As we said earlier, the op-amp has been designed to be foolproof and one of these considerations has been protection against oscillations. In order to ensure that the op-amp behaves as an amplifier and not as an oscillator, its gain is made to roll off at a very low frequency due to an internal or external *compensating capacitor*. In this way, the gain is reduced *below unity* before the additional 180° of phase shift can occur and oscillations are prevented.

A typical frequency response curve for the μA741 is shown in Fig. 11-18. Note that the open loop gain begins to roll off at only 10 Hz! From this point on, the gain decreases by 20 dB/decade such that unity gain occurs at 1 MHz. Because the gain does roll off in such a predictable manner, the following equation can be used to estimate the value of A_{OL} at any frequency.

$$A_{OL} = \frac{B_{OL}}{f} \tag{11-18}$$

where B_{OL} is the *open-loop bandwidth* or frequency where the open-loop gain is 1 (1 MHz in Fig. 11-18).

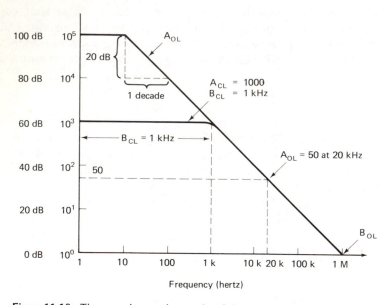

Figure 11-18 The open-loop voltage gain of the op-amp begins to fall off at 10 Hz. Also illustrated is the frequency response for a closed-loop gain of 1000. In this case, the gain (A_{CL}) is flat to 1 kHz.

Example 11-16 ———————————————————————————

Determine the open loop gain of the μA741 at $f = 20$ kHz. Assume $B_{OL} = 1$ MHz.

Solution   From Eq. 11-18,

$$A_{OL} = \frac{B_{OL}}{f} = \frac{10^6}{20 \times 10^3} = 50$$

This point is also shown in Fig. 11-18.

This example introduces an interesting point. If A_{OL} is only 50 at 20 kHz, is it still valid to assume $E_d = 0$ V? The answer is no. $E_d = V_o / A_{OL}$, and as A_{OL} *falls*, E_d must *rise*. Therefore, as frequency increases, we will find more of the input voltage dropping across the op-amp input terminals and less and less amplification taking place.

But this is only logical. We really couldn't assume $A_{CL} = -R_F / R_1$ out to some *infinite* frequency, particularly when the open-loop gain itself begins to fall off at only 10 Hz!

The frequency response for a closed-loop gain of 1000 (60 dB) is shown in Fig. 11-18. As long as A_{OL} is greater than 1000, the closed-loop gain remains constant at 60 dB. However, once $A_{OL} = A_{CL}$, both gains must roll off together at a 20dB/decade rate. We can say that the *closed-loop bandwidth* is 1 kHz (actually A_{CL} is 3 dB down at this point), meaning that A_{CL} is flat to 1 kHz and then falls off at a 20 dB/decade rate.

By studying Fig. 11-18, we can see that the *lower* the closed-loop gain, the *greater* the bandwidth ($B_{CL} = 20$ kHz, for $A_{CL} = 50$). Finally, when $A_{CL} = 1$, the bandwidth is the full 1 MHz (B_{OL}). In equation form,

$$B_{CL} = \frac{B_{OL}}{A_{CL}} \tag{11-19}$$

where B_{CL} = closed loop bandwidth
  B_{OL} = open loop bandwidth
  A_{CL} = closed loop gain.

Example 11-17

Determine the maximum closed-loop gain for a 100-kHz bandwidth. Assume a μA741 op-amp.

Solution   From Eq. 11-19,

$$A_{CL} = \frac{B_{OL}}{B_{CL}} = \frac{10^6}{100 \times 10^3} = 10$$

Another way of interpreting EQ. 11-19 is to say that *the gain bandwidth product is constant*. Multiplying A_{CL} (gain) by B_{CL} (bandwidth) always results in B_{OL}, which is constant. In a sense, you trade gain for bandwidth. The more closed-loop gain desired, the less bandwidth over which it is available. Although a voltage gain of 10,000 sounds great, it is only available to 100 Hz. On the other hand, a gain of 10 has a 100 kHz bandwidth.

The open loop bandwidth of the op-amp is set by its compensating capacitor. When this is *internal* (as with the 741), it cannot be changed. Op-amps such as the LM301A allow some leeway because the capacitor is external to the package. In any case, typical open-loop bandwidths are between 1 and 10 MHz.

In summary, *frequency response* and *slew-rate limiting* may be the most serious limitations to the nonideal op-amp. In both cases, if there are problems in meeting your ac design goals, little can be done but to select an op-amp with better specifications.

KEY TERMS

Operational Amplifier   Multistage integrated amplifier featuring differential inputs, high input resistance, low output resistance, and extremely large open-loop voltage gain (A_{OL}).

Linear Operation   Operation of the op-amp such that the output voltage is limited to values less than V_{sat} and governed by $V_o = E_d \times A_{OL}$, where E_d is the differential input voltage.

Nonlinear Operation   A digital or switching-like application in which the output voltage equals $\pm V_{sat}$. For nonlinear operation, it is *not valid* to assume $E_d = 0$ V.

Open-Loop Voltage Gain A_{OL}, gain of the op-amp without any feedback. Typically, A_{OL} is greater than 100,000 at dc.

Closed-Loop Voltage Gain A_{CL}, gain of the op-amp with feedback. In this case, resistor ratios determine the circuit gain.

Feedback   Process of feeding a portion of the output voltage back to the input. Feedback may be *negative*, which reduces overall amplifier gain but stabilizes the circuit, or *positive*, which tends to cause regeneration and oscillations.

Virtual Ground   In linear op-amp circuits, the inverting input of the op-amp is at ground potential when the noninverting input is grounded. This is because E_d is nearly 0 V due to the large open-loop gain. This point is called a virtual ground because it is not actually *connected* to ground.

Output Offset Voltage   Output voltage of the op-amp when the input is 0 V. This voltage may be due to input bias and offset currents and the input offset voltage. Careful circuit design and the use of a *nulling potentiometer* can reduce this voltage to 0 V.

Slew Rate   A measure of the rate at which the output voltage can switch between one value of V_{sat} and the other. Usually measured in V/μs. Slew rate limiting occurs when the op-amp cannot follow the input signal and a distorted output signal occurs.

Open-Loop Bandwidth   Frequency at which the open-loop gain of the op-amp has fallen to 1.

Closed-Loop Bandwidth   Frequency at which the closed-loop gain is down by 3 dB from its midband value.

──────────────── QUESTIONS AND PROBLEMS ────────────────

11-1   What determines the actual values for $\pm V_{sat}$ in a typical op-amp?

11-2   What is the maximum peak-peak output voltage for a μA741 op-amp with ± 15-V power supplies? How small a load resistance will this amplifier typically drive with a full output signal swing?

11-3   Refer to Fig. 11-19 and determine the output voltage for each circuit. Are the op-amps operating *linearly* or *nonlinearly* in these circuits? Explain. Assume $\pm V_{sat} = \pm 10$ V.

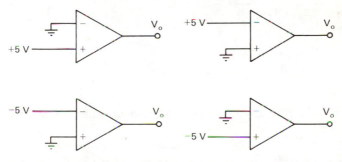

Figure 11-19

11-4   Refer to a linear data catalog and prepare a chart similar to Fig. 11-2 for the following op-amps: LM709, LM318, CA3140.

11-5   Redo Example 11-3 if $V_i = -0.2$ V. Show the direction of all currents.

11-6   Calculate currents I_1, I_2, and I_3 in Example 11-4. From this, calculate the output voltage.

11-7   Refer to Fig. 11-6; why is the inverting input *not* a virtual ground?

11-8   Show the schematic diagram of an eight-pin 741 op-amp connected as an inverting amplifier with $R_i = 5$ kΩ, and $A_{CL} = -75$. Use ± 12-V power supplies.

11-9   Determine the maximum input voltage to just cause saturation of the amplifier in Problem 11-8.

11-10   Determine the output voltage for each circuit in Fig. 11-20. Ignore offset voltages and bias currents.

11-11   If the op-amp in Fig. 11-20a is an LM301A, what is the typical minimum load resistance this circuit will drive? *Hint:* Refer to Fig. 11-2.

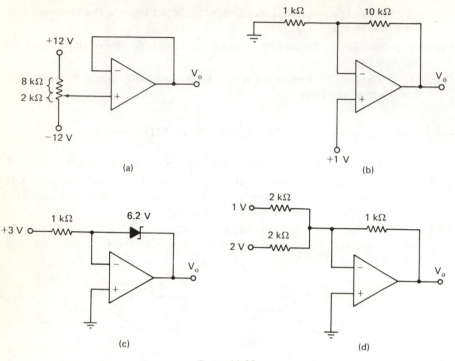

(a)   (b)

(c)   (d)

Figure 11-20

11-12 In Fig. 11-20b, determine the total op-amp source or sink current for a 1-kΩ load.

11-13 The circuit in Fig. 11-20d can be called an *averaging amplifier*. Explain why.

11-14 Why is a voltage comparator a nonlinear application of the op-amp?

11-15 Refer to Fig. 11-8; redraw the input and output waveforms (without the zener diode) if the positive and negative op-amp input connections are reversed.

11-16 Explain the operation of the circuit in Fig. 11-21.

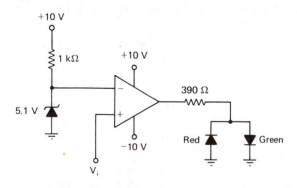

Figure 11-21

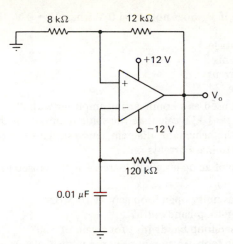

Figure 11-22

11-17 Sketch the capacitor and output voltage waveforms for the circuit in Fig. 11-22.

11-18 Why is the frequency of the astable multivibrator not affected by the magnitude of its two power supplies?

11-19 Sketch $V(+)$, $V(-)$, and V_o for the circuit shown in Fig. 11-10 if R_1 is changed to 1 kΩ and R_F to 470 kΩ.

11-20 What is the minimum trigger amplitude for the circuit described in Problem 11-19?

11-21 Why do resistors R_1 and R_2 affect the pulse width of the monostable multivibrator in Fig. 11-10?

11-22 Refer to Fig. 11-3. Calculate the maximum output voltage due to bias currents and the input offset voltage if $V_i = 0$ V, $R_1 = 22$ kΩ, and $R_F = 100$ kΩ. Assume an LM301A op-amp.

11-23 Redraw the schematic diagram for Problem 11-22 such that the effects of the bias currents are minimized. Recalculate the total output voltage.

11-24 In Problem 11-23, the output voltage is still not 0 V due to the input offset voltage. What last step must be done to null the output voltage to 0 V?

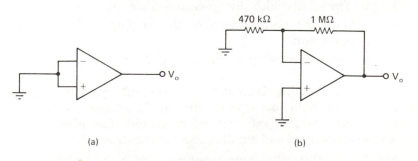

(a)   (b)

Figure 11-23   Input offset test circuits for Part III.

11-25 Refer to Fig. 11-5a; if V_o does not equal 0 V when $V_i = 0$ V, the most likely cause is
(a) Input offset voltage
(b) Input bias currents
(c) Input offset currents
(d) All of the above

11-26 A CA3130 is being used as a noninverting amplifier with $R_F = 2.2$ MΩ, and $R_1 = 100$ kΩ. If $R_G = 1$ kΩ, calculate the value required for the compensating resistor, draw the schematic diagram, and estimate the output voltage for a 0-V input due to bias currents only.

11-27 The closed-loop gain of an op-amp will increase as the closed-loop bandwidth increases. (True/False)

11-28 A certain op-amp has unity open loop gain at 10 MHz.
(a) What is the open-loop bandwidth?
(b) What is the closed-loop bandwidth for a gain of 100?
(c) Sketch the open-loop frequency response in decibels. Assume the open-loop gain becomes constant for frequencies less than 10 Hz.

11-29 A certain op-amp has a closed-loop gain of 1000 at 2 kHz. What is the open-loop bandwidth?

11-30 A certain op-amp has $V_{sat} = 10$ V, and a slew rate of 7.5 V/μs. What is the full power bandwidth?

11-31 It is desired to amplify a 20-kHz sine wave by 50 dB and obtain a 10-V peak-peak output. Determine the minimum slew rate and open-loop bandwidth required.

11-32 An op-amp has an open-loop gain of 105 dB at 10 Hz. What is the closed-loop bandwidth for a 30-dB gain.

11-33 Explain the statement, "In an op-amp with feedback, you trade gain for bandwidth."

_____ LABORATORY ASSIGNMENT 11: _____
THE OP-AMP

Objectives

1. To gain familiarity with the op-amp as a circuit element.
2. To measure the performance of the op-amp in linear and nonlinear applications.
3. To observe the ac and dc limitations of the op-amp.

Introduction In this laboratory assignment, you will assemble several linear and nonlinear op-amp circuits. The closed-loop frequency response as well as slew rate will be measured. The effects of input offsets will be observed and a nulling circuit constructed.

NOTE: In this lab, three op-amps, the μA741, LM301A, and

CA3130, will be used. All have identical pin numbers in the eight-pin package. The 301A and 3130 require a 30-pF (301A) and 50-pF (3130) *external compensating capacitor* between pins 1 and 8.

Components Required

1 μA741 op-amp

1 LM301A op-amp

1 CA3130 op-amp

Miscellaneous ¼-W resistors

Miscellaneous capacitors

Semilog graph paper

Part I: *Linear Operation*

A. *The Inverting Amplifier*

STEP 1   Assemble the inverting amplifier circuit shown in Fig. 11-3a. Use R_1 = 1 kΩ, and R_F = 10 kΩ.

STEP 2   Apply +1-V *dc* and measure the output voltage and the voltage at the inverting op-amp input. Compute the voltage gain.

Question 1   Is this circuit operating linearly? How can you tell?

STEP 3   Repeat step 1 using a 1-V peak-peak sine wave at f = 1 kHz.

STEP 4   Increase the ac input amplitude until the output saturates. Note the values of $\pm V_{sat}$ and compare to the expected voltages.

B. *Frequency Response*

STEP 1   Choose values for R_1 and R_F to obtain the following closed-loop gains at 100 Hz: 1000, 100, 1. Measure and record the frequency response from 100 Hz to 1 MHz for each of these gain values. Repeat, using the CA3130 and 50pF compensating capacitor between pins 1 and 8.

STEP 2   Use semilog graph paper and plot A_{CL} in dB versus frequency for both op-amps.

Question 2   From your graph, determine for both op-amps

(a) The rate of decrease in dB/decade of A_{CL}

(b) The closed-loop bandwidth for each voltage gain

(c) The approximate open-loop bandwidth.

C. Slew Rate

STEP 1 Modify your 741 circuit to become a voltage follower.

STEP 2 Adjust the ac signal generator to $f = 20$ kHz, and a 1-V peak-peak square wave input.

STEP 3 Slowly increase the amplitude of the input signal until the effect of slew rate limiting can be observed. Measure the rate of change of the output voltage in $V/\mu s$ and compare to the 741 specification.

STEP 4 Repeat steps 2 and 3 using the LM301A and CA3130 op-amps. Be sure to add the compensating capacitor to each.

D. The Noninverting Amplifier

STEP 1 Design and test a circuit to produce a noninverting voltage gain of 100 ±5%. Show a complete schematic diagram with pin numbers.

Part II: *Nonlinear Operation*

A. The Voltage Comparator

STEP 1 Construct the simple comparator circuit illustrated in Fig. 11-7a using a 741 op-amp.

STEP 2 Apply a 5-V peak-peak ac input at $f = 1$ kHz, and record the output waveforms (be sure to show phase relative to the input) for each case of V_{ref}: (a) 0 V, (b) +1 V, (c) –2 V, and (d) +6 V.

STEP 3 Design and test a voltage comparator to convert a 20-V peak-peak sine wave to a TTL compatible (~5 V and 0 V) 60-Hz square wave. Show your complete schematic diagram and record the input and output waveforms.

B. The Astable Multivibrator

STEP 1 Construct the 741 astable multivibrator circuit shown in Fig. 11-9a.

STEP 2 Record the waveforms across the capacitor and output terminal, showing their relative phases. Measure the frequency of oscillation and compare to calculations.

STEP 3 Try replacing R_F with a 50-kΩ potentiometer and measure the extremes in frequency of oscillation possible.

Question 3 What factors do you think limit the maximum frequency of this circuit?

C. The Monostable Multivibrator

STEP 1   Design an op-amp one-shot circuit to meet the following specifications:

1. Pulse width = 1 ms ± 10%
2. Minimum trigger amplitude = 2.5 V peak-peak

STEP 2   Assemble your circuit and test that both specifications have been met. *Note:* Use a signal generator to repeatedly trigger your circuit and observe the output waveform on an oscilloscope.

STEP 3   Record the four waveforms shown in Fig. 11-11 for your circuit.

Part III: *Input Offsets*

STEP 1   Set up the circuit shown in Fig. 11-23a. Measure the output voltage separately for all three op-amp devices.

Question 4   Explain why the output does not equal zero volts.

STEP 2   Using a 741, rewire the circuit as shown in Fig. 11-23b. Again, measure the output voltage.

Question 5   Why has the output voltage decreased compared to the circuit in step 1?

STEP 3   Calculate the value of compensating resistor required for this circuit. With this resistor in place, remeasure the output voltage.

STEP 4   Connect a 10-kΩ pot between pins 1 and 5, with the middle arm connected to –12 V. Adjust the output for 0 V.

STEP 5   With the output nulled, apply a 10-mV dc input through the 470-kΩ resistor and measure the output voltage. Compute the voltage gain. Compare to calculations (measure R_F and R_1 for optimum accuracy).

TWELVE

APPLICATIONS OF OP-AMPS

The *integrated operational amplifier* is the most widely used linear integrated circuit in production today, and for good reason. It features dc gains in excess of 100 dB, nanoamp input currents, millivolt input offset voltages, and megahertz bandwidths. Combine this with the inherent reliability and size reduction possible with integrated circuits, a typical cost of less than one dollar, and you have a circuit designer's dream come true.

Applications of op-amps are growing daily as older discrete component designs are phased out in favor of the op-amp. As indicated in Chapter 10, the op-amp can now be designed into systems in which an operational amplifier is actually a case of *overkill* but, due to its low cost and compact size, becomes a reasonable, even necessary, choice.

In this chapter, selected applications of the op-amp are covered. The examples are chosen to illustrate the wide diversity of applications possible with this component. This will by no means be a complete coverage and the reader is encouraged to continue his studies of op-amps and linear integrated circuits in any of the many texts now available.*

12.1 A REGULATED DUAL-OUTPUT TRACKING POWER SUPPLY

Various power supply circuits were discussed in Chapter 2 and the zener diode was introduced as a *voltage regulator*. Recall that a regulated

---

*For example, see Walter G. Jung, *IC OP-AMP COOKBOOK* (Indianapolis, IN: Howard W. Sams and Co., 1975); Robert F. Coughlin and Frederick F. Driscoll, *Operational Amplifiers and Linear Integrated Circuits*, (Englewood Cliffs, NJ: Prentice-Hall, 1977); Howard M. Berlin, *Design of Active Filters With Experiments*, (Indianapolis, IN: Howard W. Sams and Co., 1978).

power supply is one that has a stable (nonchanging) output voltage as the load current varies. This is an essential characteristic of any realistic power source.

The power supply shown in Fig. 12-1 provides both negative and positive regulated output voltages. The two outputs will *track* each other as the single voltage control is adjusted. This means that when V_{o1} = +5 V, V_{o2} = –5 V, and so on up to the limits of the supply.

The Unregulated Source

An *unregulated* power supply consists of a transformer to step the ac line voltage down (or sometimes up) to the value desired, a rectifier to convert the ac to pulsating dc, and a filter to smooth out the pulsations.

In Fig. 12-1, this portion of the circuit is in the top of the schematic diagram and has two outputs, labeled V^+ and V^-. Note that diodes D_1 through D_4 are connected in a *bridge* configuration but, due to the grounded transformer center tap, are actually wired as two *full-wave* rectifiers in parallel. D_2 and D_3 provide the positive output while D_1 and D_4 provide the negative output. An LED serves as a pilot light and indicates the presence of dc at the V^+ output.

Example 12-1

Assume T_1 in Fig. 12-1 is a 24 VCT 1 A transformer and that a load current of 0.5 A is drawn from each output. Determine values for C_1, C_2, and R_8. Maximum ripple voltage across either capacitor should be 0.75 V peak-peak.

Solution   Because C_1 and C_2 must work under identical load conditions, their values must be the same. Applying Eq. 2-4,

$$C_1 = C_2 = \frac{I}{\Delta V f} = \frac{0.5 \text{ A}}{(0.75 \text{ V})120 \text{ Hz}} = 5555 \ \mu F$$

The peak output voltage is $(24/2 \times \sqrt{2}) - 0.6 = 16.4$ V. A possible standard value for the two capacitors would be an electrolytic type rated at 6300 μF, 25 V.

The value of R_8 is found assuming 20 mA of LED current and a 1.6-V diode drop; R_8 = (16.4 V – 1.6 V)/20 mA = 740 Ω. Its power rating is $(20 \text{ mA})^2 \times 740$ Ω = 0.3 W. A 750-Ω ½-W resistor (closest standard value) can be selected.

Op-Amp Error Amplifier

Now that the unregulated power sources have been designed, the *regulator* circuit can be examined. This circuit, with output labeled V_{o1}, consists of a zener diode voltage reference and one half of a 747 dual

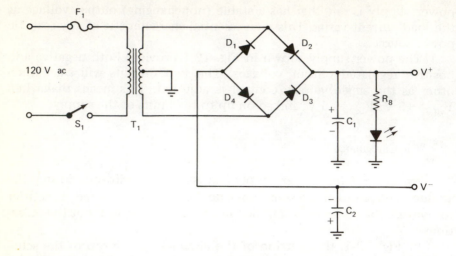

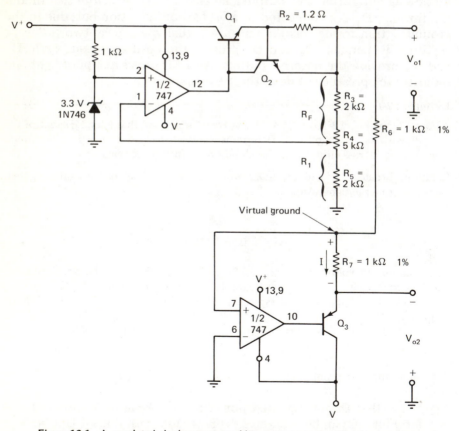

Figure 12-1 A regulated dual-output tracking power supply. The two op-amps are used as noninverting (V_{o1}) and inverting (V_{o2}) amplifiers.

op-amp integrated circuit. The 747 is similar to the 741 but contains *two* op-amps in one 14-pin package.

Because the maximum output current of the op-amp is limited to 25 mA, an external *pass transistor*, Q_1, is necessary. The actual output current absorbed by the load passes through this transistor (and resistor R_2) but *not* through the op-amp.

Note that the *error amplifier* (op-amp) is connected as a noninverting dc amplifier with R_F and R_1 as shown in Fig. 12-1.

Example 12-2

Refer to Fig. 12-1 and determine the output voltage for the two extremes of R_4 adjustment.

Solution The output voltage is governed by Eq. 11-10; $V_o = V_i(1 + R_F/R_1)$. In this case, $V_i = 3.3$ V, and R_F and R_1 vary as R_4 is adjusted. The two cases are illustrated in Fig. 12-2a and b.

$$V_o \text{ (min)} = \left(1 + \frac{2}{7}\right) \times 3.3 \text{ V} = 4.2 \text{ V}$$

$$V_o \text{ (max)} = \left(1 + \frac{7}{2}\right) \times 3.3 \text{ V} = 14.85 \text{ V}$$

Although Fig. 12-2 does not show transistors Q_1 and Q_2, it should be clear that they do not change the previous calculations. This is because the op-amp, if operating linearly, *forces* 3.3 V to appear on the sliding arm of resistor R_4. This in turn causes a specific current to flow in all three resistors (R_3, R_4, and R_5) and establishes the actual output voltage as the sum of these three voltage drops. See Fig. 12-2.

What then do these transistors accomplish? Initially, let's ignore Q_2. The base of Q_1 follows the output voltage but is approximately 0.6 V *higher* in potential. Now assume a load resistor is connected to the output. Under this condition the output voltage will have tendency to decrease due to the increased load current. This in turn means that the voltage fed back to the inverting input of the op-amp will also begin to decrease. Because the (+) input is now greater than the (–) input, the op-amp output voltage will increase. But this turns transistor Q_1 ON, pushing it towards saturation and decreasing its V_{CE} value. Because $V_o = V^+ - V_{CE}$, the output voltage will increase, *counteracting* the effect of the load.

The op-amp is actually used as an *error amplifier*, monitoring the output voltage and controlling the drive to Q_1. It employs *negative feedback* to counteract any changes in output voltage due to load variations. As long as the op-amp operates linearly, the output voltage will be regulated.

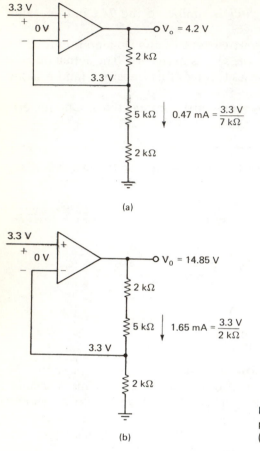

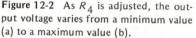

Figure 12-2 As R_4 is adjusted, the output voltage varies from a minimum value (a) to a maximum value (b).

Example 12-3

Determine the maximum possible output voltage of the power supply as limited by linear operation of the op-amp.

Solution In Example 12-1, we saw that $V^+ = 16.4$ V, but with 0.75 V of ripple. $\pm V_{sat}$ of the op-amp must then be approximately ±14.5 V. Due to Q_1, the circuit output voltage is 0.6 V below this, or 13.9 V.

Note that this result is less than the 14.85 V calculated in Example 12-2. Until saturated, the output voltage is given by $V_o = 3.3$ V $\times$ (1 + R_F/R_1). However, once the op-amp output equals V_{sat}, this equation is *no longer valid* and the op-amp no longer acts as an error amplifier.

Transistor Q_2 provides *current limiting*. As the load current flows through resistor R_2, a voltage is developed across the base-emitter junction of Q_2. When this voltage approaches 0.6 V, Q_2 begins to conduct, *stealing* base drive from Q_1. This in turn limits the output current that Q_1 can support.

Example 12-4 ———————————————————————

Determine a value for R_2 assuming a 500-mA maximum load current.

Solution When 500 mA flows through R_2, it must drop 0.6 V, or R_2 = 0.6 V/500 mA = 1.2 Ω. Its wattage rating is $(0.5 \text{ A})^2 \times 1.2 = 0.3$ W. Two 2.4-Ω ¼-W resistors in parallel could be used.

Example 12-5 ———————————————————————

Determine the maximum power dissipation of the pass transistor Q_1.

Solution The power dissipation is given by $I_C \times V_{CE}$. The worst case occurs when V_{o1} is minimum and I_C = 0.5 A, causing the maximum voltage to be dropped across the pass transistor. In this case, V_{o1} = 4.2 V, and V_{CE} = 16.4 − 4.2 = 12.2 V. P_d = 12.2 V × 0.5 A = 6.1 W. A suitable choice for Q_1 might be a 1-A, 25-V (BV_{CEO}) *npn* transistor. In order to dissipate the 6.1 W of heat, a suitable *heat sink* should be used.

Tracking Output

The last feature of Fig. 12-1 to be discussed is the *tracking output*. This circuit uses the other half of the 747 dual op-amp and transistor Q_3. Referring to the schematic diagram, assume resistors R_6 and R_7 are equal in value and closely matched. Due to the op-amp, these resistors are actually connected in *series* but their junction is at *virtual ground* (pin 7 is at 0 V).

Assume the output is adjusted to +5 V. A 5-mA current must flow through R_6 (5 V/1 kΩ) and also through R_7. Due to the virtual ground, the output voltage is the same as the drop across R_7, or V_{o2} = $-I \times R_7$ = −5 mA × 1 kΩ = −5 V. As R_4 is adjusted, V_{o2} will *track* V_{o1} but with the *opposite* polarity.

The op-amp in this circuit is actually used as an *inverting* amplifier with a gain of 1. Again, negative feedback helps to maintain a constant output voltage as the load current varies. Because this output voltage is negative, Q_3 must *sink* the load current and a *pnp* pass transistor is called for. No current limiting is shown for this output but could easily be added.

Three-Terminal Regulators

As we have just seen, the op-amp can be used to good advantage when a regulated dc output voltage is required. One disadvantage, however, is that several support components are needed to build a working circuit.

An integrated voltage regulator, called the *three-terminal regulator*, has changed this. This IC contains a stable voltage reference, error amplifier, pass transistor, and protection circuitry to prevent damage due to excessive current and heat. This device is extremely simple to use, having only an input and output terminal and ground connection. Typically, an unregulated voltage is applied to the input terminal and a regulated output voltage, dependent on the specific regulator chosen, is available at the output terminal.

Typical voltages available are 5, 6, 8, 12, 15, 18, and 24 V of either positive or negative polarity and with current ratings dependent on the package type. Figure 12-3 indicates three common packages used for these regulators. A representative type is the National Semiconductor LM78XX series of positive voltage regulators. The last two digits represent the specific output voltage. For example, the LM7805KC is a +5-V regulator in a TO-3 package capable of output currents in excess of 1.5 A. The LM7805T is the same device in the TO-220 case and is rated at 1.5 A if adequate heat sinking is provided. The LM78L05 is the low-current version of the same device packaged in the plastic TO-92 case. It is rated at 100-mA maximum output current.

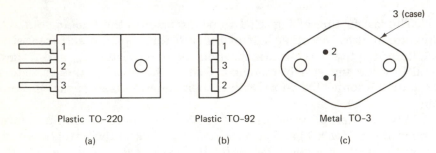

Plastic TO-220   Plastic TO-92   Metal TO-3

(a)   (b)   (c)

Figure 12-3 Three-terminal regulators are available in a variety of packages, dependent on their current rating. Pin numbers are referenced to Fig. 12-4.

Figure 12-4a illustrates the relative ease in which this regulator can be employed to construct a regulated +5-V power supply. Capacitor C_i is required only if the wire lengths between regulator and power supply filter capacitors are significant. Capacitor C_o is needed for transient (ON–OFF) response when this is important. Remember that these devices contain high-gain amplifiers subject to oscillations under certain local conditions. The capacitors help to eliminate this possibility.

Variable output voltages are also possible, as illustrated in Fig. 12-4b. In this case, the fixed 5-V output establishes a *constant current* (5 V/300 Ω), which together with the device quiescent current (I_Q) develops a voltage across the 1-kΩ potentiometer. The output voltage is then 5 V + ($I_T \times R_p$).

Three-terminal regulators have nearly eliminated the need for regulated high current power supplies in many types of electronic equip-

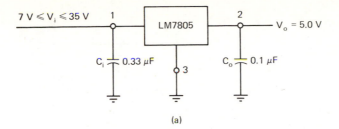

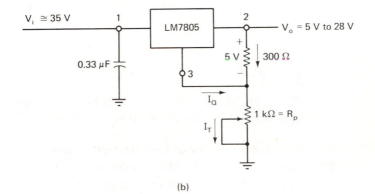

Figure 12-4 The three-terminal regulator can provide a fixed output voltage (a) or a variable output voltage (b).

ment. Instead, a high current *unregulated* supply can feed a number of smaller three-terminal regulators distributed throughout the system. For example, many computers now use one or two three-terminal regulators on each plug-in card, drawing their power from one centrally located but unregulated power source.

12.2 FIRST-ORDER ACTIVE FILTER

A *filter* is an electronic circuit designed to pass a band of frequencies while rejecting frequencies outside of this band. Filters find wide applications in communications circuits in which one frequency must typically be extracted from a wide spectrum of frequencies. Filters are also used to separate the logic 1 frequency from the logic 0 frequency produced by a modem.

A filter need not be a complex circuit. A simple RC *low-pass* filter is shown in Fig. 12-5. A rearrangement of the components in this circuit will transform it into a *high-pass* filter.

Because this circuit does not contain any active devices, it is called a *passive filter*. An *active filter*, on the other hand, uses an op-amp as

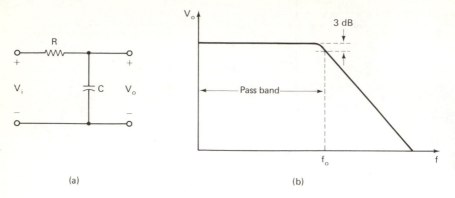

(a)   (b)

Figure 12-5 Simple RC low-pass filter. The output voltage is down by 3 dB at the break frequency (f_o) and falls of at 20 dB/decade thereafter.

part of the circuit. Active filters yield frequency response curves identical to passive component filters but offer these advantages:

1. High input resistance and low output resistance, meaning negligible *loading* effects between the external circuit and the filter.
2. No *attenuation* in the passband and, in fact, voltage gain is even possible.
3. The filter response can be achieved without inductors simplifying the circuit and lowering the overall cost.

Low-Pass Filter

The frequency response of the low-pass filter in Fig. 12-5 has been previously documented in Example 4-19. In that example, it was shown that the ratio of V_o to V_i decreased by 20 dB/decade, as is characteristic of all *single* capacitor or inductor filters.

The passive low-pass filter in Fig. 12-5 may be transformed into an active filter by placing a *voltage follower* on its output, as shown in Fig. 12-6. In this way, the load (R_L) is isolated from the filter and will not upset its frequency characteristics.

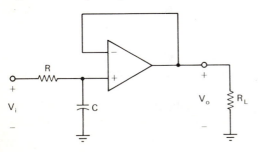

Figure 12-6 Active low-pass filter formed by adding a voltage follower to the output of the circuit in Fig. 12-5.

Example 12-6

Determine the shape of the frequency response for the circuit shown in Fig. 12-6. Assume $R = 470 \ \Omega$, and $C = 0.033 \ \mu F$.

Solution   As was shown in Example 4-19, the response of the filter will be down from its midband by 3 dB when $X_c = R = 1/(2\pi \, fC)$. Solving this for f,

$$f = f_o = \frac{1}{(2\pi \, RC)} \tag{12-1}$$

and in this case $f \cong 10$ kHz.

For frequencies above 10 kHz, the response will fall at a 20 dB/decade rate. The resulting frequency response curve is shown in Fig. 12-7, labeled $A_V = 1$.

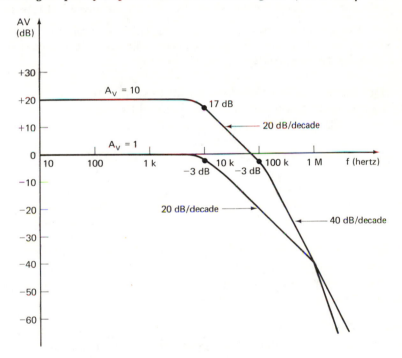

Figure 12-7 Frequency response for active filters with $A_V = 1$ and $A_V = 10$. Care must be taken to consider the frequency response of the op-amp itself. In this case, it adds an additional 20 dB/decade attenuation beginning at 100 kHz for the $A_V = 10$ case.

The circuit in Fig. 12-6 produces a frequency response *identical* to that produced by R and C alone (Fig. 12-5) but with the added advantage that the load resistor is buffered from the filter.

Often it is desired to filter a signal and amplify the result. With an active filter, these two operations can be combined in one circuit. Figure 12-8 illustrates an active filter with a gain of 10 $(1 + R_F/R_1)$.

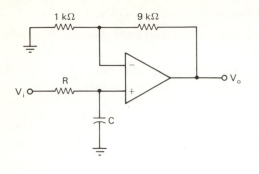

Figure 12-8 An active low-pass filter with a gain of 10. Its frequency response is graphed in Fig. 12-7.

Example 12-7

Repeat Example 12-6 for the circuit shown in Fig. 12-8. Assume a 741 type op-amp.

Solution The op-amp now acts as a noninverting amplifier, multiplying the capacitor voltage by 10 (20 dB) at each input frequency. The resulting frequency response is shown in Fig. 12-7 and labeled $A_V = 10$.

The last example points out an interesting effect due to the presence of the op-amp. Because the gain bandwidth product of the op-amp is constant (10^6 for the 741), the bandwidth (B_{CL}) for $A_{CL} = 10$ is 10^5 ($B_{CL} = B_{OL}/A_{CL}$). This means that the frequency response of the op-amp begins to fall off on its *own* beginning at 10^5 Hz (100 kHz). For this reason, the response at 100 kHz is –3 dB and a total of 23 dB down from its midband value (3 dB for the op-amp and 20 dB due to C). In addition, the gain will now decrease at a *40 dB/decade* rate (20 dB for C and 20 dB for the internal op-amp compensating capacitor) compared to 20 dB/decade for the $A_V = 1$ curve.

All of the filters discussed become high pass filters if the position of R and C is interchanged. In this case, the decrease in gain of the op-amp at higher frequencies may be undesirable, causing a *band-pass* effect rather than a high-pass effect.

Example 12-8

Assume the R and C in Fig. 12-6 are interchanged and $R = 16$ kΩ and $C = 0.01$ μF. Calculate the gain of the circuit at 100 Hz.

Solution First calculate the break frequency.

$$f = \frac{1}{[2\pi\,(16\text{ k}\Omega)(0.01\text{ }\mu\text{F})]} \cong 1\text{ kHz}$$

Because this is a high-pass filter, the response will have *fallen* by 20 dB from 1 kHz to 100 Hz. Therefore, the gain at 100 Hz is –20 dB.

Higher-Order Filters

Higher-order active filters are possible using more than one capacitor. For example, a second-order active filter has a 40 dB/decade slope for its frequency response. Second- and higher-order filters are characterized by a *damping* factor indicating the *flatness* of the response in the midband or passband region. *Butterworth* filters are considered *maximally flat* filters while other types, *Chebyshev* for example, produce ripples in the passband.

12.3 A DIGITAL LOGIC PROBE

A *digital logic probe* is a troubleshooting aid used to determine the logic status of nodes in a digital circuit. Digital circuits are allowed only two states of operation, usually called the ON and OFF or 1 and 0 states. The specific voltage levels representing these two states varies from logic family to logic family, but the most common levels are those illustrated in Fig. 12-9 for the *transistor-transistor-logic* family (TTL). In this system, a logic 1 is represented by a voltage *greater* than 2.4 V and a logic 0 by a voltage *less* than 0.4 V. Any measurement between these two levels indicates a fault condition and a defective circuit.

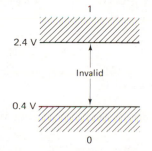

Figure 12-9 Standard TTL logic levels. Voltages between 0.4 V and 2.4 V are indicative of a faulty or open circuit.

Although a voltmeter could be used as a logic probe, this is usually not done. The reason is that the *specific* node voltage is not of concern. We only wish to know if this voltage is within the specifications shown in Fig. 12-9. A practical logic probe must indicate which of the two logic levels, a 1 or a 0, is present. Typically, this is done using two LED indicators.

The LM339

A schematic diagram for a simple logic probe is shown in Fig. 12-10. Op-amps A and B represent two of the four op-amps available in the LM339 *quad comparator*. This chip is a good example of the advantage

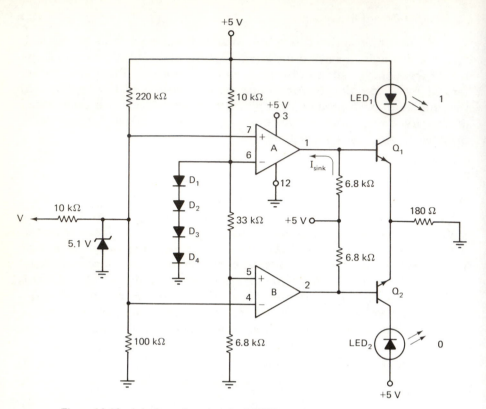

Figure 12-10 A logic probe using the LM339 quad comparator. Only two of the four op-amps are used.

of integrated circuit technology because it contains four op-amps on one silicon die!

The LM339 was specifically designed for *comparator* applications and it will operate from a single power supply. It does not have the large open-loop gain typical of the 741 type op-amps, but features a typical slew rate of 50 V/μs and a *response time* (time from application of input to the switching of the output) of 1.3 μs for a tiny 5-mV input.

The chip also features *open-collector* outputs. This means only a single transistor is used in the output stage, as shown in Fig. 12-11a. When E_d is *negative*, this transistor is *ON* and saturated and *sinks* current from the external pull-up resistor. When E_d is positive, this transistor goes *OFF*, appearing to be an *open circuit*. In this case, R pulls the output up to +V. Circuit operation can be summarized as shown in Fig. 12-11b. Using the former terminology of $\pm V_{sat}$, a $-V_{sat}$ output corresponds to a *closed* switch and $V_o = 0$ V (or V⁻ if two power supplies are used), while $+V_{sat}$ corresponds to an *open* switch in which case the output is pulled high through the external resistor.

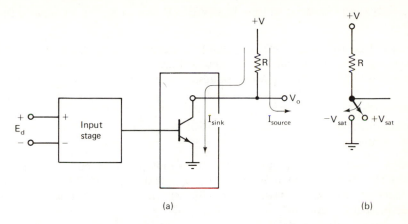

(a)   (b)

Figure 12-11   The LM339 features an open-collector output stage. This is shown in (a) as a single-output transistor with unconnected collector. When the output is in the $+V_{sat}$ condition, the transistor is OFF and V_o is pulled up to +V through R. $-V_{sat}$ corresponds to a saturated output transistor. These two cases are shown in (b).

The Window Comparator

The basic job of any logic probe is to detect voltages within certain limits or *windows*. Referring to Fig. 12-10, op-amps A and B are connected as *window comparators*. The input signal to the probe is applied through a 10-kΩ resistor and 5.1-V zener diode that protect the input circuits against excessive voltages.

Diodes D_1 to D_4 establish the 2.4-V minimum TTL high level on pin 6 of comparator A. A 1:6 voltage divider biases pin 5 of comparator B at 0.4 V. Some adjustment of the 10-kΩ resistor connected to pin 6 may be necessary to achieve exactly 2.4 V on pin 6.

Input voltages greater than 2.4 V cause op-amp A to go to $+V_{sat}$, which corresponds to an *open* switch. Transistor Q_1 is now turned ON via the 6.8 kΩ base resistor and LED_1 lights, indicating a valid logic 1 level.

While this is taking place, the output of op-amp B is saturated because its E_d is negative (pin 4 > pin 5). This holds the base of Q_2 near ground potential and LED_2 is OFF. For input voltages less than 0.4 V, Q_2 is turned on and LED_2 is activated. Note that voltages between 0.4 V and 2.4 V cause both comparators to have negative E_d's and both LEDs to be OFF, indicating an *invalid* input voltage level.

Example 12-9 ──

A simple logic level indicator was described in Example 9-5 using a 7404 logic inverter. However, the 7404 will consider an *open circuit* to be a logic 1 and erroneously light the LED. What will the logic probe shown in Fig. 12-10 indicate for an open circuit?

Solution Due to the 100-kΩ and 220-kΩ resistors, and with no input voltage applied, the two comparators see an input voltage of $100/320 \times 5$ V = 1.56 V. Because this is in the window of *invalid* values, neither LED will be lit.

Example 12-10 ────────────────────────────────────

Calculate the sink current of op-amps A and B when the output transistors are on (E_d is negative).

Solution Assuming the output transistor is saturated, the situation is as shown in Fig. 12-11b with the switch closed. If $V_{CE(sat)} \cong 0$ V, I_{sink} = 5 V/6.8 kΩ = 0.74 mA. Because the LM339 has a typical sink current of 16 mA, this current level should present no problem.

12.4 OP-AMP FUNCTION GENERATOR

A *function generator* typically generates square, sine, and triangular waveforms of variable frequency and amplitude. We have already seen how the op-amp can be used to produce square waves (the astable multivibrator) and in this section the generation of triangular waves is discussed. In addition, a fully-integrated function generator will be presented using the XR-2206.

Ramp Generator

In most RC circuits, the capacitor is charged by an *exponentially-*decreasing current. This results in a "sawtooth" waveform across the capacitor (see Fig. 11-9b for an example). If the capacitor could be charged by a *constant* current, its voltage would increase *linearly* with time, producing a *ramp* output.

Referring to Fig. 12-12a, the virtual ground at the (−) op-amp input establishes a *constant current* through the 10-kΩ input resistor and also the 100-μF capacitor. As shown in Chapter 11, the charge on this capacitor will increase such that

$$Q = I \times t$$

and the voltage will increase such that

$$V = \frac{Q}{C} = \frac{It}{C}$$

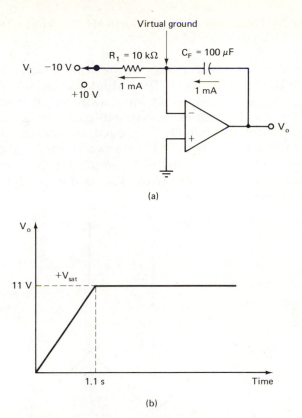

(a)

(b)

Figure 12-12   (a) The op-amp ramp generator. With the components shown, the ramp rate is 10 v/s, causing the output voltage to reach $+V_{sat}$ (11V) in 1.1 s. This is shown in (b).

The *rate* at which this voltage increases is

$$\frac{V}{t} = \frac{I}{C}$$

In Fig. 12-12a, $I = V_i/R_1$. The voltage across the capacitor must then charge at a rate given by

$$\text{ramp rate} = \frac{V}{t} = \frac{I}{C} = \frac{V_i}{R_1 C_F} \qquad (12\text{-}2)$$

Example 12-11 ───

How long will it take the output of the op-amp circuit in Fig. 12-12 to reach $+V_{sat}$ assuming ±12-V power supplies?

Solution From Eq. 12-2 the ramp rate = 10 V/(10 kΩ × 100 μF) = 10 V/s. Because V_{sat} = 11 V, t = 11 V/(10 V/s) = 1.1 s.

Imagine that the input voltage to the ramp generator was –10 V for 1 s and +10 V for the next. The output voltage would first ramp towards $+V_{sat}$ for 1 s. Then, as the input switched, it would ramp towards $-V_{sat}$ for 1 s. In both cases the ramp rate would be 10 V/s. Assuming the capacitor was initially discharged, the output voltage would ramp between 0 V and +10 V. This is shown in Fig. 12-13. Note that positive input voltages cause *negative* ramp rates and negative input voltages result in *positive* ramp rates. This is due to the direction of current flow through capacitor C_F.

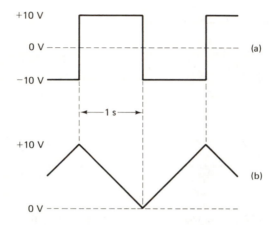

Figure 12-13 When the input voltage to the op-amp in Fig. 12-12a switches between +10 V and –10 V (a), the output waveform is a triangular wave between 0 V and 10 V (b).

Although we could use an *astable multivibrator* in place of the switch in Fig. 12-12a, one serious drawback results. The polarity of the triangular output waveform depends on the *initial* capacitor voltage and polarity of input voltage. For example, if the input to the op-amp was initially +10 V, the output voltage would be a triangular wave between 0 V and –10 V. And what if the capacitor was initially charged to +3 V? Then the ramp would be from +3 V to –7 V! These problems can be avoided if a special *dual-threshold comparator* is used.

Dual-Threshold Comparator

Figure 12-14 is the schematic diagram of a sqaure-wave generator using a 747 dual op-amp. Op-amp A functions as a ramp generator as previously discussed. Op-amp B is a *dual-threshold comparator*. It

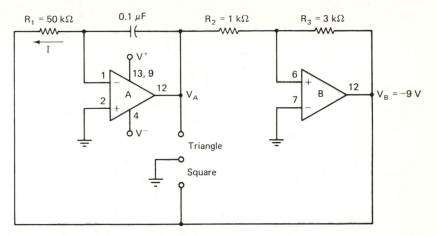

Figure 12-14 Triangular- and square-wave generator using a 747 dual op-amp. Op-amp A is a ramp generator while op-amp B is a dual threshold comparator.

operates nonlinearly ($E_d \neq 0$ V) except momentarily when it switches from one value of V_{sat} to the other.

Assume op-amp B is saturated at −9 V. This means op-amp A is ramping towards +9 V at a 1.8 V/ms (9 V/50 kΩ × 0.1 μF) rate. Because no current flows into the op-amp input pins, resistors R_2 and R_3 are in *series* with −9 V at the R_3 end (V_B) and an ever-increasing positive voltage at the R_2 end (V_A). As V_A continues to increase, a point is eventually reached where the (+) input to op-amp B becomes 0 V, causing its E_d to also be 0 V. Op-amp B is now operating *linearly* and a slight increase in the ramp output at V_A will cause V_B to switch to +9 V. Because the polarity of the input voltage to the ramp generator is now positive, its output commences to ramp *downwards* and the cycle repeats.

Example 12-12 ───────────────────────────────

Referring to Fig. 12-14, determine the value of V_A needed to cause op-amp B to switch to $+V_{sat}$ if its output is currently −9 V.

Solution The equivalent circuit is shown in Fig. 12-15. At the instant of switching, $E_d = 0$ V, placing 9 V across R_3. There must be 9 V/3 kΩ = 3 mA of current through R_2 and R_3. Because $E_d = 0$ V, $V_A = I \times R_2 = 3$ mA × 1 kΩ = +3 V. As a general rule, we can say that at the switching point $\pm V_{sat}/R_3 = -V_A/R_2$ (the current in R_2 equals the current in R_3) and

$$V_A = \frac{-R_2}{R_3} \times \pm V_{sat} \qquad (12\text{-}3)$$

Example 12-13 ───────────────────────────────

Sketch the output waveforms at V_A and V_B for the circuit in Fig. 12-13. Determine the *frequency* of oscillation.

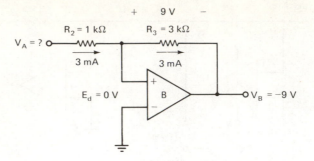

Figure 12-15 A dual threshold comparator. At the switching point, E_D = O V, placing 9 V across R_3. The resulting 3 mA of current flows through R_2 and R_3, requiring V_A to be +3 V.

Solution We have already determined from Example 12-12 that the ramp generator output will increase linearly to +3 V and, because the circuit is symmetric, decrease linearly to –3 V. The output of op-amp A is therefore a 6-V peak-peak triangular wave, while the output at V_B is a symmetric square wave between $\pm V_{sat}$. These two waveforms are shown in Fig. 12-16.

The ramp rate is given by Eq. 12-2 as 9 V/(50 kΩ × 0.1 μF) = 1.8 V/ms. It will take 6/1.8 = 3.33 ms for *one-half* period (time to charge up or down). The frequency must therefore be 1/6.66 ms = 150 Hz.

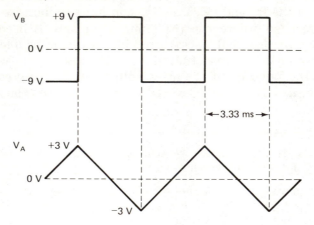

Figure 12-16 Waveforms for the function generator in Fig. 12-14.

Example 12-14 ───────────────────────────────────────

If R_1 is replaced by a 1-kΩ resistor and 100-kΩ pot, what extremes of frequency are possible for the circuit in Fig. 12-14?

Solution First, the ramp rate must be calculated for the two cases.

1. Ramp rate (max) = 9 V/(1 kΩ × 0.1 μF) = 90 V/ms
2. Ramp rate (min) = 9 V/(101 kΩ × 0.1 μF) = 0.9 V/ms

The corresponding periods are

$$T(min) = 2 \times \frac{6 \text{ V}}{90 \text{ }\frac{\text{V}}{\text{ms}}} = 133 \text{ }\mu s$$

and

$$T(max) = 2 \times \frac{6 \text{ V}}{0.9 \text{ }\frac{\text{V}}{\text{ms}}} = 13.3 \text{ ms}$$

This results in f(max) = 7.5 kHz, and f(min) = 75 Hz.

A Fully-Integrated Function Generator

It is the nature of the integrated circuit technology to put more components on a single chip of silicon. Perfectly reasonable and adequate discrete component designs become *unreasonable* and *inadequate* as their complete function (and often more) is reduced to a single chip!

An example of this is the Exar 2206 monolithic function generator. Quoting from the Exar Function Generator Data Book,

The XR-2206 is a monolithic function generator integrated circuit capable of producing high quality sine, sqaure, triangle, ramp and pulse waveforms of high stability and accuracy. The output waveforms can be both amplitude and frequency modulated by an external voltage. Frequency of operation can be selected externally over a range of 0.01 Hz to more than 1 MHz.

Perhaps its most amazing statistic is that this chip sells for less than $5! Figure 12-17 is the schematic diagram of a complete function generator system using only *one* IC, the XR-2206. It provides sine, square, and triangle wave outputs over a frequency range selectable in four positions from 1 to 100 kHz. It may be powered by a single +12-V supply or ±6-V split supplies.

The frequency of oscillation is controlled by resistors R_4 and R_{13} and the timing capacitor selected by S_1. Note that an external voltage can be applied to pin 7 via R_3, allowing a *voltage-controlled oscillator* function. A sawtooth applied to this input would cause the output to sweep through a range of frequencies. The output amplitude can also be modulated (amplitude modulation) via an external input at pin 1.

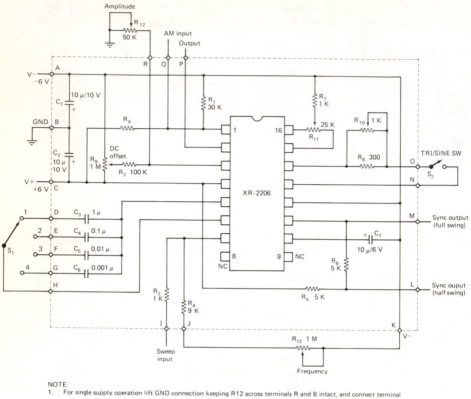

Figure 12-17 Fully-integrated function generator based on the XR-2206 integrated circuit. (Courtesy Exar Integrated Systems, Inc.)

NOTE:
1. For single supply operation lift GND connection keeping R12 across terminals R and B intact, and connect terminal

    A to GND

2. For maximum output, R_x may be open. R_x = 68 kΩ is recommended for external amplitude modulation.

12.5 CONCLUSION

The op-amp circuits discussed in this chapter should give you an idea of the wide variety of applications possible for this component. You should also note that the op-amp is often only a *building block* in more complex circuits. Circuits such as the *three-terminal voltage regulator* or *integrated-function generator* may use op-amps internally, but what is important to the user is their *external* characteristics.

The circuit designer must become proficient at reading and studying manufacturer's data sheets describing the external characteristics of these more complex ICs. Circuits such as the XR-2206 are extremely powerful but require considerable study before they can be efficiently designed into a system. As such, the traditional role of the circuit designer is changing from one of pure circuit design to the application and interface of the more complex ICs available today.

———————————————— KEY TERMS ————————————————

Pass Transistor   In a dc power source, the full load current passes through this transistor. It is usually driven by the output of the error amplifier in a negative feedback loop.

Error Amplifier   Op-amp used in a regulated power source to monitor the output voltage and adjust the base drive to the pass transistor to maintain a constant output voltage.

Three-Terminal Regulator   An integrated circuit designed to convert an unregulated voltage to some specific regulated voltage dependent on the regulator chosen. Noted for its simplicity of use and indestructability.

Active Filter   A type of electronic filter employing RC combinations and an op-amp (active device). It eliminates the need for inductors and features negligible insertion loss.

Window Comparator   A type of op-amp circuit generally using two op-amps that provide an output only if the input voltage is within a window of values.

Ramp Generator   Also called an integrator, this op-amp circuit produces a linearly rising or falling output voltage until the output is limited by V_{sat}. The circuit is similar in configuration to an inverting amplifier but with the feedback resistor replaced by a capacitor.

Ramp Rate   Similar to slew rate, this is the rate at which the output voltage of a ramp generator increases or decreases with time. Usually measured in V/s.

———————————— QUESTIONS AND PROBLEMS ————————————

12-1   Refer to Fig. 12-1. If the 747 is limited to 25-mA maximum output current, what is the minimum β for the pass transistor that will allow a 500-mA load current?

12-2   What is the smallest load resistor that can ever be connected to V_{o1} in Fig. 12-1 without causing current limiting for all values of output voltage?

12-3   If V^+ and V^- are rated at ± 18 V maximum, determine the maximum rms output voltage for transformer T_1 in Fig. 12-1 that will not exceed the supply voltage specifications for the op-amp.

12-4   For the conditions in Problem 12-3, what zener diode voltage is needed to achieve the full output voltage when R_4 is adjusted for maximum output?

12-5   Show how a 100-mA current-limiting circuit can be added to V_{o2} in Fig. 12-1.

12-6   Design a power supply with dual tracking outputs using a 12.6-V CT transformer to meet the following specifications:
(a) V_o adjustable from 1 to 6 V

(b) 150-mA current limiting on both outputs.

Hint: Bias a general purpose diode to 0.6 V for the reference.

12-7 Design a split ±12 V fixed-output power supply using one 24-V CT transformer and two three-terminal voltage regulators. Assume 500-mA maximum load currents and allow 1-V peak-peak ripple across the filter capacitors.

12-8 The LM7805 voltage regulator in Fig. 12-4a is internally protected against thermal overloads — the larger the heat sink, the more power the chip can dissipate. If the power dissipation is $I_{LOAD} \times (V_i - V_o)$ and $V_i = +10$ V, determine the maximum load current for

(a) No heat sink and power dissipation limited to 2 W

(b) Large heat sink such that $P_d = 10$ W.

12-9 Show the schematic diagrams for first-order active high- and low-pass filters with 0 dB gain and 5-kHz break frequencies. Sketch the frequency response from 10 Hz to 1 MHz for both filters.

12-10 Sketch the frequency response of the active filter in Fig. 12-18 from 10 Hz to 1 MHz.

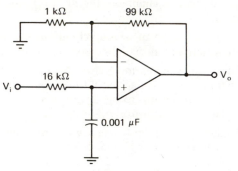

Figure 12-18

12-11 Determine the minimum βs for Q_1 and Q_2 in Fig. 12-10 if the LED current is to be 15 mA.

12-12 Design a window comparator circuit using the LM339 to detect voltages between +5 V and +10 V and light a *green* LED. Voltages outside this range should activate a *red* LED. The input should be protected against input voltages greater than +12 V.

12-13 Explain how the virtual ground forces the current in the ramp generator timing capacitor to be *constant*. Why is this same current *not constant* in the astable multivibrator shown in Fig. 11-9a?

12-14 Calculate the ramp rate of the circuit in Fig. 12-12 if C_F is changed to 10 μF and $V_i = +5$ V. Sketch V_o versus time assuming $\pm V_{sat} = 10$ V.

12-15 What is meant by a dual-threshold comparator? Show the schematic diagram of such a circuit that will switch at ±5 V. Assume $\pm V_{sat} = \pm10$ V.

12-16 Refer to Fig. 12-14; why would it be undesirable to make R_2 and R_3 the variable resistors controlling frequency?

12-17 Sketch the triangle- and square-wave outputs for the circuit in Fig. 12-14 if the ramp generator has $R_1 = 10$ kΩ, and $C = 0.05$ μF; and the comparator has $R_2 = 2.2$ kΩ, and $R_3 = 3.3$ kΩ. Assume $\pm V_{sat} = \pm9$ V.

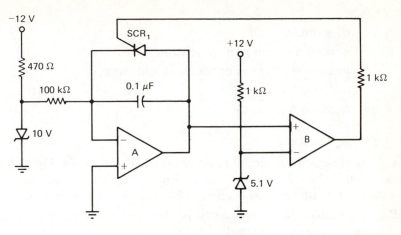

Figure 12-19

12-18 The circuit in Fig. 12-19 is a sawtooth oscillator. Sketch the output wave-form for op-amp A and estimate the frequency of oscillation.

12-19 Show that the frequency of oscillation for the triangle-wave generator in Fig. 12-14 is given by $f = 0.25 \times R_3/R_2 \times 1/RC$, independent of V_{sat}.

LABORATORY ASSIGNMENT 12:
APPLICATIONS OF OP-AMPS

Objectives

1. To gain further knowledge of the op-amp as a circuit element.
2. To measure the performance of the op-amp as a dc and ac ampli-fier, window comparator, and ramp generator.

Introduction In this laboratory assignment you will construct several of the circuits discussed in this chapter. A number of new ICs are introduced, including the LM7805 voltage regulator, LM747 dual op-amp, and the LM339 quad comparator.

Components Required

1 12.6-V CT ac source
1 741 type op-amp
1 747 type op-amp
1 339 type op-amp
1 7805 +5-V regulator
1 3.3-V and 5.1-V zener diodes

3 LEDs

5 Rectifier diodes

Miscellaneous *npn* and *pnp* transistors

Miscellaneous assortment of resistors and capacitors

Part I: *Regulated Power Source*

A. Op-Amp Regulator

STEP 1 Redesign (but do not assemble) the circuit in Fig. 12-1 to operate from a 12.6-V ac source and with V_o adjustable between 1 and 6 V. Current limiting should occur at 150 mA (see Problem 12-6).

STEP 2 Assemble the *unregulated* portion of the circuit and verify that V^+ and V^- are as calculated.

STEP 3 Add the positive output portion of the circuit and measure the output voltage extremes. Redesign the circuit if the design goals have not been met within $\pm 10\%$.

STEP 4 Determine the load regulation by setting the output to 5.0 V. Now connect a 47-Ω 1-W load resistor to the output. Calculate the load regulation as

$$\% \text{ regulation} = \frac{V_o (\text{no load}) - V_o (47 \, \Omega)}{V_o (47 \, \Omega)} \times 100\%$$

STEP 5 Record the ac waveforms at the unregulated V^+ output and the regulated dc output for the 47-Ω load condition in step 4.

Question 1 Does the regulator also tend to reduce the ripple riding on the unregulated output? Explain.

STEP 6 Momentarily short the output to ground through an ammeter. The current should limit at approximately 150 mA.

STEP 7 Add the negative output circuit to your regulator and measure and record V_{o2} as V_{o1} is adjusted between 1 and 6 V.

B. Three-Terminal Regulator

STEP 1 Construct the circuit shown in Fig. 12-4a using the positive unregulated source built in Part A. Measure V_o. It should be between 4.8 V and 5.2 V to be within the manufacturer's specifications.

STEP 2 Determine the load regulation by adding a 47-Ω load resistor and remeasure the output voltage. Calculate the load regulation as was down in step 4 of Part A.

STEP 3   Repeat step 5 of Part A and compare the ripple reduction of the two circuits.

Part II: *Active Low-Pass Filter*

STEP 1   Design and assemble an active low-pass filter that has 0 dB gain in its passband and a 5-kHz break frequency. Use a 741 op-amp.

STEP 2   Measure the voltage gain of your filter in decibels from 100 Hz to 1 MHz and plot on semilog graph paper.

STEP 3   Modify your circuit so that it has 10 dB gain in the passband and repeat step 2.

Question 2   In step 3, at what rate does the gain fall off for frequencies greater than 100 kHz? Explain.

STEP 4   Modify the 10 dB gain circuit to have a break frequency at 50 kHz and repeat step 2.

Part III: *Digital Logic Probe*

STEP 1   Assemble the logic probe circuit shown in Fig. 12-10. Replace the 10-kΩ pull-up resistor with a 1-kΩ resistor and 10-kΩ pot in series.

STEP 2   Adjust the 10-kΩ pot until pin 6 = 2.4 V.

STEP 3   Verify circuit operation by touching the input to +5 V and ground.

STEP 4   Record all node voltages in the circuit for (a) $V_i = 0$ V, (b) $V_i = 5$ V, and (c) V_i = open circuit.

STEP 5   Monitor the input voltage with a voltmeter and observe the switching points of the two comparators as V_i is varied. Check these voltages against the voltages on pins 6 and 5.

Question 3   From your data in step 4, calculate the LED current when ON.

STEP 6   Touch the logic probe input to the output of a *TTL pulse generator*. Observe the LEDs as the frequency is varied.

Question 4   If the pulse waveform was low for 80% of the cycle and high the remaining 20%, what appearance would the two LEDs have?

STEP 1   Assemble the circuit shown in Fig. 12-14. Replace the 50-kΩ resistor with a 50-kΩ pot and 1-kΩ resistor in series.

STEP 2   Calculate the minimum and maximum charging rates for the circuit. Be sure to use the proper value of V_{sat} depending on the values of V^+ and V^- used. Calculate the switching levels of the comparator and from this determine the minimum and maximum frequencies of oscillation.

STEP 3   Adjust the pot to its two extremes and measure the resulting periods. Compute the oscillation frequencies.

STEP 4   With the pot adjusted for f(max), record the waveforms at the two 747 outputs.

Question 5   What is the effect of increasing R_3? R_2?

ANSWERS TO SELECTED PROBLEMS

1-4 a. minus
     b. minus
     c. positive
     d. positive
     e. neutral

1-6 False

1-7 True

1-8 False

1-10 100,000

1-11 a. 3.5 mA
      b. 1.6 mA, 6.8 V, 10.7 mW

CHAPTER 2

2-3 D_2, D_4, D_5

2-4 2.5 mA, 2.2 mA

2-5 3.8 mA, 1.2 V

2-6 a. 0
     b. 0.1 A
     c. 0, 0.6 A
     d. 3.83 A

2-7 decrease

2-8 10.2 mA, 1.2 V

2-10 1000 μF, PIV = 33.4 V, I_{SURGE} = 9.4 A

2-11 a. 24.9 V
      b. 12.1 V
      c. 24.3 V

2-13 R_S = 8.6 Ω, C = 5313 μF, PIV = 33.4 V
2-16 1.7%
2-17 1.8%
2-18 3.2 mA, 8.2 V
2-22 0.57 V, 4.5 mA
2-23 0.64 V, 19 mA

CHAPTER 3

3-7   25 V
3-8   a. active
      b. saturated
      c. active
      d. cutoff
      e. saturated
3-9   a. 32 μA
      b. 1.3 V
      c. 0.9 mA
      d. 0 mA
      e. 0.2 mA
3-10 a. 25
      b. 75
      c. 96
      d. 96
3-12 0.99, 120
3-14 a. 35
      b. 6 V
      c. 75 V
      d. 1.0 V
      e. 800 mA
      f. 10 mA
      g. 25°C
3-16 40 μA, 8 mA, 4 V
3-17 15 μA, 0.6 V, 1.4 mA, 10 V
3-19 120
3-20 92, 100
3-21 0.014 W
3-22 0.13 V peak-peak
3-23 6.3 V, 1.4 mA
3-24 75 mA, 300 mA
3-25 940 Ω, 1.9 kΩ
3-26 520 Ω, 8.3 kΩ
3-28 150

CHAPTER 4

4-2 1.77, no
4-3 –30 dB
4-4 10 μW
4-5 26%
4-6 106 dB
4-7 10 kHz
4-9 3.1 W
4-11 34.8 dB, 26 dB, 55.6 dB
4-12 117 Ω, 750 Ω
4-17 2089 W
4-18 15.9 kHz, 45°
4-19 2.2 dBm
4-20 0.55 V

CHAPTER 5

5-5 0.03 V peak-peak
5-10 167
5-11 r_{be} = 1.7 kΩ, β_{ac} = 313, β_{dc} = 333, r_o = ∞
5-12 793 kΩ, 1.5 kΩ
5-13 R_i = 1.7 kΩ, A_v = –276, R_o = 1.5 kΩ
5-14 36 mV peak-peak
5-15 –110, 3 V peak
5-17 26.6 μA, 4 mA, –8 V
5-18 R_i = 939 Ω, R_o = 2.3 kΩ, A_v = –368
5-19 4 V peak-peak
5-20 3 μF, 1 μF
5-21 1.9 μF, 0.7 μF
5-22 4.1 mA, –7.7 V
5-23 0.4 V, 0.76 V

CHAPTER 6

6-1 V_{CE} = 5.5 V
6-2 V_{CE} = 5.1 V
6-3 V_{CE} = 4.9 V
6-4 R_i = 3.5 kΩ, R_o = 1.5 kΩ, k = –3.1, V = 8.6 V peak-peak
6-8 A_V: –172, –12.7, –5.3, –2.7
     R_i : 635 Ω, 2.8 kΩ, 3.3 kΩ, 3.5 kΩ

6-10 $V_{CE} = 4.5$ V

6-11 $V_{CE} = -13.5$ V

6-12 58

6-13 0 V, 16 V

6-14 $R_i = 634\ \Omega$, $R_o = 2\ k\Omega$, $A_V = -93$

6-15 4.7 V peak-peak

6-17 $V_{CE} = 7.1$ V

6-18 $V_{CE} = 2.55$ V, $R_i = 501\ \Omega$, $R_o = 488\ \Omega$, $A_V = -208$

6-19 $V_{CE} = 1.8$ V

6-20 $V_{CE} = 6.6$ V, $R_i = 9.5\ k\Omega$, $R_o = 56\ \Omega$, $A_V = 0.99$, $A_I = 18.7$

6-21 1.7 μF, 32 μF

6-22 3.6 V peak-peak

6-23 12.7 dB

CHAPTER 7

7-1 0.97 V, 1.94 mA

7-2 24.5 dB, -4.7 dB

7-3 1.26 V

7-4 $V_{CE1} = 5.8$ V, $V_{CE2} = 4.65$ V

7-5 $R_i = 1.5\ k\Omega$, $R_o = 7\ \Omega$, $A_V = -7.8$

7-6 19.5 dB

7-7 6.5 mW

7-8 $C_1 = 11\ \mu$F, $C_2 = 40\ \mu$F, $C_3 = 4\ \mu$F, $C_4 = 16\ \mu$F

7-9 5

7-11 $V_{CE} = 3$ V

7-12 $R_i = 3.2\ k\Omega$, $R_o = 4\ k\Omega$, $A_{VT} = 83$

7-13 41 μF

7-16 $V_{CE1} = 3.9$ V, $V_{CE2} = -9.5$ V

7-17 357

7-19 20%

7-20 10:1, 0.4 V peak-peak

7-21 (a) 20 mA
(b) 625 Ω
(c) 25 V peak-peak, 5 V peak-peak
(d) 42%

7-23 $V_{B1} = 0.6$ V, $V_{B2} = -0.6$ V

7-24 $V_{B4} = 5.3$ V, $V_{B3} = 6.5$ V

CHAPTER 8

8-6 -4 V

8-11 $R_D = 5.8\ k\Omega$

8-13 V_{DS} = 5.3 V, 9.8 V

8-15 9.75 V, 3.5 mA

8-16 V_{DS} = 4.2 V

8-17 750 Ω

8-18 V_{GS} = +2 V, I_D = 5 mA

8-19 1800 μS

8-20 89.8 mA/V

8-21 A_V = −2.3

8-22 V_{DS} = 2.5 V, V_{CE} = 4.6 V

8-23 R_i = 680 kΩ, R_o = 2 kΩ, A_V = 131

8-24 R_1 = 20 kΩ, R_2 = 30 kΩ, R_D = 1.2 kΩ

8-26 NOR

CHAPTER 9

9-5 10.5 V

9-7 13.5 V

9-8 a. 90°-180°
b. 270°-360°
c. 0-360°

9-9 146°

9-11 I⁺, III⁻

9-14 88°, 168°

9-15 142.6 kΩ

9-16 high, low

9-17 160 Ω, 6 kΩ

9-26 23 mA

9-27 9.5 kΩ

9-28 26 mA

CHAPTER 10

10-7 69 mils

10-8 5.65×10^{-3} cm²

10-11 62%

10-13 V_{C1} = V_{C2} = 7.25 V

10-14 200

10-15 500, 0.04, 82 dB

10-16 1 mV

10-17 V_{C1} = V_{C2} = 5 V, V_{B3} = −4.9 V

10-18 V_{C1} = V_{C2} = 6.67 V, V_o = 6.1 V

10-19 Q_1

11-2   28 V peak-peak

11-3   +10 V, –10 V, +10 V, –10 V

11-5 V_o = +4 V, I = 4.4 mA

11-6   0.6 mA, 0.3 mA, –0.4 mA

11-8 R_F = 375 kΩ

11-9   147 mV

11-10 a. –7.2 V
           b. +11 V
           c. –0.6 V
           d. –1.5 V

11-11 277 Ω

11-12 12 mA, source

11-17 f = 500 Hz

11-19 55 ms

11-20 7.6 V

11-22 67 mV

11-23 47 mV

11-25 a

11-26 94.7 kΩ, 66 μV

11-27 False

11-28 a. 10 MHz
           b. 100 kHz

11-29 2 MHz

11-30 119 kHz

11-31 6.3 MHz, 0.63 V/μs

11-32 56.3 kHz

CHAPTER 12

12-1   20

12-2   28

12-3   26.3 VCT

12-4   3.5 V

12-8   0.4 A, 2 A

12-10 f_{break} = 100 kHz, A_V = 40 dB

12-11 60

12-14 50 V/s

12-17 f = 750 Hz

12-18 196 Hz

INDEX